# Topics in Artificial Intelligence Applied to Industry 4.0

*Edited by*

*Mahmoud Ragab AL-Refaey*
*Information Technology Department, Faculty of Computing and Information Technology (FCIT),*
*King Abdulaziz University (KAU), Jeddah, Saudi Arabia*
*Mathematics Department, Faculty of Science, Al-Azhar University, Naseir City, Cairo, Egypt*

*Amit Kumar Tyagi*
*Department of Fashion Technology, National Institute of Fashion Technology, New Delhi, India*

*Abdullah Saad AL-Malaise AL-Ghamdi*
*Information Systems Department, Faculty of Computing and Information Technology (FCIT), King*
*Abdulaziz University (KAU), Jeddah, Saudi Arabia*
*Information Systems Department, School of Engineering, Computing and Design, Dar Al-Hekma University,*
*Jeddah, Saudi Arabia*

*Swetta Kukreja*
*Department of Computer Science and Engineering, Amity University, Mumbai, Maharashtra, India*

*Registered Offices*
John Wiley & Sons, Inc., 111 River Street, Hoboken, NJ 07030, USA
John Wiley & Sons Ltd, The Atrium, Southern Gate, Chichester, West Sussex, PO19 8SQ, UK

For details of our global editorial offices, customer services, and more information about Wiley products visit us at www.wiley.com

Wiley also publishes its books in a variety of electronic formats and by print-on-demand. Some content that appears in standard print versions of this book may not be available in other formats.

**Library of Congress Cataloging-in-Publication Data applied for:**

Hardback ISBN: 9781394216116

Cover Design: Wiley
Cover Image: © VicenSanh/Adobe Stock Photos

Set in 9.5/12.5pt STIXTwoText by Straive, Pondicherry, India

# Contents

# About the Editors

**Mahmoud Ragab AL-Refaey** obtained his Ph.D. degree from the Faculty of Mathematics and Natural Sciences at the Christian-Albrechts-University in Kiel (CAU), Schleswig-Holstein, Germany.

He received his B.SC. degree in Statistics Computer Science from Mansoura University in Mansoura, Egypt. He is a professor of data science at the Department of Information Technology, Faculty of Computing and Information Technology, King Abdulaziz University in Jeddah, Saudi Arabia and the Mathematics Department, Faculty of Science, Al Azhar University in Cairo, Egypt. He worked in different research groups at various universities such as the Combinatorial Optimization and Graph Algorithms Group (COGA), Faculty of Mathematics and Natural Sciences, Berlin University of Technology in Berlin, Germany; Faculty of Informatics and Computer Science at the British University in Egypt BUE; Integrated Communication Systems Group at Ilmenau University of Technology TU Ilmenau, in Thüringen, Germany. Now he is a researcher at various centers such as: University of Oxford Centre for Artificial Intelligence in Precision medicines; Center of Research Excellence in Artificial Intelligence and Data Science; Center of Excellence in Smart Environment Research at King Abdulaziz University, Jeddah, Saudi Arabia. He has published over 100 papers in refereed high-impact journals, books, and patents. His research focuses on: AI, Deep learning, Optimization, Mathematical Modeling, Data Science, Neural Networks, Time series analysis, and decision support systems.

**Amit Kumar Tyagi** is working as an assistant professor at the National Institute of Fashion Technology, New Delhi, India. Previously, he worked as an assistant professor (Senior Grade 2) and senior researcher at Vellore Institute of Technology (VIT), Chennai Campus, Chennai, Tamil Nadu, India, for the period of 2019–2022. He received his PhD degree (full-time) in 2018 from Pondicherry Central University, Puducherry, India. Regarding his academic experience, he joined Lord Krishna College of Engineering, Ghaziabad (LKCE) for the periods of 2009–2010 and 2012–2013. He was an assistant professor and head of research at Lingaya's Vidyapeeth (formerly known as Lingaya's University), Faridabad, Haryana, India, for the period of 2018–2019. His supervision experience includes more than 10 master's dissertations and one PhD thesis. He has contributed to several projects such as AARIN and P3- Block to address some of the open issues related to privacy breaches in vehicular applications (such as parking) and medical cyber-physical systems (MCPS). He has published over 100 papers in refereed high-impact journals, conferences, and books, with some of his articles receiving best paper awards. Also, he has filed more than 20 patents (nationally and internationally) in the areas of deep learning, the Internet of Things, cyber-physical systems, and computer vision. He has edited more than 20 books for IET, Elsevier, Springer, CRC Press, and so on. Furthermore, he has authored three books on the Internet of Things, intelligent transportation systems, and vehicular ad hoc networks with BPB Publication, Springer, and IET publisher, respectively. He is a winner of the Faculty Research Award for the years 2020, 2021, and 2022 (consecutively three years), given by Vellore Institute of Technology, Chennai, India. Recently, he has received the best paper award for a paper titled "A Novel Feature Extractor Based on the Modified Approach of Histogram of Oriented Gradient" at ICCSA 2020 in Italy, Europe. His current research focuses on next-generation machine-based communications, blockchain technology, smart and secure computing, and privacy. He is a regular member of the ACM, IEEE, MIRLabs, Ramanujan Mathematical Society, Cryptology Research Society, and Universal Scientific Education and Research Network, CSI, and ISTE.

**Abdullah Saad AL-Malaise AL-Ghamdi** is a professor of software & systems Engineering and AI, associated with Faculty of Computing and Information Technology (FCIT) at King Abdulaziz University (KAU), Jeddah, Saudi Arabia. He is a professor at the Information Systems Department, School of Engineering, Computing and Design, Dar Al-Hekma University, in Jeddah, Saudi Arabia. He received his PhD. degree in computer science from George Washington University, USA, in 2003. He is a member of the Scientific Council and holds the position of secretary general of the scientific council at KAU. In addition, he is working as the head of Consultant's unit at the Vice-President for Development Office, as a consultant to the vice-president for Graduate Studies & Scientific Research at KAU. Previously, he has worked as the head of the IS Department, vice dean for Graduate Studies and Scientific Research, and head of the Computer Skills Department at FCIT. Recently he is a researcher at various centers such as: University of Oxford Centre for Artificial Intelligence in Precision Medicines; Center of Research Excellence in Artificial Intelligence and Data Science; Center of Excellence in Smart Environment Research at King Abdulaziz University, Jeddah, Saudi Arabia. He has supervised many MSc & PhD students who are now successful in and outside academia. He has published many papers in refereed high-impact journals, books, and patents. His main research areas are software engineering and systems, artificial intelligence, data analytics, business intelligence, and decision support systems.

**Swetta Kukreja** is working as an associate professor in the Department of CSE at Amity University, Mumbai. She has more than 10 years of teaching and research experience. She has completed her PhD from Lingaya's University, Faridabad. She had served as an editor for many international conferences and journals. She has many publications (including patents) in national and international conferences and journals and has also served as a reviewer for the same. She is a member of ACM and IEEE.

# List of Contributors

**Monika Agarwal**
Computer Science and Engineering
Dayananda Sagar University
Bengaluru, Karnataka, India

**Abdullah Saad AL-Malaise AL-Ghamdi**
Information Systems Department, Faculty of Computing
and Information Technology (FCIT)
King Abdulaziz University (KAU)
Jeddah, Saudi Arabia

Information Systems Department
School of Engineering, Computing and Design
Dar Al-Hekma University
Jeddah, Saudi Arabia

**K. Annamalai**
VIT University
Chennai, Tamil Nadu, India

**Nusrat J. Ansari**
Computer Science Department
Vivekanand Education Society's Institute of Technology
Mumbai, Maharashtra, India

**Dinesh Kumar Atal**
Department of Biomedical Engineering
Deenbandhu Chhotu Ram University of Science and
Technology
Sonipat, Haryana, India

**Ananda K. Behera**
Artificial Intelligence and Machine Learning Programme
Liverpool John Moores University
Liverpool, UK

**Rajiv Kumar Berwer**
Department of Computer Science and Engineering
Deenbandhu Chhotu Ram University of Science and
Technology
Sonipat, Haryana, India

**Biswajit R. Bhowmik**
BRICS Laboratory
Department of Computer Science and Engineering
National Institute of Technology Karnataka
Mangalore, Karnataka, India

**Sovers Singh Bisht**
Noida Institute of Engineering and Technology
Greater Noida, Uttar Pradesh, India

**Priyanka Chandani**
Noida Institute of Engineering and Technology
Greater Noida, Uttar Pradesh, India

**Mani D. Choudhry**
Department of Information Technology
KGiSL Institute of Technology
Coimbatore, Tamil Nadu, India

**Jyoti Dabass**
DBT Centre of Excellence Biopharmaceutical
Technology, IIT
New Delhi, India

**Manju Dabass**
EECE Department
The Northcap University
Gurugram, Haryana, India

**José P.G. de Oliveira**
Polytechnic School of Pernambuco
University of Pernambuco
Recife, Pernambuco, Brazil

**M.K. Dharani**
Department of AI
Kongu Engineering College
Erode, Tamil Nadu, India

**N. Ethiraj**
Dr. M.G.R. Educational and Research Institute
Chennai, Tamil Nadu, India

**Carmelo J.A.B. Filho**
Polytechnic School of Pernambuco
University of Pernambuco
Recife, Pernambuco, Brazil

**S. Geetha**
Dr. M.G.R. Educational and Research Institute
Chennai, Tamil Nadu, India

**K.K. Girish**
BRICS Laboratory
Department of Computer Science and Engineering
National Institute of Technology Karnataka
Mangalore, Karnataka, India

**V.M. Gobinath**
Rajalakshmi Institute of Technology
Chennai, Tamil Nadu, India

**M. Gunasekar**
Department of Information Technology
Kongu Engineering College
Erode, Tamil Nadu, India

**Sanjeev Indora**
Department of Computer Science and Engineering
Deenbandhu Chhotu Ram University of Science and Technology
Sonipat, Haryana, India

**Garima Jain**
Noida Institute of Engineering and Technology
Greater Noida, Uttar Pradesh, India

**Guru Akaash N. Janthalur**
Computer Science Department
Vivekanand Education Society's Institute of Technology
Mumbai, Maharashtra, India

**A.S. Jayasurya**
Department of Electrical and Electronics
Universiti Teknologi Petronas
Perak, Malaysia

**G. Belshia Jebamalar**
Department of Computer Science and Engineering
S.A. Engineering College
Chennai, Tamil Nadu, India

**Manoj Joshi**
Department of ECE
JSS Academy of Technical Education
Noida, Uttar Pradesh, India

**Igor Jurcic**
Telecommunications and Informatics Department
HT ERONET
Mostar, Bosnia and Herzegovina

**S.K. Rajesh Kanna**
Rajalakshmi Institute of Technology
Chennai, Tamil Nadu, India

**Vinod M. Kapse**
Noida Institute of Engineering and Technology
Greater Noida, Uttar Pradesh, India

**Manigandan Kashimani**
Computer Science Department
Vivekanand Education Society's Institute of Technology
Mumbai, Maharashtra, India

**A. Kathirvel**
Panimalar Engineering College
Chennai, Tamil Nadu, India

**M. Keerthika**
Department of Computer Science and Engineering
Rajalakshmi Engineering College
Chennai, Tamil Nadu, India

**Utku Köse**
Computer Engineering
Suleyman Demirel University
Kaskelen, Kazakhstan

**Swetta Kukreja**
Department of Computer Science and Engineering
Amity University
Mumbai, Maharashtra, India

**Ambeshwar Kumar**
Computer Science and Engineering
GITAM University
Visakhapatnam, Andhra Pradesh, India

**K. Pradheep Kumar**
Department of CSIS
BITS Pilani
Pilani, Rajasthan, India

**K.R. Prasanna Kumar**
Department of Information Technology
Kongu Engineering College
Erode, Tamil Nadu, India

**Sunil Kumar**
BRICS Laboratory
Department of Computer Science and Engineering
National Institute of Technology Karnataka
Mangalore, Karnataka, India

**R. Lalitha Priya**
Computer Science Department
Vivekanand Education Society's Institute of Technology
Mumbai, Maharashtra, India

**K. Logeswaran**
Department of AI
Kongu Engineering College
Erode, Tamil Nadu, India

**M. Manjula**
Computer Science and Engineering
Dayananda Sagar University
Bengaluru, Karnataka, India

**Bireshwar D. Mazumdar**
Department of Computer Science and Engineering
Faculty of Engineering and Technology
United University Prayagraj
Allahabad, Uttar Pradesh, India

**K. Mehata**
Dr. M.G.R. Educational and Research Institute
Chennai, Tamil Nadu, India

**Rodrigo de Paula Monteiro**
Polytechnic School of Pernambuco
University of Pernambuco
Recife, Pernambuco, Brazil

**Sundarrajan Munusamy**
Department of Networking and Communications
SRM Institute of Science & Technology
Chennai, Tamil Nadu, India

**Parimala D. Muthusamy**
Department of Electronics and Communication
Engineering
Velalar College of Engineering and Technology
Erode, Tamil Nadu, India

**Ajay R. Nair**
Computer Science Department
Vivekanand Education Society's Institute of Technology
Mumbai, Maharashtra, India

**Kanchan Naithani**
School of Computing Science and Engineering
Galgotias University
Greater Noida, Uttar Pradesh, India

**Sérgio C. Oliveira**
Polytechnic School of Pernambuco
University of Pernambuco
Recife, Pernambuco, Brazil

**Ts. Sundaresan Perumal**
Universiti Sains Islam Malaysia
Bandar Baru Nilai
Negeri Sembilan, Malaysia

**Pooja**
Computer Science and Engineering
Dayananda Sagar University
Bengaluru, Karnataka, India

**M. Pragadeesh**
Department of Information Technology
Rajalakshmi Engineering College
Chennai, Tamil Nadu, India

**Soniya Priyatharsini**
Dr. M.G.R. Educational and Research Institute
Chennai, Tamil Nadu, India

**Mahmoud Ragab AL-Refaey**
Information Technology Department, Faculty of
Computing and Information Technology (FCIT)
King Abdulaziz University (KAU)
Jeddah, Saudi Arabia

Mathematics Department
Faculty of Science
Al-Azhar University
Naseir City, Cairo, Egypt

**R. Rahul**
Dr. M.G.R. Educational and Research Institute
Chennai, Tamil Nadu, India

**R. Rajadevi**
Department of AI
Kongu Engineering College
Erode, Tamil Nadu, India

**Manikandan Ramachandran**
School of Computing
SASTRA Deemed University
Thanjavur, Tamil Nadu, India

**M. Santhiya**
Department of Computer Science and Engineering
Rajalakshmi Engineering College
Chennai, Tamil Nadu, India

**V. Saravanan**
Department of Computer Science
College of Engineering and Technology
Dambi Dollo University
Dambi Dollo, Oromia Region, Ethiopia

**S. Savitha**
Department of CSE
K.S.R. College of Engineering
Tiruchengode, Tamil Nadu, India

**Juergen Seitz**
Duale Hochschule Baden-Württemberg
Wirtschaftsinformatik, Heidenheim, Germany

**S. Sendilvelan**
Dr. M.G.R. Educational and Research Institute
Chennai, Tamil Nadu, India

**Neha Sharma**
Tata Consultancy Services
Pune, Maharashtra, India

**Arun K. Singh**
Department of Computer Science & Engineering
Greater Noida Institute of Technology
Greater Noida, Uttar Pradesh, India

**Mohan Singh**
Department of ECE
G.L. Bajaj Institute of Technology and Management
Greater Noida, Uttar Pradesh, India

**Jeevanandham Sivaraj**
Department of Information Technology
Sri Ramakrishna Engineering College
Coimbatore, Tamil Nadu, India

**P. Suresh**
Department of Database Systems
School of Computer Science and Engineering
Vellore Institute of Technology
Vellore, Tamil Nadu, India

**Shrikant Tiwari**
School of Computing Science and Engineering
Galgotias University
Greater Noida, Uttar Pradhesh, India

**Varun D. Tripathy**
Computer Science Department
Vivekanand Education Society's Institute of Technology
Mumbai, Maharashtra, India

**Amit Kumar Tyagi**
Department of Fashion Technology
National Institute of Fashion Technology
New Delhi, India

**Kapil D. Tyagi**
Department of ECE
Jaypee Institute of Information Technology
Noida, Uttar Pradesh, India

**Vaibhav B. Tyagi**
Department of ECE
ISBAT University
Kampala, Uganda

**Harish Venu**
Institute of Sustainable Energy (ISE)
Universiti Tenaga Nasional
Putrajaya Campus, Malaysia

**Virendra K. Verma**
Department of Industrial & Production Engineering
Institute of Engineering and Rural Technology (IERT)
Allahabad, Uttar Pradesh, India

**Ramesh S. Wadawadagi**
Department of Information Science and Engineering
Nagarjuna College of Engineering
Bengaluru, Karnataka, India

**Vivek Yadav**
Expresslending Pty Ltd
Melbourne, Victoria, Australia

# Preface

Industries, as we all know, are the ones that produce goods and services for society. Workers in the textile industry design, fabricate, and sell cloth. The tourist industry includes all the commercial aspects of tourism. The automobile industry makes cars and car parts. The food service industry prepares food and delivers it to hotels, schools, and other big facilities. "Industry" comes from the Latin "*industria*," which means "diligence, hard work," and the word is still used with that meaning. Generally, the industry has been through various evolutions during the last three decades. The industry started in the eighteenth century; that is, in 1784, the first power loom was developed. Hence, Industry 1.0 was all about mechanization with water and steam. In the next phase of the industry revolution, that is, Industry 2.0, the electrification of the industry took place. It started from 1900 to 1950. During this revolution, the "Assembly Line was developed." Further, Industry 3.0 is about the automation of data. During this revolution, the adoption of computers and automation, enhanced by smart and autonomous systems, is fueled by data and machine learning. All the data that was available manually began to be stored disks. Industry 3.0 was a major revolution in terms of automating things, and even operational technologies came into existence, but there was still a felt need to merge information technology with operational technology to truly digitize the world. This would be called "the Digital Transformation" in the true sense, and the resolution happening in Industry 4.0 is to move toward that direction.

Later, Industry 4.0 is to improve manufacturing efficiency; it is about transforming the way your entire business operates and grows. It is associated with cyber-physical system, in which digital technologies can create virtual versions of real-world installations, processes, and applications. This can then be robustly tested to make cost-effective, decentralized decisions. These virtual copies can then be created in the real world and linked via the Internet of Things (IoT), allowing cyber-physical systems to communicate and cooperate with each other and human staff to create a joined-up real-time data exchange and automation process for Industry 4.0 manufacturing. This should allow for digital transformation and automated and autonomous manufacturing with joined-up systems that can cooperate with each other. This technology will help solve problems and track processes while increasing productivity. It also primarily focuses on the use of large-scale machine-to-machine communication and IoT deployments to provide increased automation, improved communication, and self-monitoring, as well as smart machines that can analyze and diagnose issues without the need for human intervention. The idea of connected manufacturing or smart factories is becoming increasingly ubiquitous. Factories and their machines across the globe are getting smarter as connected products and systems operate as part of a larger, more responsive, and agile information infrastructure. The aim is to harvest benefits and improvements in efficiency and profitability, increased innovation, and better management of safety, performance, and environmental impact. This book will provide a complete experience of industrial revolution and its progress toward emerging technology.

*Mahmoud Ragab AL-Refaey*
*Amit Kumar Tyagi*
*Abdullah Saad AL-Malaise AL-Ghamdi*
*Swetta Kukreja*

# Acknowledgment

First of all, we would like to extend our gratitude to our family members, friends, and supervisors, who stood with us as advisors in completing this book. Also, we would like to thank our almighty God, who inspires us to write this book. We also thank Wiley Publishers, who have provided their continuous support all the time, and our colleagues, authors with whom we have worked together inside the college/university and others outside of the college/university, who have provided their continuous support towards completing this book on *Topics in Artificial Intelligence Applied to Industry 4.0*.

Further, the authors also gratefully acknowledge the support provided by the Faculty of Computing and Information Technology (FCIT) and King Abdulaziz University (KAU), Jeddah, Saudi Arabia, to produce this book. Furthermore, we thank the School of Engineering, Computing and Design, Dar Al-Hekma University, Jeddah, Saudi Arabia, for their support in performing this book.

We also acknowledge the support provided by the Department of Fashion Technology, National Institute of Fashion Technology, New Delhi, and the Department of Computer Science and Engineering, Amity University Mumbai, India.

Lastly, we would like to thank our respected madam Prof. G. Aghila, Prof. Siva Sathya, Manisha Kinnu (IRS), our respected sir Prof. N Sreenath, and Prof. Aswani Kumar Cherukuri for giving their valuable inputs and helping us in completing this book with Wiley Publisher.

Once again, thanks to all.

*Mahmoud Ragab AL-Refaey*
*Amit Kumar Tyagi*
*Abdullah Saad AL-Malaise AL-Ghamdi*
*Swetta Kukreja*

# 1

# Introduction to Industry 4.0 and Its Impacts on Society

*Shrikant Tiwari[1], Kanchan Naithani[1], Arun K. Singh[2], Virendra K. Verma[3], Ramesh S. Wadawadagi[4], and Bireshwar D. Mazumdar[5]*

[1] School of Computing Science and Engineering, Galgotias University, Greater Noida, Uttar Pradhesh, India
[2] Department of Computer Science & Engineering, Greater Noida Institute of Technology, Greater Noida, Uttar Pradesh, India
[3] Department of Industrial & Production Engineering, Institute of Engineering and Rural Technology (IERT), Allahabad, Uttar Pradesh, India
[4] Department of Information Science and Engineering, Nagarjuna College of Engineering, Bengaluru, Karnataka, India
[5] Department of Computer Science and Engineering, Faculty of Engineering and Technology, United University Prayagraj, Allahabad, Uttar Pradesh, India

## 1.1 Introduction

The ongoing Fourth Industrial Revolution (4IR) is characterized by the continuous transformation of society and the economy through technological advancements [1, 2]. This revolution encompasses breakthroughs in artificial intelligence (AI), robotics, the Internet of Things (IoT), and other digital technologies. What sets it apart from previous revolutions is not only the creation of new machines or processes but also the integration of these technologies into existing systems and the development of previously unimaginable systems [3].

The potential impacts of the 4IR on society and the economy are extensive and profound [4]. While these new technologies have the potential to improve productivity, efficiency, and quality of life for many, they also raise important questions regarding the equitable distribution of benefits and the potential exclusion of certain groups.

This chapter serves as an introduction to the 4IR and its societal impact. It explores the technological advancements driving this revolution, examines its potential effects on the economy and society, and addresses the ethical and governance considerations that arise in this era of technological progress.

The objective of this exploration is to deepen our understanding of the 4IR and its implications for society and the economy. Additionally, it explores how individuals, businesses, and governments can collaborate to shape this revolution in a way that maximizes its potential benefits while mitigating any negative consequences.

The 4IR refers to the current phase of technological advancements, encompassing AI, robotics, the IoT, and other digital technologies [5]. Coined by Klaus Schwab in his 2016 book *The Fourth Industrial Revolution*, it builds upon the transformative changes initiated by previous industrial revolutions [6].

The First Industrial Revolution introduced mechanization and steam power, while the Second Industrial Revolution brought electricity and mass production [7]. The Third Industrial Revolution, known as "the digital revolution," introduced computers and digital technology. However, the 4IR is distinctive because it integrates and converges technologies across all aspects of life [8]. It blurs the boundaries between physical, digital, and biological systems, enabling unprecedented levels of automation, connectivity, and data analysis. This integration has the potential to revolutionize industries, boost productivity, and create new avenues for economic growth.

The anticipated impact of the 4IR on society and the economy is profound. It is crucial for us to understand its implications and foster collaboration to ensure the equitable distribution of its benefits.

*Topics in Artificial Intelligence Applied to Industry 4.0*, First Edition. Edited by Mahmoud Ragab AL-Refaey, Amit Kumar Tyagi, Abdullah Saad AL-Malaise AL-Ghamdi, and Swetta Kukreja.

### 1.1.1 Overview of the Major Technological Advancements Driving the Revolution

The ongoing 4IR is characterized by remarkable technological advancements that are profoundly reshaping our lifestyles and work dynamics [9]. These advancements include:

- Artificial Intelligence: AI encompasses machines capable of performing tasks that traditionally require human intelligence, such as speech recognition, decision-making, and experiential learning. It continues to evolve and finds applications in various fields, including autonomous vehicles, personalized medicine, and intelligent virtual assistants.
- Robotics: Robotics involves the use of robots and automated systems to perform tasks that typically require human intervention. Advances in robotics enable increased levels of automation in industries such as manufacturing, logistics, and more.
- Internet of Things: The IoT is a network that connects physical objects embedded with sensors, software, and other technologies, enabling data exchange and collection. It promotes connectivity and data analysis, leading to valuable insights and efficiencies in sectors such as health care, agriculture, and transportation.
- Big Data Analytics: Big data refers to vast amounts of data generated by the IoT, social media, and other sources. Big data analytics involves employing advanced analytical techniques to extract insights and value from this data. It empowers organizations to make informed decisions and optimize their operations.
- 3D Printing: 3D printing involves the layer-by-layer creation of physical objects. Advances in 3D printing technology allow to produce intricate and precise objects, unlocking new possibilities in health care, aerospace, manufacturing, and other industries.

These advancements in the 4IR are revolutionizing various sectors and presenting exciting opportunities for innovation and growth (refer to Figure 1.1). They have the potential to reshape our society and economy in profound ways, driving us toward a more interconnected and technologically advanced future.

### 1.1.2 Importance of Studying and Understanding the Impacts of the Fourth Industrial Revolution

Studying and understanding the impacts of the 4IR is critical for several reasons [6, 10]:

- Economic Growth: The 4IR has the potential to drive substantial economic growth by fostering innovation, creating new employment opportunities, and enhancing productivity. A deep understanding of the opportunities presented by these technological advancements enables businesses and governments to capitalize on them, leading to sustainable economic growth.

**Figure 1.1** Major technological advancements driving the revolution.

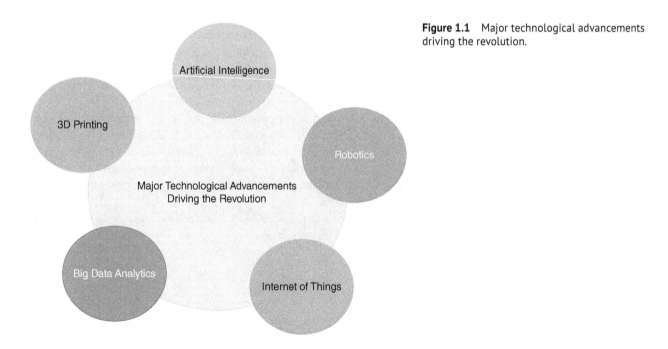

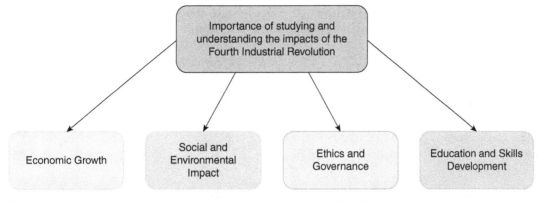

**Figure 1.2** Importance of studying and understanding the impacts of the Fourth Industrial Revolution.

- Social and Environmental Impact: The 4IR carries the potential for significant social and environmental consequences. It may exacerbate income inequality, contribute to job displacement through automation, and have adverse environmental effects. By comprehending the potential negative impacts of these technologies, we can develop policies and strategies to address and mitigate these challenges effectively.
- Ethics and Governance: The 4IR raises crucial ethical and governance considerations. Privacy, security, and accountability become paramount concerns in a technologically advanced era. A thorough understanding of these considerations empowers us to establish ethical frameworks and governance structures that ensure responsible development and use of these technologies.
- Education and Skills Development: The 4IR transforms the skill sets and educational requirements needed for the workforce of the future. Understanding the evolving demands and identifying the necessary skills equip us to prepare individuals and communities for the changing nature of work, fostering adaptability and lifelong learning.

In summary, studying and comprehending the impacts of the 4IR are vital to optimize the benefits of these technological advancements while minimizing their negative consequences. By doing so, we can create a more equitable, sustainable, and prosperous future for all (refer to Figure 1.2).

## 1.2 The Technological Advancements of the Fourth Industrial Revolution

The 4IR represents a wave of transformative technological advancements that are fundamentally reshaping our lives and work dynamics [7, 11]. These advancements encompass a range of key technologies, which include the following:

- Artificial Intelligence: AI enables machines to perform tasks that traditionally require human intelligence, such as speech recognition, decision-making, and experiential learning. Ongoing advancements in AI have led to its application in various domains, including autonomous vehicles, personalized medicine, and intelligent virtual assistants.
- Robotics: Robotics involves the use of robots and automated systems to perform tasks that typically require human intervention. Advances in robotics have facilitated increased levels of automation in industries such as manufacturing, logistics, and beyond.
- Internet of Things: The IoT consists of a network of physical objects embedded with sensors, software, and other technologies for data collection and exchange. It enables enhanced connectivity and data analysis, leading to valuable insights and efficiencies in sectors like health care, agriculture, and transportation.
- Big Data Analytics: Big data refers to vast amounts of data generated by the IoT, social media, and other sources. Big data analytics involves employing advanced techniques to extract insights and value from this data. Organizations leverage these insights to make informed decisions and optimize their operations.
- 3D Printing: 3D printing is an additive manufacturing process that constructs physical objects layer by layer. Advancements in 3D printing technology enable the production of intricate objects with exceptional precision, unlocking new possibilities in health care, aerospace, manufacturing, and more.

- Blockchain: Blockchain is a distributed ledger technology that enables secure and transparent transactions without intermediaries. It has the potential to transform various industries, including finance, supply chain management, and real estate.
- Augmented Reality (AR) and Virtual Reality (VR): AR and VR technologies provide immersive experiences that blend the physical and digital realms. These technologies find applications in fields such as education, entertainment, and retail.

These technological advancements are reshaping industries, creating new opportunities for growth. However, they also give rise to critical ethical and governance considerations. Understanding these advancements and their potential impact is crucial for individuals, businesses, and governments as we navigate the 4IR and strive to maximize its benefits while addressing its challenges.

### 1.2.1 Discussion of the Potential Benefits and Drawbacks of Each Technology

The key technologies driving the 4IR present both potential benefits and drawbacks [12]:

- Artificial Intelligence
  Benefits: AI has the potential to enhance decision-making, improve efficiency, and enable better understanding of complex data. It offers applications in various fields, including health care and finance.
  Drawbacks: Ethical concerns regarding bias and discrimination in AI decision-making processes exist. There are also concerns about job displacement as AI and automation become more prevalent.
- Robotics
  Benefits: Robotics can increase efficiency, reduce costs, and improve safety in industries like manufacturing and logistics. It enables the completion of tasks that are dangerous or difficult for humans.
  Drawbacks: Job displacement is a concern as robotics and automation advance. Safety and ethical considerations in the development and deployment of robots require attention.
- 3D Printing
  Benefits: 3D printing enables faster, cheaper, and more customizable production of products. It has the potential to reduce waste and improve sustainability.
  Drawbacks: 3D printing is still relatively costly compared to traditional manufacturing methods. The quality of printed products may not always meet the standards of traditional manufacturing.
- Internet of Things
  Advantages: The IoT offers real-time insights and improved efficiency across industries. It enhances resource monitoring and management, promoting energy and water conservation.
  Disadvantages: Security and privacy concerns arise as more devices become interconnected, necessitating robust data protection measures. Increased energy usage associated with the IoT raises environmental concerns.
- Blockchain
  Advantages: Blockchain enhances transparency, security, and efficiency in industries like finance, supply chain management, and real estate. It enables peer-to-peer transactions, reducing reliance on intermediaries.
  Disadvantages: Energy consumption, especially with cryptocurrencies, poses environmental challenges. Scalability and interoperability of blockchain systems require attention.

Understanding the potential benefits and drawbacks of these technologies is crucial as we navigate the 4IR (refer to Figure 1.3). It allows for informed decision-making regarding their development and deployment, enabling us to harness their advantages while addressing the associated challenges.

### 1.2.2 Examples of How These Technologies Are Already Being Used in Various Industries

The key technologies of the 4IR are already being applied in various industries, leading to transformative outcomes [12]:

- Artificial Intelligence
  Health Care: AI assists in disease diagnosis, treatment development, and improving patient outcomes. It analyzes medical images, predicts high-risk individuals, and enables proactive interventions.
  Finance: AI enhances fraud detection, risk management, and customer service in the financial industry. AI-powered chatbots offer personalized financial advice and efficient customer support.

**Figure 1.3** Potential benefits.

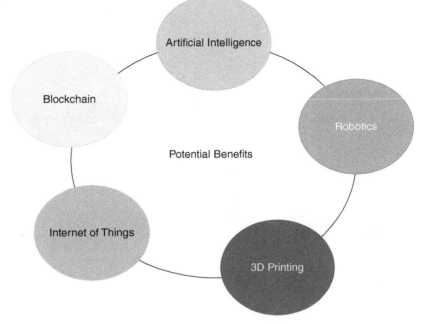

- Robotics
  Manufacturing: Robotics automates manufacturing processes, increasing efficiency and reducing costs. Robots assemble products, handle materials, and perform quality control checks.
  Health Care: Robotics assists in surgeries and medical procedures, enabling precision and reducing risks.
- 3D Printing
  Manufacturing: 3D printing creates prototypes, custom parts, and complete products. Automotive companies use it to produce lightweight, high-performance vehicle components.
  Health Care: 3D printing creates customized prosthetics, implants, and surgical instruments, improving patient outcomes and cost-effectiveness.
- Internet of Things
  Agriculture: IoT sensors monitor soil moisture, temperature, and environmental factors, optimizing crop yields while conserving water resources.
  Transportation: IoT sensors enable vehicle performance monitoring, inventory tracking, and route optimization in logistics, enhancing efficiency and reducing costs.
- Blockchain
  Finance: Blockchain ensures secure and transparent transactions, facilitating peer-to-peer payments and cross-border transfers, enhancing efficiency and transparency in international money transfers.
  Supply Chain Management: Blockchain improves transparency and efficiency by enabling reliable tracking and verification of goods throughout the supply chain.

These examples illustrate how these technologies are already being utilized in various industries. Their potential applications are extensive and diverse, promising further advancements and transformations.

## 1.3 Impacts on the Economy

The 4IR has the potential to bring significant impacts to the global economy, as indicated by research [13, 14]. Consider the following key aspects:

- Enhanced Productivity: The integration of automation, robotics, and AI in industries like manufacturing, transportation, and logistics can boost productivity and efficiency. This can lead to reduced production costs and improved profitability for businesses.

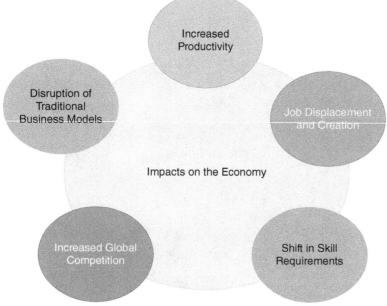

**Figure 1.4** Impacts on the economy.

- Job Transformation: While certain industries may experience job displacement due to technological advancements, new job opportunities can arise in other sectors. For instance, the implementation of AI in customer service may reduce the need for human operators while creating new roles in AI development and maintenance.
- Shift in Skills: The increasing prevalence of automation and AI across industries may require workers to acquire new skills. Proficiency in programming, data analysis, and other tech-related skills can become crucial for individuals to remain competitive in the evolving job market.
- Heightened Global Competition: The 4IR can foster increased global competition as companies leverage technology to enhance their products and services. This drive for innovation and efficiency can bring benefits, but it may also intensify competition for jobs and market share.
- Disruption of Business Models: New technologies have the potential to disrupt traditional business models. For example, the rise of e-commerce has disrupted brick-and-mortar retail, leading to store closures, while simultaneously creating opportunities in online retailing.

In summary, the 4IR offers significant economic benefits (refer to Figure 1.4), but it also presents challenges and disruptions. Governments, businesses, and individuals should comprehend these potential impacts to effectively navigate this transformative era. Collaboration is vital to maximize the benefits while mitigating any negative effects.

### 1.3.1   The Potential Impacts of the Fourth Industrial Revolution on the Economy

The 4IR has the potential to significantly impact the economy, including changes in employment, productivity, and industry structure [15, 16]. Consider the following key points:

- Employment Changes: The 4IR is expected to result in the displacement of jobs that are repetitive or routine, particularly in sectors like manufacturing and transportation. However, it can also create new opportunities in emerging fields such as data analysis, AI, and robotics. Acquiring specialized skills will be crucial to capitalize on these new roles, potentially leading to a skills gap and limited opportunities for workers without these skills.
- Increased Productivity: Automation, AI, and robotics are anticipated to enhance productivity in various industries, particularly manufacturing, logistics, and transportation. This can lead to lower production costs and increased profitability for businesses.
- Industry Structure: The 4IR has the potential to disrupt traditional industry structures and give rise to new business models. For example, e-commerce has disrupted brick-and-mortar retail, while ride-sharing platforms have transformed transportation. These disruptions may reshape industry structures and introduce new players across multiple sectors.

**Figure 1.5** Potential impacts of the Fourth Industrial Revolution on the economy.

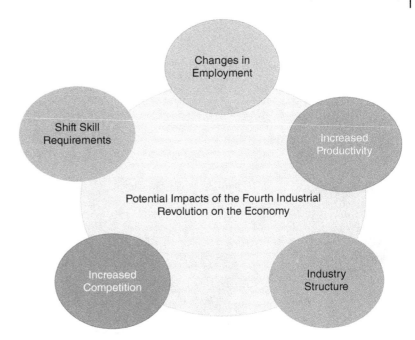

- Heightened Competition: The 4IR is projected to intensify global competition as companies leverage technology to improve their products and services. This can drive innovation and efficiency but may present challenges for smaller businesses and potentially lead to industry consolidation.
- Shift in Skill Requirements: The 4IR will require workers to possess new and specialized skills, particularly in areas such as data analysis, AI, and robotics. This shift in skill requirements may create a skills gap and limit employment opportunities for individuals lacking these proficiencies. However, it also presents an opportunity for individuals to acquire new skills and remain competitive in the evolving job market.

In summary, the 4IR has the potential to bring significant transformations to the economy (refer to Figure 1.5). It is crucial for businesses, governments, and individuals to understand and adapt to these potential impacts by investing in education and training, as well as implementing policies and regulations that ensure a fair and equitable distribution of the benefits generated by the 4IR across society.

### 1.3.2 Analysis of Fourth Industrial Revolution Is Changing the Nature of Work and Engages in Economic Activities

The 4IR is bringing about significant transformations in work and economic activities, resulting in various implications [17, 18]. Consider the following examples:

- Automation: The integration of automation, robotics, and AI is automating routine and repetitive tasks, particularly in manufacturing and logistics. While this may lead to the displacement of low-skill or repetitive jobs, it also creates new employment opportunities in fields such as data analysis and AI development.
- Gig Economy: The 4IR has fueled the growth of the gig economy, allowing individuals to engage in freelance or contract work through digital platforms. While it offers flexibility, it also presents challenges such as job insecurity and limited benefits.
- Remote Work: Technological advancements have made remote work more feasible, with the COVID-19 pandemic further accelerating its adoption. Remote work offers flexibility and reduces commuting time, but it also brings challenges like social isolation and managing work-life balance.
- Skill Requirements: The 4IR demands new and specialized skills, particularly in areas such as data analysis, AI, and robotics. This may result in a skills gap, limiting employment opportunities for individuals without these specific skills.
- New Business Models: The 4IR has enabled the emergence of innovative business models like the sharing economy and subscription-based services. While these models create new opportunities for workers and consumers, they also disrupt traditional industries and pose challenges for businesses that fail to adapt.

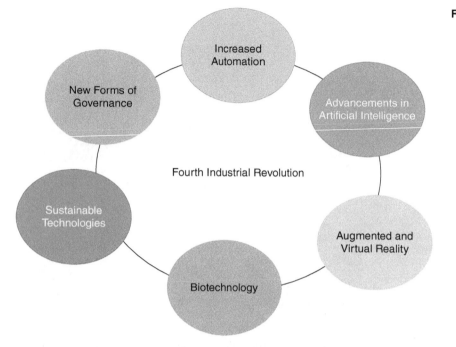

**Figure 1.6** Fourth Industrial Revolution.

In summary, the 4IR is reshaping work and economic activities in various ways (refer to Figure 1.6). While it offers flexibility and efficiency, it also presents challenges such as job displacement, evolving skill requirements, and the need to adapt to new business models. It is crucial for businesses, governments, and individuals to understand and embrace these changes to ensure that the benefits of the 4IR are equitably shared across society.

### 1.3.3 Discussion of How Businesses and Governments Can Prepare for and Adapt to These Changes

To effectively prepare for and adapt to the transformative effects of the 4IR [19, 20], businesses and governments can undertake several proactive measures. Consider the following examples:

- Invest in Training and Education: Businesses can allocate resources to training programs that enhance the skills of their workforce in areas such as data analysis, AI, and robotics. Simultaneously, governments can invest in education and training initiatives to equip individuals with the necessary skills demanded by the evolving job market.
- Foster Innovation: Businesses can foster innovation by dedicating resources to research and development, collaborating with startups and other innovative entities, and exploring new business models. Governments can support innovation by implementing policies that encourage it, such as providing tax incentives and research grants.
- Promote Entrepreneurship: Governments can establish policies that promote entrepreneurship and offer support to small businesses, including access to funding and streamlined regulatory processes. This can stimulate the emergence of new enterprises and industries in response to the evolving economic landscape.
- Embrace Digital Transformation: Businesses can embrace digital transformation by adopting new technologies and business models, such as those found in the sharing economy and subscription-based services. Governments can facilitate digital transformation by investing in digital infrastructure and creating regulatory frameworks that foster innovation and digitalization.
- Address Social and Economic Inequalities: The 4IR has the potential to exacerbate social and economic disparities. To tackle this, businesses and governments can invest in programs that provide training and education opportunities to underserved communities, support initiatives that promote diversity and inclusion, and implement policies to assist workers displaced by automation.

In summary, collaboration between businesses and governments is crucial to effectively prepare for and adapt to the changes brought about by the 4IR. By investing in education and training, fostering innovation, promoting entrepreneurship, embracing digital transformation, and addressing social and economic inequalities, they can ensure that the benefits of the 4IR are distributed equitably throughout society.

## 1.4 Impacts on Society

The 4IR is anticipated to have profound impacts on society, as suggested by various studies [21, 22]. Here are some potential impacts to consider:

- Increased Connectivity: The 4IR has the capacity to enhance connectivity among individuals, communities, and nations. This can foster the exchange of information, ideas, and resources, promoting global cooperation and collaboration.
- Changes in Social Structures: The 4IR may bring about changes in social structures, influencing how people work, socialize, and interact with one another. For instance, remote work and the sharing economy could reshape the traditional employer-employee relationship, while social media and VR may alter the dynamics of socialization and relationship-building.
- Disruption of Existing Industries: The 4IR has the potential to disrupt established industries, leading to workforce displacement and shifts in the economy. Automation and robotics could replace human workers in certain sectors, while 3D printing might disrupt traditional manufacturing processes.
- New Economic Opportunities: The 4IR is also expected to create fresh economic opportunities, such as the emergence of new industries and jobs in fields like AI, robotics, and biotechnology. This could contribute to economic growth and the generation of employment opportunities.
- Ethical and Social Implications: The 4IR introduces ethical and social considerations that must be addressed. These include concerns related to privacy, security, and the impact of technology on human well-being. It is crucial to tackle these issues proactively to ensure the fair distribution of benefits across society.

In summary, the 4IR will have multifaceted effects on society (refer to Figure 1.7). While it offers significant potential benefits, it is essential to recognize and mitigate any potential adverse impacts. By addressing ethical concerns and striving for equitable distribution of benefits, we can navigate this revolution in a way that maximizes its positive outcomes for society.

### 1.4.1 Discussion of How the Fourth Industrial Revolution Is Impacting Society

The 4IR is ushering in profound societal changes, impacting key areas such as health care, education, and lifestyle [23, 24]. Here are notable examples of its effects:

- Health Care: The 4IR is revolutionizing health care through the integration of new technologies. Telemedicine enables remote diagnosis and treatment, enhancing access to medical services. Wearable devices provide real-time health data, enabling proactive monitoring and personalized care. Additionally, personalized medicine leverages genetic data to tailor treatments to individuals, improving patient outcomes.

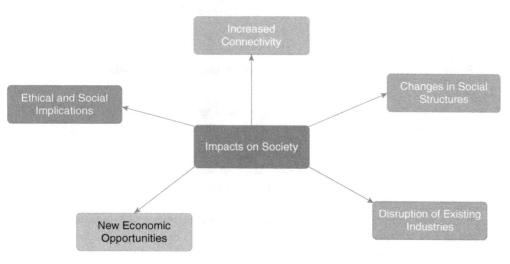

**Figure 1.7** Impacts on society.

- Education: The 4IR is transforming education by introducing innovative teaching methods and technologies. VR and AR create immersive learning experiences, enhancing student engagement and understanding. Online courses and e-learning platforms offer flexible and accessible education opportunities, reaching learners globally.
- Lifestyle: The 4IR is redefining lifestyles, influencing how people work, shop, socialize, and entertain themselves. The sharing economy has transformed travel and transportation, offering convenient and sustainable alternatives. Social media and VR have reshaped social interactions, facilitating global connections and changing the way relationships are formed. These changes have a profound impact on how individuals engage with the world around them.

Overall, the 4IR is driving significant societal shifts, with advancements in technology and innovative solutions reshaping health care, education, lifestyle, and entertainment. While these changes bring numerous benefits, it is crucial to address potential negative impacts and ensure equitable access to the advantages for all members of society.

### 1.4.2 Analysis of How Individuals and Communities Can Best Adapt to These Changes

Preparing for the transformative changes brought about by the 4IR can be challenging, but there are strategies individuals and communities can employ to navigate this new era [25]. Here are some recommended steps:

- Lifelong Learning: With rapid technological advancements and evolving industries, individuals must commit to continuous learning. Engaging in lifelong learning initiatives, such as online courses, vocational training, and professional development programs, allows individuals to acquire new skills and stay relevant in the changing job market.
- Community Building: Building strong communities that support and empower one another is crucial in navigating the 4IR. Sharing knowledge, resources, and experiences within the community can help individuals explore entrepreneurial opportunities, discover new career paths, and collectively adapt to the changing economic landscape.
- Embracing Technology: Embracing technology and cultivating digital literacy are essential in this era. Being open to learning new tools, software, and emerging technologies can enhance individual productivity, expand career possibilities, and enable active participation in the digital economy.
- Resilience and Adaptability: The 4IR brings unprecedented changes and uncertainties. Developing resilience and adaptability is key to navigating this landscape. Being open to new opportunities, being willing to acquire new skills, and being adaptable to changing circumstances will enable individuals to thrive amidst the evolving economic and technological landscape.
- Advocacy: Individuals and communities have an important role to play in advocating for policies and regulations that ensure fairness and equity in the 4IR. This includes advocating for access to quality education and training, promoting worker protection, and championing the ethical and responsible use of emerging technologies.

In summary, preparing for the 4IR requires individuals and communities to adopt a proactive and adaptable mindset. By embracing lifelong learning, fostering community building, adopting technology, cultivating resilience and adaptability, and advocating for equitable policies, individuals and communities can navigate the challenges and seize the opportunities presented by this transformative era.

### 1.4.3 Examples of How the Fourth Industrial Revolution Is Being Used to Address Social Challenges and Promote Social Good

The 4IR is not only reshaping economies and industries but also playing a crucial role in addressing social challenges and promoting social good [3, 26]. Here are some examples of how the 4IR is being utilized to tackle these challenges:

- Health Care: AI is revolutionizing health care by improving disease diagnosis and treatment. AI-powered diagnostic tools enable early detection and accurate diagnoses, while robots assist in surgical procedures and rehabilitation. Wearable technologies and sensors facilitate real-time health monitoring, empowering individuals to take proactive measures for preventing and managing chronic diseases.
- Education: The 4IR is transforming education by introducing innovative learning methods and enhancing educational accessibility. Online learning platforms, VR and AR, and personalized learning tools make education more affordable, flexible, and engaging. These technologies also support teacher training, resulting in improved educational outcomes for students.

- Sustainable Development: The 4IR contributes to sustainable development by generating innovative solutions for energy, water, and waste management. IoT sensors and data analytics optimize energy and water systems, promoting efficiency and conservation. 3D printing utilizes recycled materials, reducing waste and enabling the creation of sustainable products. Blockchain technology ensures transparent and ethical supply chains, fostering sustainable finance models and responsible sourcing.
- Social Justice: The 4IR empowers social justice initiatives through tools for advocacy and civic engagement. Social media and digital platforms raise awareness of social issues, mobilizing communities, and facilitating dialogue. AI algorithms are used to identify and address biases and discrimination in domains like criminal justice and employment, promoting fairness and equity.

In summary, the 4IR presents opportunities to address social challenges and promote social good in various domains such as health care, education, sustainable development, and social justice. By harnessing the power of new technologies and innovative approaches, we can work toward creating a more equitable and sustainable future for all.

## 1.5 Ethics and Governance

The 4IR presents a wide range of ethical and governance challenges that must be addressed to harness the benefits of new technologies responsibly and sustainably [27–29]. It is crucial to consider the following key issues:

- Privacy and Data Protection: The vast amount of data generated during the 4IR raises concerns about privacy and data protection. Clear regulations regarding data ownership, usage, and protection are essential to safeguard individuals' privacy rights and prevent the misuse of personal information.
- Bias and Discrimination: Algorithms and AI systems driving 4IR technologies can inadvertently perpetuate bias and discrimination. Establishing guidelines for their development and deployment is crucial to mitigate bias and ensure fairness and inclusivity in decision-making processes.
- Cybersecurity: With increased reliance on digital systems and networks, cybersecurity becomes paramount. Businesses and governments must prioritize investments in robust cybersecurity infrastructure and provide training to protect systems and data from cyber threats.
- Job Displacement and Retraining: Automation within the 4IR may result in job displacement. It is vital to provide adequate support and training to affected workers, enabling them to transition into new roles or industries successfully and ensuring inclusive economic growth.
- Ethical Use of Technology: The 4IR introduces ethical dilemmas, such as the use of autonomous weapons and gene editing. Establishing comprehensive ethical guidelines and regulations is necessary to ensure responsible and ethical development and use of these technologies, prioritizing human well-being and societal benefit.

In conclusion, addressing the ethical and governance challenges of the 4IR requires proactive measures to protect privacy, minimize bias, enhance cybersecurity, support workers, and uphold ethical standards (refer to Figure 1.8). Collaboration among governments, businesses, and civil society organizations is essential in establishing clear rules and guidelines to navigate the transformative impact of new technologies in a responsible and sustainable manner. By doing so, we can ensure that the benefits of the 4IR are harnessed for the betterment of society.

### 1.5.1 Discussion of the Ethical and Governance Challenges Posed by the Fourth Industrial Revolution

The 4IR introduces a host of ethical and governance challenges that must be addressed to ensure the responsible and sustainable use of new technologies [28]. These challenges include the following:

- Privacy and Data Protection: The proliferation of data raises concerns about privacy and the safeguarding of personal information. It is crucial to establish clear regulations on data ownership, usage, and protection to protect individuals' privacy rights.
- Cybersecurity: As reliance on digital systems and networks grows, cybersecurity becomes increasingly important. Businesses and governments should prioritize investments in robust cybersecurity infrastructure and provide training to mitigate cyber threats and protect systems and data.

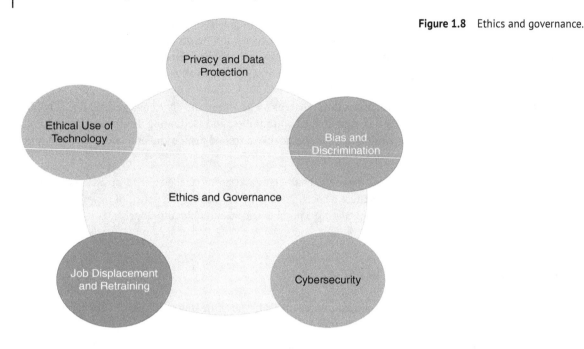

**Figure 1.8** Ethics and governance.

- Bias and Discrimination: Algorithms and AI systems can inadvertently perpetuate bias and discrimination. Guidelines for the development and deployment of these technologies are necessary to minimize bias and ensure fairness and inclusivity in decision-making processes.
- Job Displacement and Retraining: Automation in the 4IR may result in job displacement. Adequate support and training must be provided to help affected workers transition into new roles or industries, ensuring inclusive economic growth.
- Inequality: The 4IR has the potential to widen existing inequalities within and between countries. Access to technology and education can create disparities, emphasizing the importance of equitable distribution of benefits and ensuring broad participation in the digital economy.

Addressing these ethical and governance challenges requires collaboration among governments, businesses, and civil society organizations. By working together to establish clear rules and guidelines, we can ensure that the 4IR is harnessed responsibly and sustainably, benefiting society.

### 1.5.2 Analysis of the Role of Government, Businesses, and Individuals in Addressing These Challenges

Addressing the ethical and governance challenges posed by the 4IR requires a collective effort involving governments, businesses, and individuals [30]. Each stakeholder has a unique role to play in promoting responsible and sustainable use of emerging technologies.

- Government: Governments play a crucial role in establishing regulatory frameworks and guidelines that ensure the ethical utilization of 4IR technologies. They should allocate funding for research and development initiatives that benefit society as a whole and establish regulatory bodies to oversee compliance with ethical standards. For example, the European Union's General Data Protection Regulation sets guidelines for data protection and privacy.
- Businesses: Businesses have a responsibility to develop and deploy 4IR technologies in an ethical and sustainable manner. This entails adopting ethical guidelines during the development and implementation of new technologies, investing in robust cybersecurity infrastructure and training, promoting transparency and accountability within their systems and processes, and minimizing bias in algorithms and AI systems. Additionally, businesses should provide support and training to workers affected by automation, ensuring a just transition.
- Individuals: Individuals can contribute to the promotion of responsible and ethical use of 4IR technologies. They should stay informed about the risks and benefits associated with these technologies and advocate for their ethical and transparent application. Individuals can also safeguard their own privacy and data while supporting policies that protect privacy rights for everyone. Additionally, endorsing initiatives that provide training and support for workers displaced by automation is important.

To effectively address the ethical and governance challenges stemming from the 4IR, transparent collaboration among all stakeholders is vital. By working together, we can ensure that emerging technologies are developed and deployed in a manner that benefits society while upholding ethical standards.

### 1.5.3 Examples of Best Practices in Ethical and Responsible Technology Development and Deployment

Ethical and responsible technology development and deployment are exemplified by several noteworthy initiatives and principles [31]. Here are some examples:

- The IEEE Global Initiative for Ethical Considerations in AI and Autonomous Systems: This collaborative effort brings together stakeholders to create ethical guidelines for AI and autonomous systems. The guidelines prioritize transparency, accountability, and social responsibility.
- Microsoft's AI Principles: Microsoft has established a set of principles guiding the development and deployment of AI systems. These principles emphasize transparency, accountability, and fairness. The company also has an AI ethics review process to ensure responsible practices.
- The Partnership on AI: This organization unites technology companies, civil society organizations, and academic institutions to promote responsible AI development and deployment. They have developed ethical guidelines for AI systems and provide resources and support to organizations working toward ethical AI practices.
- The Montreal Declaration for Responsible AI: AI researchers collaborated to create this declaration, aiming to establish ethical principles for AI development and deployment. The declaration emphasizes respect for human autonomy, prevention of harm, and transparency.
- The OpenAI Charter: OpenAI, a nonprofit research company, has developed this charter to guide their work in safe and beneficial AI development. The charter emphasizes principles such as transparency, accountability, and ethical use of data.

These examples highlight the importance of collaboration among stakeholders, including technology companies, researchers, civil society organizations, and governments, in driving ethical and responsible technology development. By jointly establishing ethical guidelines and principles, we can ensure that emerging technologies are developed and deployed in a way that benefits society.

## 1.6 Future Directions

The 4IR is still in its early stages, and its future trajectory holds a multitude of possibilities [18, 32]. Here are some potential directions it could take:

- Enhanced Automation: Ongoing advancements in automation technology may lead to a significant increase in the number of jobs that become automated, resulting in transformative changes within the labor market.
- Advancements in AI: While AI is already widely utilized, continuous improvements may result in more sophisticated and powerful AI systems capable of performing tasks that are currently unimaginable.
- Augmented and Virtual Reality: AR and VR technologies, although still in early stages, have the potential to revolutionize various aspects of life, including entertainment, gaming, education, and health care, by altering how we perceive and interact with our surroundings.
- Biotechnology: The 4IR encompasses more than just digital technologies; advancements in biotechnology could have a profound impact on society, ranging from personalized medicine to bioengineering.
- Sustainable Technologies: As the consequences of climate change become increasingly apparent, there is a growing demand for sustainable technologies that mitigate our carbon footprint and protect the environment.
- New Models of Governance: The 4IR might give rise to novel forms of governance and decision-making. Technologies like blockchain and decentralized systems offer the potential for more distributed and transparent decision-making processes.

These examples provide a glimpse into potential future directions for the 4IR. As technology continues to progress rapidly, making precise predictions about its evolution remains challenging. However, it is certain that the 4IR will have a significant impact on society and the world at large.

### 1.6.1 Potential Future Directions of the Fourth Industrial Revolution

The 4IR, still in its early stages, holds immense potential to bring about profound societal transformations through transformative technologies [33, 34]. Here are some possible future directions for the 4IR:

- Quantum Computing: Quantum computers, utilizing qubits instead of traditional bits, have the potential to revolutionize fields such as cryptography and drug discovery with their ability to perform faster calculations in certain domains.
- Edge Computing: Processing data closer to its source, known as "edge computing," can reduce latency and enhance data-processing efficiency. This approach is particularly relevant for applications like autonomous vehicles and smart cities.
- Brain-Computer Interfaces: Brain-computer interfaces (BCIs) enable direct communication between the brain and computers or devices. BCIs have profound implications for health care and accessibility, empowering individuals with disabilities to control prosthetic limbs or other devices through their thoughts.
- Nanotechnology: Manipulating materials at the nanoscale, nanotechnology opens up possibilities for novel materials and devices with unique properties. This technology finds applications in medicine, electronics, energy, and other fields.
- 5G Networks: 5G networks offer significantly faster data-transfer speeds and reduced latency compared to previous wireless network generations. This advancement unlocks possibilities such as remote surgery and autonomous vehicles.

While these technologies hold tremendous promise, it is crucial to consider their ethical, social, and economic implications as they continue to evolve. Responsible and inclusive development and deployment of these technologies will be essential to ensure equitable access to their benefits. For instance:

- Privacy and Security: As 4IR technologies generate and process increasing amounts of data, concerns about privacy and security grow. Establishing robust security protocols and safeguards is crucial to effectively protect user data.
- Employment Displacement: Ongoing advancements in automation technologies pose the risk of significant job displacement, resulting in substantial labor market changes. Mitigating this requires providing workers with opportunities for adaptation through access to training and education programs that equip them with the skills needed in the evolving job landscape.
- Inequality: The 4IR has the potential to exacerbate existing inequalities if certain individuals and communities are left behind by technological advancements. Ensuring equitable access to these technologies and developing them in ways that benefit all of society are imperative to reduce disparities.
- Ethical Concerns: As AI and other advanced technologies progress, ethical concerns arise regarding their societal impact. Prioritizing the ethical and responsible development and deployment of these technologies, while considering potential consequences and implications, is vital.

In conclusion, the future trajectory of the 4IR presents both opportunities and challenges. Addressing these challenges proactively is crucial to harnessing and utilizing these technologies in ways that benefit society. By fostering responsible and inclusive development, we can navigate the transformative impact of the 4IR while ensuring its benefits are shared by all.

### 1.6.2 Analysis of the Fourth Industrial Revolution

The 4IR is still in its early stages, and the rapid evolution of technology makes it challenging to predict its precise trajectory and long-term consequences [35, 36]. Nevertheless, several potential directions emerge for the future of this revolution.

One direction involves the further integration and growth of AI and machine learning across various industries and societal domains. This advancement could result in increased automation and productivity, although it also raises ethical concerns regarding AI's role in decision-making and the potential displacement of jobs.

Another potential direction is the advancement and widespread adoption of biotechnology, including gene editing and personalized medicine. These developments have the potential to bring about significant improvements in health care, but they also prompt ethical considerations regarding the use and equitable distribution of such technologies.

The expansion of the IoT represents another potential direction, leading to increased connectivity and data collection. This expansion could have both positive and negative effects on society, including privacy concerns and potential security risks associated with the vast amount of data being generated.

Furthermore, the future trajectory of the 4IR will be influenced by the actions and decisions of governments, businesses, and individuals in addressing the ethical and governance challenges that arise. The establishment of regulations and policies around data privacy and security, for instance, will likely play a crucial role in shaping the course of this revolution.

In conclusion, the 4IR will undoubtedly continue to have significant impacts on society. It is imperative to approach these developments with careful consideration and strategic planning to ensure that the impacts are positive, sustainable, and in line with ethical principles in the long term. As we navigate this revolution, a proactive approach is necessary to harness its potential while addressing the challenges it presents.

### 1.6.3 Discussion of the Fourth Industrial Revolution in a Positive and Equitable Way

Achieving a positive and equitable future for the 4IR necessitates the collaborative efforts of individuals, businesses, and governments [37, 38]. Each group can contribute in the following ways:

- Individuals can become informed and engaged citizens by staying updated on technological advancements and advocating for ethical and responsible technology development and deployment. Investing in education and training will ensure they are prepared for evolving work dynamics and can contribute meaningfully to the economy.
- Businesses play a vital role in prioritizing ethical and responsible technology practices. They should ensure that their products and services have a positive impact on society and the environment. Investing in employees through training and education will help them adapt to changing work requirements.
- Governments have the responsibility to create policies and regulations that promote the well-being of all members of society in the 4IR. This includes addressing concerns related to privacy, security, and inequality. Additionally, governments should invest in education and training programs to equip individuals with the skills needed for the evolving job market.
- Collaboration between these groups is crucial for ensuring a positive and equitable future. For instance, businesses can work alongside governments to develop policies and regulations that support ethical and responsible technology practices. Individuals can engage with both businesses and governments to voice their concerns and advocate for their priorities.

Ultimately, shaping the future of the 4IR in a positive and equitable manner requires a shared commitment to values such as sustainability, equity, and transparency. Through concerted efforts, individuals, businesses, and governments can ensure that the revolution benefits all members of society, fostering an inclusive and prosperous future. Together, we can navigate the challenges and opportunities of the 4IR and create a future that is both technologically advanced and socially responsible.

## 1.7 Conclusion

In summary, the 4IR represents a transformative era characterized by rapid technological advancements. It holds immense potential to revolutionize various aspects of our lives, but it also poses notable challenges related to ethics, governance, and social impact.

Throughout this chapter, we have delved into the key technological advancements driving the 4IR, exploring their potential benefits and drawbacks, as well as their present and future applications across diverse industries. Additionally, we have examined the revolution's impacts on the economy, society, and the nature of work.

To ensure a positive and equitable future within the 4IR, it is crucial for individuals, businesses, and governments to collaborate in addressing its challenges. This entails giving precedence to ethical and responsible technology development and deployment, making investments in education and training programs to equip individuals for the evolving work landscape, and establishing policies and regulations that guarantee societal-wide benefits.

While significant challenges lie ahead, the 4IR also presents an opportunity to construct a better and more equitable future for all. By working collectively and upholding values such as sustainability, equity, and transparency, we can shape the course of the revolution in a manner that advances the well-being of all members of society.

In conclusion, the 4IR is a transformative force that requires proactive and inclusive approaches to harness its potential for the benefit of humanity. By embracing collaboration and prioritizing ethical considerations, we can navigate the challenges and seize the opportunities presented by this revolution, ultimately shaping a future that is technologically advanced, socially responsible, and equitable for all.

### 1.7.1 Reflection on the Importance of Studying the Fourth Industrial Revolution and Its Impacts on Society

Understanding the 4IR and its impact on society is crucial for individuals, businesses, and governments to navigate the future effectively. The rapid technological advancements associated with this revolution have the potential to bring about transformative changes in various industries and improve lives globally. However, these advancements also pose significant challenges, including ethical considerations, governance issues, and societal implications.

To effectively navigate the 4IR, it is important for individuals and organizations to acquire knowledge about the potential benefits and risks associated with the technologies driving this revolution. This knowledge will enable informed decision-making regarding the integration of these technologies into work and daily lives. Additionally, understanding the broader impacts of the revolution on the economy, society, and the environment is essential for collaboration and ensuring inclusivity, equity, and sustainability in the long term.

In essence, studying the 4IR and its societal impacts is vital for shaping a future that is not only technologically advanced but also ethically sound, socially inclusive, and environmentally responsible. By actively engaging with this revolution, we can work toward a positive and equitable future that benefits all members of society.

### 1.7.2 Fourth Industrial Revolution in a Way That Benefits All Members of Society

The 4IR presents immense potential for societal benefit, but it also brings notable challenges that demand our attention. Each of us, as individuals, businesses, and governments, plays a crucial role in shaping the future of this revolution to ensure its inclusive benefits.

As individuals, it is important to stay informed about the latest technological advancements and their potential impact on society. Supporting businesses and governments that prioritize ethical and responsible technology development is a meaningful way to contribute.

Businesses can prioritize sustainability and ethics in their technology strategies. By investing in employee training and development, we can equip our workforce with the skills needed to thrive in a rapidly evolving technological landscape.

Governments hold the power to establish policies and regulations that promote ethical and responsible technology practices. By investing in education and workforce development programs, we can ensure that everyone has access to the skills and knowledge required to participate in the 4IR.

In conclusion, the 4IR presents a range of opportunities and challenges. Through collaboration among individuals, businesses, and governments, we can ensure that this revolution benefits everyone and remains sustainable in the long run.

# References

**1** Penprase, B.E. (2018). The fourth industrial revolution and higher education. *Higher Education in The Era of the Fourth Industrial Revolution* 10 (1): 978–981.

**2** Philbeck, T. and Davis, N. (2018). The fourth industrial revolution. *Journal of International Affairs* 72 (1): 17–22.

**3** Li, G., Hou, Y., and Wu, A. (2017). Fourth industrial revolution: technological drivers, impacts and coping methods. *Chinese Geographical Science* 27: 626–637.

**4** Brand, J.E. (2015). The far-reaching impact of job loss and unemployment. *Annual Review of Sociology* 41: 359–375.

**5** Scepanovič, S. (2019). The fourth industrial revolution and education. In: *2019 8th Mediterranean Conference on Embedded Computing (MECO)*, 1–4. IEEE.

**6** Schwab, K. (2017). *The Fourth Industrial Revolution*. Currency.

**7** Xu, M., David, J.M., and Kim, S.H. (2018). The fourth industrial revolution: opportunities and challenges. *International Journal of Financial Research* 9 (2): 90–95.

**8** Park, S.C. (2018). The fourth industrial revolution and implications for innovative cluster policies. *AI & Society* 33: 433–445.

**9** Marr, B. (2020). *Tech Trends in Practice: The 25 Technologies that Are Driving the 4th Industrial Revolution*. Wiley.

**10** Lin, H.F. (2007). Knowledge sharing and firm innovation capability: an empirical study. *International Journal of Manpower* 28 (3/4): 315–332.

**11** Dash, D., Farooq, R., Panda, J.S., and Sandhyavani, K.V. (2019). Internet of things (IoT): the new paradigm of HRM and skill development in the fourth industrial revolution (Industry 4.0). IUP. *Journal of Information Technology* 15 (4): 7–30.

**12** Schäfer, M. (2018). The fourth industrial revolution: how the EU can lead it. *European View* 17 (1): 5–12.

**13** Bonciu, F. (2017). Evaluation of the impact of the 4th industrial revolution on the labor market. *Romanian Economic and Business Review* 12 (2): 7–16.

**14** Shaturaev, J. (2022). Economies and management as a result of the fourth industrial revolution: an education perspective. *Indonesian Journal of Educational Research and Technology* 3 (1): 51–58.

**15** Caruso, L. (2018). Digital innovation and the fourth industrial revolution: epochal social changes? *AI & Society* 33 (3): 379–392.

**16** Manda, M.I. and Ben Dhaou, S. (2019). Responding to the challenges and opportunities in the 4th Industrial revolution in developing countries. In: *Proceedings of the 12th international conference on theory and practice of electronic governance*, 244–253.

**17** Chalmers, D., MacKenzie, N.G., and Carter, S. (2021). Artificial intelligence and entrepreneurship: implications for venture creation in the fourth industrial revolution. *Entrepreneurship Theory and Practice* 45 (5): 1028–1053.

**18** Hirschi, A. (2018). The fourth industrial revolution: issues and implications for career research and practice. *The Career Development Quarterly* 66 (3): 192–204.

**19** Schiølin, K. (2020). Revolutionary dreams: future essentialism and the sociotechnical imaginary of the fourth industrial revolution in Denmark. *Social Studies of Science* 50 (4): 542–566.

**20** Skilton, M. and Hovsepian, F. (2018). *The 4th Industrial Revolution*. Springer Nature.

**21** Colombo, A.W., Karnouskos, S., Kaynak, O. et al. (2017). Industrial cyberphysical systems: a backbone of the fourth industrial revolution. *IEEE Industrial Electronics Magazine* 11 (1): 6–16.

**22** Cowie, P., Townsend, L., and Salemink, K. (2020). Smart rural futures: will rural areas be left behind in the 4th industrial revolution? *Journal of Rural Studies* 79: 169–176.

**23** Cockerham, W.C. (2005). Health lifestyle theory and the convergence of agency and structure. *Journal of Health and Social Behavior* 46 (1): 51–67.

**24** Irwin, T. (2015). Transition design: a proposal for a new area of design practice, study, and research. *Design and Culture* 7 (2): 229–246.

**25** Facer, K. (2011). *Learning Futures: Education, Technology and Social Change*. Routledge.

**26** Oztemel, E. and Gursev, S. (2020). Literature review of Industry 4.0 and related technologies. *Journal of Intelligent Manufacturing* 31: 127–182.

**27** Crane, A., Matten, D., Glozer, S., and Spence, L.J. (2019). *Business Ethics: Managing Corporate Citizenship and Sustainability in the Age of Globalization*. USA: Oxford University Press.

**28** Epstein, M.J. and Buhovac, A.R. (2014). *Making Sustainability Work: Best Practices in Managing and Measuring Corporate Social, Environmental, and Economic Impacts*. Berrett-koehler publishers.

**29** Kibert, C.J. (2010). The Ethics of Sustainability. In: *In Theory and In Fact, In reshaping the built Environment* (ed. C.J. Kibert). Washington, D.C.: Island Press.

**30** Morrar, R., Arman, H., and Mousa, S. (2017). The fourth industrial revolution (Industry 4.0): a social innovation perspective. *Technology Innovation Management Review* 7 (11): 12–20.

**31** Von Schomberg, R. (2012). Prospects for technology assessment in a framework of responsible research and innovation. In: *Technikfolgen abschätzen lehren* (ed. M. Dusseldorp and R. Beecroft). für Sozialwissenschaften: VS Verlag.

**32** Bloem, J., Van Doorn, M., Duivestein, S. et al. (2014). The fourth industrial revolution. *Things Tighten* 8 (1): 11–15.

**33** Elheddad, M., Benjasak, C., Deljavan, R. et al. (2021). The effect of the fourth industrial revolution on the environment: the relationship between electronic finance and pollution in OECD countries. *Technological Forecasting and Social Change* 163: 120485.

**34** Moll, I. (2021). The myth of the fourth industrial revolution. *Theoria* 68 (167): 1–38.

**35** Baller, S., Dutta, S., and Lanvin, B. (2016). *Global Information Technology Report 2016*. Geneva: Ouranos.

**36** Grübler, A. (2003). *Technology and Global Change*. Cambridge University Press.

**37** Schwab, K. and Davis, N. (2018). *Shaping the Future of the Fourth Industrial Revolution*. Currency.

**38** Xing, B. and Marwala, T. (2017). Implications of the fourth industrial age on higher education. *arXiv preprint* arXiv:1703.09643.

**2**

# Digital Transformation Using Industry 4.0 and Artificial Intelligence

*M. Keerthika[1], M. Pragadeesh[2], M. Santhiya[1], G. Belshia Jebamalar[3], and Harish Venu[4]*

[1] *Department of Computer Science and Engineering, Rajalakshmi Engineering College, Chennai, Tamil Nadu, India*
[2] *Department of Information Technology, Rajalakshmi Engineering College, Chennai, Tamil Nadu, India*
[3] *Department of Computer Science and Engineering, S.A Engineering College, Chennai, Tamil Nadu, India*
[4] *Institute of Sustainable Energy (ISE), Universiti Tenaga Nasional, Putrajaya Campus, Malaysia*

## 2.1 Introduction

Ever since the dawn of the industrial era in the seventeenth century, contemporary industry has endured several transformations. Commodity manufacture, such as tools, clothes, and armament, stayed manual till the introduction of industrial processes in the latter decades of the eighteenth century. The shift from the industrial revolution, or Industry 1.0, to the soon-to-be industrial years of age, referred to as "Industry 4.0," was quick. Technology's rapid growth, on the other hand, and economic conditions, on the contrary, have led to the establishment of Industry 4.0. The industrial revolution brought about the transition from an economy focused on farming and crafting to one centered on machinery and manufacturing for industry. The industrial revolution is the name given to the paradigm shift in technology for manufacturing [1]. As a result, we now have societies to do. People's employment and living arrangements have changed because of current industrial advances. This is how the history of the industrial revolution is explained: between 1760 and 1820, there was the first industrial era, sometimes known as "Industry 1.0." It heralded a shift away from manually labor-intensive industrial methods and toward machines driven by steam and water. Industry 2.0, also referred to as the subsequent revolution of industry or the technological revolution, spans the years 1870–1914. This revolution was mostly brought about by the development of electrically powered equipment. The inaugural assembly line was utilized to develop the streamlining of the mass production process, which subsequently spread like wildfire. The third revolution in industry, often referred to as "Industry 3.0," occurred in the latter part of the twentieth century. It is frequently defined as the "digital revolution" since communication and computing technologies are used so extensively in the industrial process. Germany is where the term "Industries 4.0" or "Industry 4.0," as the phrase was initially used, originated [2]. Industry 4.0 is a national strategic initiative that is being driven by the Ministries of Economic Affairs and Energy (BMWI) and Education and Research (BMBF) in Germany. It aims to enhance the manufacturing process by increasing digitization and the interconnection of products, chains of value, and business models. Industry 4.0 marks the beginning of the digital age. Traditional production will come to an end with the arrival of Industry 4.0. The merging of the barriers between the virtual and physical worlds resulted in cyber-physical systems [3]. A possibility exists to change how the company responds to societal expectations, thanks to Industry 4.0, which stands for connection. Unlike prior industrial revolutions, which were propelled by innovations in manufacturing processes and systems, Industry 4.0 advancements are driven by a smart, networked, pervasive environment. Figure 2.1 illustrates this. During the first industrial age, water and steam power were utilized when mass production began, whereas electricity was employed during the second. Electronics and computers were used more often in the industry during the Third Industrial Revolution; the fourth, dubbed Industry 4.0, will include automated procedures and technology for digital twins. Therefore, Industry 4.0 is being described as "the emerging field of the automation and electronic information transfer for manufacturing and similar technologies that includes Internet of Things (IoT), cyber-physical networks, cloud-based computing, integrating systems, and big-data

*Topics in Artificial Intelligence Applied to Industry 4.0*, First Edition. Edited by Mahmoud Ragab AL-Refaey, Amit Kumar Tyagi, Abdullah Saad AL-Malaise AL-Ghamdi, and Swetta Kukreja.

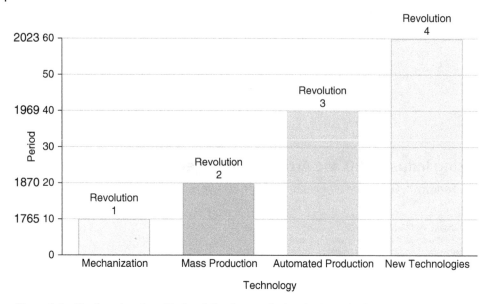

**Figure 2.1** The four decades of industrialization are depicted schematically.

analytics that serve in establishing the smart [7, 8] sectors and factories." Industry 4.0 refers to an interconnected system of intelligent machines, tools, and systems for many different sectors that leverage technology and communication. The advancement of AI and the use of deep learning (DL) and machine learning (ML)-focused methodologies are enabling Industry 4.0. Applications of AI have made progress in tackling the difficulty of automatically detecting patterns in data. AI-powered solutions can assist experts in the field as they conduct evaluations in the context of their duties at work when it comes to complex information and strategies. Focused implementation of data-driven decision-making in the industrial sectors may lead to the successful integration of analytical viewpoints on business operations. A complex management strategy that promotes trust in the broader AI-based systems' activities, inference procedures, and outputs is required for such integration. But for artificial intelligence (AI) centered systems to be effectively implemented and accepted by professionals, choices and consequences must be understandable or, in other words, "explainable."

In this context, the purpose of this research is to give a comprehensive evaluation of AI, XAI-based techniques, and their applications in the field of Industry 4.0. We began by discussing Industry 4.0 technology before rapidly describing many AI- and XAI-based methodologies. Following that, we will go through many Industries 4.0 enablers as well as the reason why and where AI is being employed. In addition, we classified Industry 4.0 applications into categories. Finally, we discussed the difficulties and trends that will impact AI's digital transition in Industry 4.0. The following is a list of survey contributions. The remainder of this article is structured as follows. Section 2.1 will give an overview of the many technologies that will allow Industry 4.0. Sections 2.3 and 2.4 will take us through various AI-based methods or methodologies used in the context of Industry 4.0. Section 2.5 discusses the many applications of the suggested methods in Industry 4.0. Section 2.6 elaborates on Industry 4.0 case studies, and Section 2.7 shows existing barriers to employing AI-based technologies in industrial applications, as well as recommendations for additional research. This chapter concludes with a consideration of potential developments.

## 2.2 Industry 4.0 Technologies

"Industry 4.0" is a prevalent idea in the manufacturing and industrial sectors. The more complex goals concentrate on developing, integrating, and evolving multiple intelligent machines or models, with data and teamwork being essential. Automated operations, improved processes, and increased productivity are frequently the initial targets. Achieving a high level of automation while also achieving exceptional operational performance and productivity is the main objective of Industry 4.0. Providing autonomous, intelligent, actual-time, and interoperable factories is another goal of Industry 4.0 [9]. The big data [1], system integration [2], cyber-physical structure [3], augmented and virtual reality [4], cloud-based computing [5], IoT [6], and the use of additive technology are just a few of the advanced and cutting-edge data and information extraction technologies that are combined to achieve this goal, as shown in Figure 2.2. Table 2.1 outlines the linked technologies utilized across a range of use cases in the industry and is organized in the following manner: the chapter's contribution is outlined in the fourth column: after the first column includes the referred-to article and emphasizes the relevant technology, the second column indicates the specific industry or implementation, the next column provides a brief overview of the technology that was used, and the last column determines the work's sources.

**Figure 2.2** Support of key technologies in Industry 4.0.

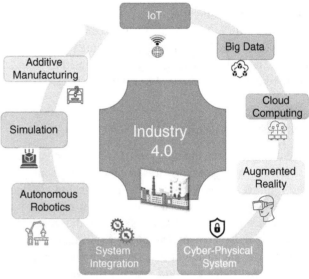

**Table 2.1** An overview of recent cutting-edge technologies, which enable the Fourth Industrial Revolution.

| Technology | Application | Technology description | Main contribution summary |
|---|---|---|---|
| Big data [1] | Business | Selection and evaluation of extensively available datasets, applying a set of methods to clean and record the data. Presents observations during data processing with various varieties and higher velocities in greater volumes of data. | Investigated the operational and necessary impacts of big data. Presented systematic analysis and case study conclusions. Studied the applications and highlighted future directions. |
| System integration [2] | Manufacturing | Establishment of a standard data network system. Allows various organizations and departments to be integrated and linked, where a smooth collaboration and computerized value chains are feasibly formed. | Reviewed important aspects of additive manufacturing new improvements in process development and material science. Analyzed modern science and technological trends and highlighted its possible applications. |
| Cyber-physical system [3] | Industry 4.0 | Set of advanced technologies; links the processes of physical resources and computational capacities and controls physical system, while designing a virtual model. | Reviewed current research trends of cyber-physical systems and their applications in industries and identified challenges. |
| Augmented reality [4] | Industry | Collection of HCI methods can insert virtual objects. Collaborates in the physical environment. | Presented an overview of the importance of AR. |
| Cloud computing [5] | Industry 4.0 | System for establishing online storage functions (data applications), models, and programs in a virtual server. | Explored emerging IT trends: IoT, big data, and cloud computing. Investigated their industrial implementation. |
| IoT [6] | Manufacturing | Networking of smart physical objects (sensors, devices, machines, cameras, vehicles, buildings). Allows exchange and collection of data, communication, and collaboration of objects. | Presented a brief summary of Industry 4.0 and associated technologies. |

## 2.3 AI Features in Industry 4.0

Advances in technology, which enable software, networks, machines, and products to detect, perceive, grow, comprehend, and acquire knowledge from their own experiences or to expand human activities, are combined to create AI. AI enables industrial production systems to accomplish exceptional tasks better than people. AI can also enable robots to carry out activities that humans would not, such as handling delicate or hazardous commodities or microscopic components. This puts into perspective the fact that many industrial robots already in use are not as intelligent as people. Even if they have limited programming, they can nonetheless do a variety of tasks skillfully and in a variety of settings. This technology is advancing with Industry 4.0, thanks to continuous innovation. The following list of Figure 2.3 depicts and discusses AI-related endeavors in Industrial 4.0: a summary of several AI-related topics, such as DL, computer vision (CV) ML, and natural language processing (NLP), is presented in Table 2.2. We also examine alternative approaches and algorithms that are used in various Industry 4.0 enabling systems.

### 2.3.1 Machine Learning

One of the fundamental ideas of AI is described by the subfield of ML. Instead of merely following directions, it gains knowledge through experiences or datasets. By training, ML-based techniques, such as those in [10], automatically learn and improve system performance. These techniques look at each identifiable pattern's result and attempt to reverse-engineer elements to produce an output. It creates a framework for forming judgments and decisions based on prior experiences [11]. According to Figure 2.3, the three primary subcategories of ML techniques are unsupervised, supervised, and reinforcement learning. For statistical services and strict prognostication, qualitative, quantitative, and investigators used the fusion of data, ML methods, and techniques like support vector machine (SVM), conversion of DWT, rapid Fourier transformation, which is the evaluation of principal components, Gaussian combination models, K-nearest neighbor (KNN), and neural network models (ANNs) [12]. ANN was used by Atici-Ulusu et al. [13] to examine product faults in Industry 4.0. A built-in system created using ML was described by Naseraldin etal. [14], who also looked at the impacts at the tactical, operational, and strategic levels. To recognize and forecast malfunctions of products across Industry 4.0, Tavakolizadeh et al. [15] employed ANN, multiplayer perception (MLP), gradient augmentation, and random forest. J.-C. Hong et al. [16] used clustering, auto-encoding, regression, linear K-means, random forests, and spatial clustering based on the density of applications with noise (DBSCAN) approach to construct a context-aware framework for detection.

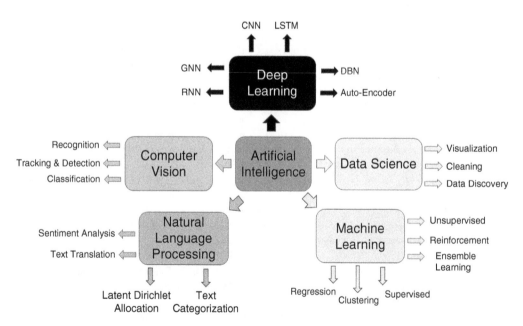

**Figure 2.3** Areas of Industry 4.0 with a focus on AI.

**Table 2.2** An overview of the various AI with XAI techniques applied in different industrial applications.

| Broad area | Application | Technology | Method/models | Main contribution |
|---|---|---|---|---|
| ML [10] | Design, manufacturing operation | Simulation, cyber-physical system, IoT, big data analysis | Generic ML algorithm | Data analysis and processing for design improvements of ships manufacturing. |
| ML [11] | Tracking in production | Cloud computing | Ontology-based interpretable model | Product-in-use assessment in Industry 4.0 |
| ML [22] | Automotive industry | IoT, cyber psychosocial system | Not specified | Customized and advanced service Industry 4.0 |
| ML DL [12] | Manufacturing industry | System integration, IoT | SVM, DWT Transformation, Fast Fourier Transform, FPCA, Gaussian mixture models, GRNN, Hidden Markov Model, KNN, RNN, ANNs. | Overview of prescriptive, predictive, and prescriptive prognostic maintenance and analytics in Industry 4.0. |
| ML [14] | Assembly systems | IoT, cloud computing, additive manufacturing | Autonomous decision using ML. | Explore the impact of Industry 4.0 Strategic, tactical, operational levels of assembly lines. |
| ML [15] | Production line failure | IoT, simulation | Random forest, gradient boosting, MLP, ANN. | Presented case study, a real-time early detection and product failure system for Industry 4.0. |
| ML [16] | Manufacturing process | IoT, cyber-physical system | Clustering, auto-encoder, linear regression, K-mean, random forest, and DB scan. | Presented a context-aware intrusion detection system for Industry 4.0 |
| ML CV [23] | Tracking human | IoT | Feature and blob-based method. | Presented a features-based person tracker model for the industrial environment. |
| ML CV [24] | Manufacturing industry | Big data, IoT | SVM, decision trees, random forests, logistics regression, KNN | Identification and classification of materials in the context of Industry 4.0. |
| ML [25] | Maintenance industry | IoT, cyber-physical system | K-means, Gaussian mixture | Presented a real-world implementation cycle, for knowledge discovery using machining learning. |
| NLP [26] | Human resource management | IoT | Text mining | Presented a tool to automatically determine industry 4.0 impact on human resource management. |
| NLP [27] | Supply chain | IoT | Latent Dirichlet allocation | Multi Tier supply chain in Industry 4.0 |

### 2.3.2 Deep Learning

A sort of combination of both ML and AI is DL. It gives instructions to an electronic device or system to layer-process information, classify data, assess data, and predict outcomes. As shown in Figure 2.3, a few of the most popular DL approaches are convolutional neural networks (CNN), neural networks with recurrent connections (RNN), and generative neural networks (GNN). The networks of neural networks and human brain cells operate according to almost comparable principles. A group of methods or algorithms establishes the connection between multiple underlying elements and arranges the data in the same way as the brain of a person does. Because these devices or models can provide cutting-edge accuracy, DL is gaining popularity. Diez et al. [12] employed DL-based approaches for forecasting, descriptive in nature and prescriptive evaluation (including RNN, hidden Markov model, reverse propagation neural network, as well as deep belief networks (DBN)). In Li et al. [17], for example, and Luo et al. [18], writers presented a CNN model for a rapid quality assurance system and early fault detection in industrial industries, respectively. The lifting net methodology, which is based on CNN, was utilized by researchers to categorize problems and identify characteristics flexibly utilizing noisy data with no prior knowledge [19]. A DL-based method for quick defect identification in time-varying circumstances was created by Wang et al. [18]. A multilayered recurrent unit with gated circuitry (MGRU) approach was created by R.-V. Sánchez et al. [20] to diagnose gear spur failure. SVM and long short-term memory (LSTM) are used to assess the method's classification accuracy. A forecasting maintenance method developed by Zhu et al. [21] employs LSTM-RNN and adaptive kernel spectral clustering (AKSC) to classify unusual behaviors on the basis of various degradation criteria.

### 2.3.3 Natural Language Processing

NLP is an art that enables systems or machines to comprehend, interpret, and read languages. A system operates effectively once it understands what the user is trying to say. A device or machine's capability to learn, recognize, and comprehend spoken and written human language is a component of AI or NLP. The objective of NLP aims to recognize and comprehend human language in order to convey a result in a convincing manner. Nowadays, most NLP approaches employ ML and DL methodologies to glean information from human language. For instance, Awasthi et al. [27] investigated the Latent Dirichlet Allocation approach, which is based on NLP, for applications in the industry including multi-tier distribution networks. Ferreira et al. [28] described an attention mechanism–based neural network model for translating English conversational language to a structured query language (SQL). It is used to store data from multiple devices and sensors across an SQL-based database designed for Industry 4.0.

### 2.3.4 Computer Vision

CV develops technology that allows computers to "see," "interpret," and "understand" digitized images and movies. Its objective is to draw inferences from visual cues and adapt them to real-world issues. CVs already have numerous applications, but there will be a lot more in the years to come. A wide range of industrial organizations, for instance, use CV for identifying flaws, fraud protection, allowing mobile deposits, and information visualization [19, 30, and 18]. In the framework of Industry 4.0, Karumbu et al. [24] used computational vision-based item recognition and categorization. A DL-based automated control of quality technique for picture classification in the publishing sector was presented by Silva et al. [29]. For the manufacturing sector, Lee [30] introduced a defect identification and categorization system based on CV and DBN-DL.

## 2.4 Industry 4.0 and XAI

AI-powered techniques are improving in accuracy, giving professionals additional reasons or information to explain the need of certain choices and instructions [31–38]. The choice must be transparent and understandable in order for systems based on AI to be adopted smoothly and for experts to approve or support them. It is common knowledge that explain ability might increase consumers' belief in these models. Despite "explain ability" being rarely a well-defined concept, it does encompass a variety of traits, objectives, and purposes. For instance, the nature and adequacy of an opinion strongly depend on its specific setting and the user's traits. It is important for data specialists and scientists to understand the fundamentals of ML and sophisticated ML models, but doing so is also associated with significant cognition stress beyond the extent of extreme applications, according to suggestions from the cognitive study. Numerous techniques for examining explain ability frameworks have been developed in the literature [46, 47]. The categorization approaches are usually not perfect; they can differ widely and be categorized into several types, such as overlapped or non overlapping categories, based on the components of the methods. Several of the methods in Figure 2.4 are examined in depth by Arri et al. [48]. Thanks to XAI, humans can now recognize, comprehend, interpret, and explain the decision-making process used by an AI architecture or system. The technology supporting Industry 4.0 is described in Table 2.3, along with several XAI-based approaches or tactics employed in diverse sectors of industry.

## 2.5 Industry 4.0 Integration Using an XAI-Based Methodology with AI

When Industry 4.0 technologies are combined with cutting-edge XAI based on AI (Figure 2.4), the results in multiple software applications are remarkably successful, accurate, and high quality. By using large data samples, comprehensible and interpretable architectures may attain a high degree of abstraction, garnering significant interest across all areas of Industry 4.0. Table 2.4 explains a few uses of XAI- and AI-based methodologies. The table provided an overview of the main enabling technologies, different approaches, and applications in the designated areas.

**Figure 2.4** An Outline of XAI-based techniques.

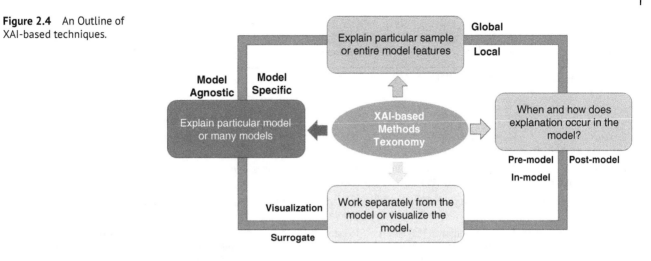

**Table 2.3** An outline of the various AI with XAI-based techniques in different sectors.

| Broad area | Application | Technology | Method | Main contribution |
|---|---|---|---|---|
| XAI ML [39] | Business | Internet of Things | Visualization and ML {Shapley values, XGBoost predictive classification algorithm.} | Proposed an XAI model that can be applied to justify why a customer buys or leaves nonlife insurance coverage. |
| XAI ML [40] | Industrial machinery | Internet of Things | Visualization and ML techniques, local and global explanations, random forest, ELI5, and LIME. | Implementation and explanations of a residual life estimator design based on machine learning employed to industrial data. |
| XAI ML [41] | Manufacturing | Internet of Things | Visualization Nonlinear modeling with SHAP values, data-driven decision model. | Proposed a data-driven decision model to improve process quality in manufacturing by combining nonlinear modeling and SHAP values from the field of explainable AI. |
| XAI ML [42] | Manufacturing | Internet of Things | XAI methods Smoothed integrated gradients, guided gradient class activation. Mapping deep SHAP. | Presented a DL-based classification method for fiber layup fault identification in the automatic composite manufacturing. |
| XAI DL [43] | Prediction techniques | Internet of Things | Localized a post-hoc justification DL, substituting "Decision Tree." | Introduced a new localized post-hoc explanation method for monitoring problems in the predictive process. |
| XAI [44] | Diagnostic troubleshooting | Internet of Things | Feature importance of isolation forest (Local-DIFFI). | Presented a fault diagnosis and anomaly detection techniques in rotating machinery to interpret black-box models. |
| XAI DL [45] | Manufacturing | Internet of Things | Lightweight online detector of anomalies. Minimum covariance determinant PCA, statistic. | Presented an XAI-based method for fault analysis and penetration harvesting for steel plate manufacturing. |

**Table 2.4** Industry 4.0 in conjunction with extensions of AI and XAI-based techniques.

| Application | Specific area | Technology | Contribution summary |
|---|---|---|---|
| Human-computer interaction | Manufacturing and industry [62] health care [63] | IoT, cyber-physical system, AR, IoT | Provided state-of-the-art human-computer interaction applications in Industry 4.0. AI (DL, CV) and XAI for human-computer interaction. |
| Smart assistance | Demand forecasting in manufacturing [67]. | IoT | Demonstrated a system for demand forecasting use in manufacturing. |
| Industrial robots | AR [71] | IoT, cyber-physical system, AR | Implementation of AR to analyze and estimate the usage of a production and industrial robot. |

*(Continued)*

**Table 2.4** (Continued)

| Application | Specific area | Technology | Contribution summary |
|---|---|---|---|
| Autonomous vehicles | Cyberattacks [77] | IoT | Autonomous vehicles in the era of Industry 4.0. Analyze threat classification in autonomous vehicles, utilizing accountability, authentication, and service availability. |
| Chatbots | Human chatbots communication [78] | Robotics | AI for human to human to chatbot conversation. |
| Autonomous resource exploration | Autonomous exploration and mapping system [79] | Emerging technologies | Autonomous exploration and mapping systems. |
| Augmented reality | Human-robot collaboration [4] | IoT, AR, autonomous robotics | AR applications in Industry 4.0. |
| Smart transportation | Traffic surveillance [52] smart lighting and smart parking transportation [75, 76] | IoT, IoT, IoT | CV, AI application of Industry 4.0. Presented a multi camera solution for traffic monitoring. AI and video analytics-based systems for smart transportation. Reviewer of ML and IoT for smart transportation. |

### 2.5.1 Intelligent Communities

In order to meet the increasing needs of urbanization, metropolitan management and sustainable development procedures are formed and carried out in intelligent cities, Figure 2.5 explains the structures are primarily made up of interaction, knowledge, and technologies. IoT applications are built on a big data–supported intelligent network of interconnected equipment, sensors, systems, and objects that exchange data via the cloud and wirelessly [52]. These

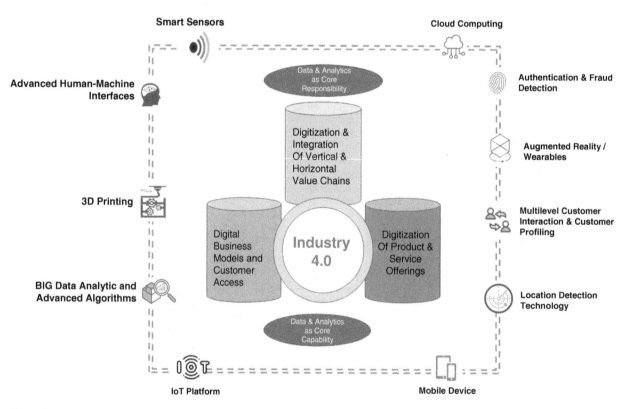

**Figure 2.5** Industry 4.0 framework and contributing digital technologies.

data are used by smart cities to establish guidelines that raise the living conditions and productivity of their citizens. Smart environments, smart systems, or smart solutions for regulating environmental factors, such as irrigation, photovoltaics, garbage, lighting, water supply, and weather stations, are all parts of the application of smart cities [53]. It strives to improve both the environment and energy performance in cities. Additionally, it plays a crucial part in smart mobility, which includes smart monitoring techniques for walkers, spaces for parking, bike paths, capacity management, control of traffic, charging points, and tourist saturation. Additionally, it tries to eliminate city noise so that people may move properly. Smart living is another component, which includes smart air conditioning, surveillance cameras, and personal sports facilities. Its goal is to improve people's quality of life. Additionally, it carries a crucial responsibility for residents as intelligent individuals, which entails optimizing their services. To improve communication between the many stakeholders, these systems integrate informative digital signage (mupis), citizen applications, citizen credit cards, and social Wi-Fi.

### 2.5.2 Smart Industries

As the most recent technological advancement guarantees to provide a more vigorous, flexible, connected, and adaptable production along with related sectors, Industry 4.0 is currently transforming factories [56]. When marketing "smart factories," there are plenty of novel aspects to consider, including the IoT and AI [57, 58]. By creating and promoting their datasets, smart factories may integrate XAI, which uses AI to create predictive studies of equipment performance to enhance control of quality, coherence, and servicing, and entirely simplify industrial boundaries. Smart factories have effectively achieved a digital transition. While the benefits of AI in relation to production processes are well known, it is important to recognize that manufacturers should have a plan for developing AI and XAI as well as a strategy for the kind of modernization and automated platform to use. The most effective way to transition factories into the goal of smart manufacturers for Industry 4.0 aims to achieve IoT gadgets, big data, cloud-based computing, and cyber-physical infrastructure.

### 2.5.3 Adaptive Production

By building and delivering independent, self-regulating smart devices, Industry 4.0 makes manufacturing plants smarter, more adaptable, and more productive [81]. Consequently, high standards of industrialization and self-optimization are produced by machines, systems, and equipment. Additionally, as envisioned, the production method includes the capacity to meet additional difficult and effective product criteria and circumstances. Therefore, Industry 4.0's primary goals are innovative manufacturing and smart production. AI and XAI, when implemented properly, have many advantages for the manufacturing sector, including cost reduction, decreased error rates (intelligent computations can execute obligations that decrease vulnerability to mistakes during procedures accomplished by human beings), and cost savings (many banks and e-commerce sites use robots to start providing customer service). By doing this, businesses may cut back on staff costs or transfer employees to more crucial and challenging positions that will increase productivity and allow them to focus more intently on their work.

### 2.5.4 Intelligent Health Care

The health-care sector now can manufacture newly tailored systems, cutting-edge tools, and gadgets, thanks to Industry 4.0. It offers a unique type of digital health-care service and a faultless tracking system that perfectly suits the demands of the healthcare and pharmaceutical industries in terms of cost and conditions. Big data, online computing, and IoT usage enable the development of a modern, smart health-care industry. It enhances data communication by using the Web of Things, advanced manufacturing methods, hardware, firmware, robots, detectors, and other smart technologies for information [59]. By changing traditional health-care institutions into innovative health-care entities, technological and communication advancements improve the quality of health care. Smart health care is a framework for preserving health that connects individuals, organizations, and components involved in health care; obtains data automatically; and then actively governs and efficiently responds to requirements derived from the pharmaceutical companies' ecosystem [60, 61]. It does this by utilizing sensors on clothes, the IoT, and adaptable Internet. Computerized record tracking, noninvasive services, and intelligent, linked pharmacological equipment and sensors are the fundamental concepts underpinning smart health care.

### 2.5.5 Human-Computer Interaction

The methods by which individuals can communicate with machines and systems are covered in this section. How can humans engage with technologies like gadgets, automated machinery, data, and services? That is the topic that better and evolved human-machine interaction or human-machine interaction (HCI), seeks to address in Industry 4.0. It focuses primarily on the development, assessment, and use of technology for communication and information, with a particular focus on enhancing user behavior, task performance, and experience quality [62]. Utilizing AI, more XAI applications are emerging; at the same time, HCI is now developing and expanding [83, 63], paving the way for the rapid emergence of novel and exciting study topics. AI and XAI seek to offer, establish, encourage, and discover fresh approaches and empirical understanding of HCI-related activities, encompassing yet not confined to mobile and e-commerce, employment opportunities, businesses, relationships between people and intelligent machines and equipment, human-robot communications, novel interface designs for virtual environments, and augmented reality (AR) but also the assessment of HCI problems in relation to neurophysiological devices, tools, as well as gadgets (e.g. GSR, EEG, MRI, and CT scan).

### 2.5.6 Prevention-based Maintenance

Technology for communication and information advancement has forced the industrial sector to adopt a more intelligent strategy. Performance data is regularly analyzed to identify actionable penetrations that foretell product failure, enhance uptime, and boost asset productivity [64]. In order to manage the state of equipment, real-time data is obtained during predictive maintenance. The goal is to find patterns that can help detect and eventually foresee problems; usually, AI techniques [65] are used in training procedures to achieve this goal. Systems become increasingly crucial for assessing device needs and determining if administration should be carried out when scheduled upkeep is computerized. Undoubtedly, the use of AI, as well as XAI-based technologies [66], leads to considerable cost reductions, improved predictability, and increased system accessibility.

### 2.5.7 Intuitive Assistance

AI and XAI give computers and gadgets the ability to comprehend speaking, speech, or text input and reply appropriately using a suitable initiative. These systems are preprogrammed and work with ML, DL, which is NLP, AI, and XAI. Intelligent tools, robotics, networks, and other technologies are designed to do specific tasks requiring human intellect or cognitive ability [67].

### 2.5.8 Smart Gadgets

IoT, computing via the cloud, cyber-physical networks, AI, and XAI-based techniques, are used in Industry 4.0 smart devices to enable human interaction [68]. Industry 4.0, which enables cutting-edge technology for product advancement, must be blended with the present manufacturing techniques. It incorporates real-time applied production systems, knowledge, and sophisticated, upgraded automated experience. Therefore, it is essential to use cutting-edge technology in integrated digital and environmental design while creating intelligent goods. Industry 4.0 is being advanced by a number of operators, including the IoT, big data, cloud-based computing, mass customization, and the development of production time [69, 70].

### 2.5.9 Commercial Robotics

Industrial robots can autonomously run and regulate robots in many industries, thanks to advanced XAI and AI techniques. Examples of common duties include welding, ironing garments, applying paint, apparatus, place and pick, packaging, evaluating products, and experimenting, which are all carried out with great strength, velocity, and precision. Industry 4.0-related technologies, including AR [71, 72].

### 2.5.10 Digital-Privacy and Security

New difficulties arise when information and communication technologies are combined, particularly in cybersecurity. The IoT has significantly changed how many cyber dangers there are today. Technical challenges, major reasons for cyberattacks, requirements for cybersecurity, and strategies with a global view, including the divide between the public and private sectors, are all brought about by security risks and vulnerabilities of IoT. In Industry 4.0's smart network systems, AI is used to decide cybersecurity and privacy elements like cyber threat detection [73, 74]. Analyzing the network structure and spotting potential cyberattack risks in real time are often included. It also frequently incorporates malware detection, endpoint identification, acknowledgment, and network traffic analysis.

### 2.5.11 Intelligent Transportation

With the advent of Industry 4.0, the transportation sector also experiences significant advancement [52]. With the use of autonomous or self-driving cars, sensors, alternative route planning, traffic sign detection, and other smart technologies, transit times can be reduced or recommendations for more efficient modes of transportation are provided. Using sophisticated, computerized vision-based systems for automatic vehicle recognition and tracking, AI-based technologies [75] have also improved transportation capacities. The growth of big data, cloud computing, and IoT also enhance transportation services [76, 77].

## 2.6 Case Studies for Industry 4.0

Numerous technological innovations are used to power Industry 4.0. Today, they have already been effectively included in the working procedures of several firms and organizations. Here are the main Industry 4.0 innovations that help modern businesses with production and commercial operations.

### 2.6.1 Analytics of Big Data

There has been a huge growth in the amount of raw data generated by different IoT devices and embedded sensors as a result of the widespread adoption of IoT technology across sectors and private residences. Businesses and manufacturers may better understand past trends, recognize human behavior, and understand how technology works by using big data and deep analytics technologies to the collected data. As a result, businesses are better equipped to decide how to expand their businesses and produce new products. Companies may study sales data and how it relates to manufacturing, for instance, and then optimize their production processes and warehouses. A complete digital picture of operational procedures can notify companies of potential production line faults in the future or provide more details on how to enhance certain operations.

### 2.6.2 Intelligent Robotics

AI completely alters how businesses acquire data and make judgments. Large-scale data processing is done by AI algorithms, which then forecast consumer behavior, social shifts, or manufacturing procedures. Numerous modern technologies associated with Industry 4.0, including data mining, big data processing, business software, and others, are built on the algorithms [78–80]. AI technology may replicate human learning processes in addition to having predictive capabilities. By working with well-known patterns, the system can automatically train and learn from datasets while producing less and fewer errors each time. With the aid of AI technology, equipment can generate high-quality products faster and with less waste while maintaining improved productivity and uptime.

### 2.6.3 Cloud-Based Computing

Another essential technology without which the fourth industrial age would not be conceivable is cloud computing. The technology serves as the nucleus for data processing, system integration, and networking. Businesses may combine diverse activities from engineering for supply chain management (SCM) to manufacturing to distribution and sales into one cohesive system by using the cloud to build a totally digital environment. Additionally, because SME manufacturers do not have to spend a lot of money on building and maintaining hardware, cloud systems can drastically reduce expenses. As their enterprises grow, they may scale up their information processing and storing requirements.

### 2.6.4 Digital Industry 4.0 and IoT

With 9 billion gadgets, including cell phones, laptops, RFID readers, and sensors, linked to the Internet, the term "Internet of Things" (IoT) is no longer just a marketing gimmick. The experts anticipate a dramatic increase of up to 1 trillion linked items in the next 10 years. This data demonstrate a significant paradigm change, which makes it necessary to stay current with shifting customer needs and operational environments. Only by developing devices with embedded software to manage their ongoing improvement and updating can this problem be solved. "4th Industrial Revolution" ("Digital Industry 4.0"), which is made possible by IoT, may occasionally be used to describe entirely digital value chains being created by highly intelligent linked systems.

### 2.6.5 Standard Product Lifecycle Management Packages and Industry 4.0

The tangible things we see around us have a specific touch of a product lifecycle management PLM system in their design and production. Each firm has a unique approach for developing products. Some people just need to produce 3D models using CAD, whereas others run large PLM platforms with millions of items and hundreds of clients. Most international organizations demand that PLM systems meet the IoT reality, regardless of the extent of their use. When we conceive smart Industry 4.0 as it relates to PLM engineering, we envision integrated sensors, processing units, and instructional programs in goods linked to the cloud through the Internet and analyzing and storing data on product usage. The question of whether to deal with the item itself or the information obtained from the product is still up for debate. Any PLM development business should be aware of the significance when deciding how to proceed with the deployment of IoT software and take into account the digital interconnection among mechanical, then computerized, and intellectual goods in addition to the cross-enterprise interaction from manufacturing to assistance, operation, or administration. It's challenging to change most old PLM solutions into IoT-enabled digital ones. Over 50% of PLM system providers assert that their systems do not support cross-disciplinary cooperation. Another noteworthy aspect and barrier to the deployment of IoT solutions is that, generally, only a small number of users may access the system over the extended organizational value chain. With their PLM systems, just 21% of worldwide manufacturers could accommodate over 1,000 people. Finally, 75% of suppliers admit that it is challenging for their PLM systems to keep up with the evolving business needs. The fragmented procedures in the current state of PLM systems pose a significant risk of becoming a major problem. Data and technology gaps, unresolved business procedures, a lack of data exchange, and a general lack of flexibility are all brought to light, which has the potential to undermine new product efforts. For true digitalization along with IoT deployment to succeed, the PLM platform has to blend in any holes in order to unite system barriers into an integrated approach that suits current as well as potential organizational demands [83].

### 2.6.6 Internet of Things and Product Lifecycle Management

Engineers from Scandinavia have appropriate experience in the creation and maintenance of sizable Microsoft.NET-based PLM platforms using AML/XML meta-modeling interface. We know exactly how to adapt outdated PLM technologies into digitally enhanced ones. The topic of conversation in this specific instance is inter-engineering collaboration. Some businesses produce sensors, while others have the know-how to program them. Others are knowledgeable about business and marketing trends as well as the best ways to employ sensors. Finally, some businesses, like ours, are adept at effectively processing sensor data with the assistance concerning cloud-based filing along with huge data analytics.

### 2.6.7 Leveraging 3D Printing

Mass production with 3D printing is still in its infancy. Though technology is still evolving, more businesses are beginning to reach the 3D printing market. It is effectively used, for instance, in the manufacture of automobiles. By 2025, Volkswagen expects to print in 3D 100,000 auto-components yearly. Meanwhile, a growing number of businesses that specialize in three-dimensional printing are appearing. By 2030, it's anticipated that 21.5 million 3D printers will have been delivered worldwide.

### 2.6.8 Technologies for Cybersecurity

New requirements for cybersecurity standards at businesses are being put forth by an increased reliance on cloud connection and networked IoT devices. As a result, businesses undergoing an Industry 4.0 digital transition should also think about the safety of the Internet of their operating machinery. The industrial automation's cybersecurity protocols can be an examination of risks and threats, product testing, network security assessment, security certification, and staff training, enhancing digital infrastructure with the most up-to-date and reliable security tools, including blockchain-based applications, multi-factor authentication, role-based system access, and others.

## 2.7  Challenges of Industry 4.0

Industry 4.0 has a lot of chances waiting for it to develop. All industries and academic fields will be impacted by Industry 4.0; hence, it is important to solve the problems outlined here. In the words of McKinsey, the idea of Industry 4.0 disrupts the value chain, requiring firms to rethink the way they execute trade. To flourish in the new environment, their organization needs to oversee the digital resurgence of their industry.

The following are the primary difficulties encountered in implementing Industry 4.0:

1) Major Investment: It is necessary for a seamless digital change and evolution since sector 4.0 has made the functioning of the sector more process-driven in educating the current workforce as well.
2) Desiring to Take into Account Novel Business Models: In the era of Industry 4.0, businesses are in an industrial era where it is necessary to change how we communicate with customers, comprehend business cases, change manufacturing operations, and change how consumers interact and use goods.
3) Adaptation to Change: It is important to assess the corporate culture and improve the ability to change for the better. Launching the radical digitization strategy lacks boldness.
4) Process Reorganization: To implement Industry 4.0, vertical and horizontal value chains must be automated and integrated. Therefore, it's critical to pinpoint the areas where action is required. Additionally, doing pilot research is necessary to achieve better results.
5) Workforce: Due to a shortage of internal talent, current developments in technology are bringing about change at an unheard velocity. The present workforce must be developed and trained in order to increase quality and efficiency because the skill sets they now possess are becoming outdated. There will be a requirement to hire a new generation of tech-savvy workers as new job categories arise. In accordance with the World Economic Forum's 2018 Global Trends study, 76% of hiring managers and recruiters believe that sector 4.0 would significantly affect the recruiting sector.
6) Standardization: The entire globe is presently undergoing a digital change that will transform every sector of business. Industries need guidelines and requirements to guarantee that the various parts are compatible and work together.
7) Management of Data: Industries produce a lot of manufacturing and quality data in real time. Data must be made easily accessible and available.
8) Competitiveness: Growing competition is promoting consumer and manufacturer interaction at various product phases.
9) Information Security: Industry 4.0 works with a lot of data; thus, data security is important. Security must be a top priority if Industry 4.0 is to live up to its full promise. End-to-end encryption must be used in order to prevent attacks such as phishing and vulnerability.

## 2.8  Advantages of Intelligent Factory

The industrial sector has been significantly impacted by the digital transformation, and intelligent factories have grown to be quite essential. Industrial businesses were formerly on the edge of IT and digitalization. However, because of the digitization of industrial processes, intelligent factories now provide tremendous benefits. Table 2.5 Explains the Challenges of Industry 4.0 versus traditional industry. The great quantity of connected data allows for a more personalized approach, as well as acquiring a very broad capacity to adapt to demand.

1) Production procedures have been optimized for 24-hour manufacturing capacity and are repetitive and error free.
2) Every automated manufacturing process improves accuracy. Downtime is prevented.

**Table 2.5**  Challenges of Industry 4.0 versus traditional industry.

| Industry 4.0 | Traditional industry |
| --- | --- |
| Flexibility in work organization | Work rigidity |
| Use of the product as a service | Product purchase |
| Customization according to customer requirements | Mass production |
| Smart factories with flexible production at a competitive cost | Large factories to manufacture big volumes of a specific product |
| Maximization ROCE: profitablity/capital used | Minimization of cost |
| Dynamic production according to demand | Orderly planning based on the anticipation with a stock |

3) A significant improvement in the product's manufacture and design.
4) Quicker production processes. A rise in sales.
5) Larger efficiency from exhaustive controls results in larger cost savings.
6) Automated operations use less labor, make fewer mistakes, and use more raw materials, energy, etc.

### 2.8.1 Positive Effects of Industrial Revolution 4.0

The industrial sector has been significantly impacted by the digital revolution. Businesses were able to obtain a lot of information about the manufacturing processes and automate many of their operations as a result. However, there was still a very low degree of data analysis and communication within divisions, as well as with clients and business partners. The problems are efficiently solved by Industry 4.0 technology. Each manufacturing process is network connected, which enhances data sharing and workflow monitoring, enables remote job management, and allows for real-time process optimization. The following are the primary advantages businesses experience when deploying Industry 4.0 solutions:

a) Manufacturing Flexibility: Businesses may efficiently manage their manufacturing procedures by keeping an eye on the supplies available for manufacture and client requests. This enables companies to focus just on producing the commodities that are actually desired by customers and to avoid making unnecessary resource purchases for their products.
b) Enhanced Supply Chain Transparency and Logistics: Companies may increase the transparency of their purchasing and procurement chains for their internal departments and all chain participants by using 4.0 industry technology. By doing this, businesses may drastically save the expenses associated with purchasing and shipping the necessary ingredients as well as marketing and delivering the finished goods.
c) Smart Resource Allocation: The resource supply operations may be automated using a variety of enterprise-level platforms, including enterprise resources planning systems (ERPs), PLMs, SCMs, warehouse management systems (WMSs), and others. The numerous gadgets integrated into the workflow of the organization may automatically make orders while keeping track of the materials that are in stock and choosing the most reputable suppliers at the most competitive prices.
d) Improved Client Experience: Industry 4.0 technologies help marketing become more client focused. Utilizing a variety of tools and devices, businesses may adapt their marketing materials and product offers to the wants and needs of their target audiences. Customers may easily locate and order the goods they desire that will be uniquely tailored for them in this way.
e) Effective Utilization of Data: The volume of data generated by both humans and robots has greatly increased during the past 10 years. Before the advent of big data technologies and advanced analytics, a large portion of this data had never been handled and analyzed. Companies that use these technologies to strengthen their digital ecosystems receive a greater understanding of how to increase their processes and customer interactions, making it simpler for them to achieve a competitive edge in the market.

## 2.9 Discussion and Emerging Trends

The data, knowledge, and information revolution began midway through the twentieth century, and a summary of the digital transition and how AI fueled it is provided. One area that is being impacted by digital transformation is business, where digital technology has been incorporated into all areas of operation with an eye on how you function and provide value to consumers. The idea behind digital transformation is to reevaluate outdated business models. The emphasis is on how technological advances can be used to employ smarter ways of functioning using digital devices like mobile phones, laptops, tablets, videoconferencing, home offices, and the use of application platforms made possible by the use of cloud computing, Internet services, and solutions that are cloud based. The adoption of paper-fewer working modes in the majority of commercial activities is another factor. This all points to the term "Industry 4.0" being synonymous with the fourth industrialization [49], which is the concept of manufacturing facilities with machines improved by AR and wireless access, as well as a plant environment in which detectors connected to computers visualize the entire manufacturing process and make automated choices [50]. IoT, in addition to the growth of cyber-physical computer networks, is also covered during the age of industrial IoT, computing via the cloud, and AI, as well as the trend toward robotics and data interchange in industrial technologies and processes [51, 52]. Cultural shifts are required in order for new business models to be assimilated, particularly for generations of "digital migrants" or those who were born before the dawn of the Internet.

But the digital revolution is not limited to the commercial world; it is permeating almost every element of human life, including the field of education [53–55]. According to the literature analysis, approaches based on XAI and AI and Industry 4.0 have made significant strides in a variety of domains. According to this viewpoint, AI may be viewed as one of the most important aspects of the industry 4.0 revolution. However, there are still numerous problems with AI techniques. In fact, the majority of algorithmic procedures, networks, and techniques built around AI are greedy for power and require a growing number of cores and GPUs to function properly. The uncertain nature of AI-based techniques must also be taken into account. How are models built with ML as well as DL able to predict the outcomes? Assessing the accuracy of these systems at the individual level is challenging for researchers. Additionally, AI approaches frequently require the following: big datasets, hyper-parameters optimizing, fine-tuning, reliable processing ability, and persistent information training. As a result, a crucial component of all models that use DL as well as ML is the availability of resources and data. These data are often gathered and created from countless sources, gadgets, and users; there is also a possibility that cyberattacks might have an impact on sensitive or personal data. The efficiency of the AI-based system for applications that operate in real time is also impacted by the bias in the data. As a result, the evident justification that XAI may provide for a mechanism that enables intellect on par with that of humans is necessary to explain the fundamental nature of AI. Explain ability is a new and powerful field of AI support; it is a crucial collection of approaches that can offer insights that standard linear models were unable to. The confidence and transparency of the simulations can be increased by using XAI-based models; however, there are still many issues to be solved. Since the existing models do not ensure that the tools have been trained on correct or impartial sets of data (how the set of data is acquired and generated), the training method, design, model, and objective function are susceptible. We have talked about a number of XAI-based techniques, but we are unable to guarantee the level of confidentiality for any model that poses a security concern. Similar to created systems, algorithms, and models, occasionally, they are exceedingly complicated while also being widely understood. So a layperson or industry requires an illogically clear, exact knowledge. It is a situation in which using XAI approaches may be advantageous since it may create simpler systems, models, and algorithms. The generated algorithms and models occasionally employ true and accurate data to make predictions and judgments that are inconsistent, unreasonable, biased, or occasionally out of order.

Reasonability is important and offers a different perspective that is dependent on the specific data provided as input to the AI algorithms and models. If utilized by an AI system, it does not guarantee that the output forecasts and choices are sufficient. In certain cases, we can describe how a model, algorithm, or system operates, but we need to be clear about how the system complies with moral and legal standards and is consistent with them. We have always been impacted by technological developments and innovations; now, the impact of AI and the Fourth Industrial Revolution is the most momentous development in human history. Industry 4.0's growth rate is currently outpacing all other technical breakthroughs, including the Internet, business, manufacturing, and digitalization. This development will continue to expand as an industry pioneer in the anticipated future. As a result, XAI and AI, in general, become major industrial problems. Several global sectors are emerging with significant advancements in XAI and AI. They are also being developed as new technologies including big data, cities that are smart, production, education, manufacturing facilities, health care, AR, robots, and IoT, which have an influence on the future of all sectors and people. The social environment and standard of existence in the future will be impacted by these technologies as they develop. By integrating predictive maintenance techniques into production processes and replacing visual inquiries with machines or co-bots that conduct quality checks more precisely and efficiently, XAI systems can boost automated acquisition of methods that could improve the end product's quality.

Applications may be made better by XAI and Industry 4.0, and embracing digital transformation enables businesses to adapt to changing customer needs in the future. With automated production acting as the cornerstone of the world's economic system, the installation of modern technology is all about unlocking the usual potential for items and reasons for prospects in the decades to come in this area. Statistics and IoT have a crucial role to play in the Fourth Industrial Revolution because they can recognize trends and behaviors including relay instantaneous fashion data to linked sectors and producers' biometrics. The health surveillance application can additionally use significant technological advances and XAI-based methodologies, for instance, enabling the investigation and predictions of situations of pandemics such as COVID-19 [84].

## 2.10 Conclusion

Manufacturing and associated businesses can benefit from the current mechanization and data communications paradigm known as "Industry 4.0." It links several big data along with the Internet of gadgets for analytical techniques, computing in the cloud, and many more, and creates clever, useful applications. This chapter offered a thorough analysis in accordance with XAI as well as AI-based techniques used in various Industrial 4.0 situations and addressed essential technologies enabling the Fourth Industrial Revolution in addition to providing an extensive study of multiple AI-based approaches matched with Industrial 4.0. Evaluation analytics drawn from large data, which enhances data collecting for analysis in many industrial applications, can also assist such a paradigm. New cutting-edge systems have been created throughout this revolution, and they can continually adapt to the changes in industry. We concluded that these intelligent applications and systems may be implemented automatically and in real time, thanks to AI and XAI. Consequently, AI is the key element of the industrial revolution that enables intelligent robots to carry out activities independently, whereas XAI provides a collection of processes that may offer explanations that are intelligible by humans. The industry 4.0 revolutions have been examined in terms of their prospects, problems, and potential future research areas.

## References

**1** Oztemel, E. and Gursev, S. (2020). Literature review of Industry 4.0 and related technologies. *Journal of Intelligent Manufacturing* 31 (1): 127–182.

**2** Wamba, S.F., Akter, S., Edwards, A. et al. (2015). How 'Big Data' can make big impact: findings from a systematic review and a longitudinal case study. *International Journal of Production Economics* 165: 234–246.

**3** Kim, J.H. (2017). A review of cyber-physical system research relevant to the emerging it trends: Industry 4.0 IoT big data and cloud computing. *Journal of Industrial Integration and Management.* 2 (3).

**4** De Pace, F., Manuri, F., and Sanna, A. (2018). Augmented reality in Industry 4.0. *American Journal of Computer Science Technology* 6 (1): 17.

**5** Lu, Y. (2017). Cyber physical system (CPS)-based Industry 4.0: a survey. *Journal of Industrial Integration and Management* 2 (3).

**6** Dilberoglu, U.M., Gharehpapagh, B., Yaman, U., and Dolen, M. (2017). The role of additive manufacturing in the era of Industry 4.0. *Procedia Manufacturing* 11: 545–554.

**7** Bahrin, M.A.K., Othman, M.F., Azli, N.H.N., and Talib, M.F. (2016). Industry 4.0: a review on industrial automation and robotic. *Jurnal Teknologi* 78: 6–13.

**8** Ghobakhloo, M. (2018). The future of manufacturing industry: a strategic roadmap toward Industry 4.0. *Journal of Manufacturing Technology Management* 29 (6): 910–936.

**9** Qin, J., Liu, Y., and Grosvenor, R. (2016). A categorical framework of manufacturing for Industry 4.0 and beyond. *Procedia CIRP* 52: 173–178.

**10** Ang, J.H., Goh, C., Saldivar, A.A.F., and Li, Y. (2017). Energy-efficient through-life smart design manufacturing and operation of ships in an Industry 4.0 environment. *Energies* 10 (5): 610.

**11** Bougdira, A., Akharraz, I., and Ahaitouf, A. (2019). A traceability proposal for Industry 4.0. *Journal of Ambient Intelligence and Humanized Computing* 1–15.

**12** Diez-Olivan, A., Del Ser, J., Galar, D., and Sierra, B. (2019). Data fusion and machine learning for industrial prognosis: trends and perspectives towards Industry 4.0. *Information Fusion* 50: 92–111.

**13** Kucukoglu, I., Atici-Ulusu, H., Gunduz, T., and Tokcalar, O. (2018). Application of the artificial neural network method to detect defective assembling processes by using a wearable technology. *Journal of Manufacturing Systems* 49: 163–171.

**14** Cohen, Y., Naseraldin, H., Chaudhuri, A., and Pilati, F. (2019). Assembly systems in Industry 4.0 era: a road map to understand assembly 4.0. *International Journal of Advanced Manufacturing Technology* 105 (9): 4037–4054.

**15** Soto, J.C., Tavakolizadeh, F., and Gyulai, D. (2019). An online machine learning framework for early detection of product failures in an Industry 4.0 context. *International Journal of Computer Integrated Manufacturing* 32 (4/5): 452–465.

**16** Park, S.-T., Li, G., and Hong, J.-C. (2020). A study on smart factory-based ambient intelligence context-aware intrusion detection system using machine learning. *Journal of Ambient Intelligence and Humanized Computing* 11 (4): 1405–1412.

**17** Li, L., Ota, K., and Dong, M. (2018). Deep learning for smart industry: efficient manufacture inspection system with fog computing. *IEEE Transactions on Industrial Informatics* 14 (10): 4665–4673.

**18** Luo, B., Wang, H., Liu, H. et al. (2019). Early fault detection of machine tools based on deep learning and dynamic identification. *IEEE Transactions on Industrial Electronics* 66 (1): 509–518.

**19** Pan, J., Zi, Y., Chen, J. et al. (2018). LiftingNet: a novel deep learning network with layerwise feature learning from noisy mechanical data for fault classification. *IEEE Transactions on Industrial Electronics* 65 (6): 4973–4982.

**20** Tao, Y., Wang, X., Sánchez, R.-V. et al. (2019). Spur gear fault diagnosis using a multilayer gated recurrent unit approach with vibration signal. *IEEE Access* 7: 56880–56889.

**21** Cheng, Y., Zhu, H., Wu, J., and Shao, X. (2019). Machine health monitoring using adaptive kernel spectral clustering and deep long short-term memory recurrent neural networks. *IEEE Transactions on Industrial Informatics* 15 (2): 987–997.

**22** Fraga-Lamas, P. and Fernández-Caramés, T.M. (2019). A review on blockchain technologies for an advanced and cyber-resilient automotive industry. *IEEE Access* 7: 17578–17598.

**23** Ahmed, I., Ahmad, A., Piccialli, F. et al. (2018). A robust features-based person tracker for overhead views in industrial environment. *IEEE Internet of Things Journal* 5 (3): 1598–1605.

**24** Penumuru, D.P., Muthuswamy, S., and Karumbu, P. (2019). Identification and classification of materials using machine vision and machine learning in the context of Industry 4.0. *Journal of Intelligent Manufacturing* 1–13.

**25** Diaz-Rozo, J., Bielza, C., and Larrañaga, P. (2017). Machine learning-based CPS for clustering high throughput machining cycle conditions. *Procedia Manufacturing* 10: 997–1008.

**26** Fareri, S., Fantoni, G., Chiarello, F. et al. (2020). Estimating Industry 4.0 impact on job profiles and skills using text mining. *Computers in Industry* 118.

**27** Zhou, R., Awasthi, A., and Stal-Le Cardinal, J. (2020). The main trends for multi-tier supply chain in Industry 4.0 based on natural language processing. *Computers in Industry*.

**28** Ferreira, S., Leitão, G., Silva, I. et al. (2020). Evaluating human-machine translation with attention mechanisms for Industry 4.0 environment SQL-based systems. *Proceedings of IEEE International Workshop on Metrology for Industry 4.0 & IoT*, 229–234.

**29** Villalba-Diez, J., Schmidt, D., Gevers, R. et al. (2019). Deep learning for industrial computer vision quality control in the printing Industry 4.0. *Sensors* 19 (18).

**30** Lee, H. (2017). Framework and development of fault detection classification using IoT device and cloud environment. *Journal of Manufacturing Systems* 43: 257–270.

**31** Ahmed, I., Anisetti, M., and Jeon, G. (2021). An IoT-based human detection system for complex industrial environment with deep learning architectures and transfer learning. *International Journal of Intelligence Systems*.

**32** Gade, K., Geyik, S., Kenthapadi, K. et al. (2020). Explainable AI in industry: Practical challenges and lessons learned. *Proceedings of Companion Proceedings of the Web Conference 2020*, 303–304.

**33** Rehse, J.-R., Mehdiyev, N., and Fettke, P. (2019). Towards explainable process predictions for Industry 4.0 in the dfki-smart-lego-factory. *KI-Künstliche Intelligenz* 33 (2): 181–187.

**34** Carletti, M., Masiero, C., Beghi, A., and Susto, G.A. (2019). Explainable machine learning in Industry 4.0: Evaluating feature importance in anomaly detection to enable root cause analysis. *Proceedings of IEEE International Conference on Systems, Man and Cybernetics (SMC)*, 21–26.

**35** Christou, I.T., Kefalakis, N., Zalonis, A., and Soldatos, J. (2020). Predictive and explainable machine learning for industrial Internet of Things applications. *Proceedings of 16th International Conference on Distributed Computing in Sensor Systems (DCOSS)*, 213–218.

**36** Le, D.D., Pham, V., Nguyen, H.N., and Dang, T. (2019). Visualization and explainable machine learning for efficient manufacturing and system operations.

**37** Langone, R., Cuzzocrea, A., and Skantzos, N. (2020). Interpretable anomaly prediction: predicting anomalous behavior in Industry 4.0 settings via regularized logistic regression tools. *Data & Knowledge Engineering* 130.

**38** Daglarli, E. (2021). Explainable artificial intelligence (xAI) approaches and deep meta-learning models for cyber-physical systems. *Proceedings of Artificial Intelligence Paradigms for Smart Cyber-Physical Systems*, 42–67.

**39** Gramegna, A. and Giudici, P. (2020). Why to buy insurance? An explainable artificial intelligence approach. *Risks* 8 (4): 137.

**40** Serradilla, O., Zugasti, E., Cernuda, C. (2020). Interpreting remaining useful life estimations combining explainable artificial intelligence and domain knowledge in industrial machinery. *Proceedings of IEEE International Conference on Fuzzy Systems (FUZZ-IEEE)*, 1–8.

**41** Senoner, J., Netland, T., and Feuerriegel, S. (2021). Using explainable artificial intelligence to improve process quality: evidence from semiconductor manufacturing. *Management Science* 224 (1).

**42** S. Meister, M. Wermes, J. Stüve and R. M. Groves, "Investigations on explainable artificial intelligence methods for the deep learning classification of fibre layup defect in the automated composite manufacturing", Composites Part B: Engineering, 2021.

**43** Mehdiyev, N. and Fettke, P. (2021). Explainable artificial intelligence for process mining: a general overview and application of a novel local explanation approach for predictive process monitoring. *Interpretable Artificial Intelligence: A Perspective of Granular Computing* 937: 1.

**44** Brito, L.C., Susto, G.A., Brito, J.N., and Duarte, M.A. (2022). An explainable artificial intelligence approach for unsupervised fault detection and diagnosis in rotating machinery. *Mechanical Systems and Signal Processing* 163.

**45** Kharal, A. (2020). Explainable artificial intelligence based fault diagnosis and insight harvesting for steel plates manufacturing.

**46** Stiglic, G., Kocbek, P., Fijacko, N. et al. (2020). Interpretability of machine learning-based prediction models in healthcare. *Wiley Interdisciplinary Reviews: Data Mining and Knowledge Discovery* 10 (5).

**47** Arya, V. et al. (2019). One explanation does not fit all: A toolkit and taxonomy of ai explainability techniques.

**48** Arrieta, A.B. et al. (2020). Explainable artificial intelligence (XaI): concepts taxonomies opportunities and challenges toward responsible AI. *Information Fusion* 58: 82–115.

**49** Ying, R., Bourgeois, D., You, J. et al. (2019). Gnnexplainer: generating explanations for graph neural networks. *Advances in Neural Information Processing Systems* 32.

**50** Ribeiro, M.T., Singh, S., and Guestrin, C. (2016). 'why should i trust you?' explaining the predictions of any classifier. *Proceedings of the 22nd ACM SIGKDD International Conference on Knowledge Discovery and Data Mining*, 1135–1144.

**51** Lundberg, S.M. and Lee, S.-I. (2017). *A unified approach to interpreting model predictions in Advances in Neural Information Processing Systems 30*, 4765–4774. New York, NY, USA: Curran Associates, Inc http://papers.nips.cc/paper/7062-a-unified-approach-to-interpreting-model-predictions.pdf.

**52** Lom, M., Pribyl, O. and Svitek, M. (2016). Industry 4.0 as a part of smart cities. *Proceedings of Smart Cities Symposium Prague (SCSP)*, 1–6.

**53** Pellicer, S., Santa, G., Bleda, A.L. et al. (2013).A global perspective of smart cities: A survey. *Proceedings of 7th International Conference on Innovative Mobile and Internet Services in Ubiquitous Computing*, 439–444.

**54** Allam, Z. and Dhunny, Z.A. (2019). On big data artificial intelligence and smart cities. *Cities* 89: 80–91.

**55** Thakker, D., Mishra, B.K., Abdullatif, A. et al. (2020). Explainable artificial intelligence for developing smart cities solutions. *Smart Cities* 3 (4): 1353–1382.

**56** Shrouf, F., Ordieres, J. and Miragliotta, G. (2014). Smart factories in Industry 4.0: A review of the concept and of energy management approached in production based on the Internet of Things paradigm. *Proceedings of IEEE international conference on industrial engineering and engineering management*, 697–701.

**57** Grabowska, S. Smart factories in the age of Industry 4.0. *Management Systems in Production Engineering* 28 (2): 90–96.

**58** Wan, J., Yang, J., Wang, Z., and Hua, Q. (2018). Artificial intelligence for cloud-assisted smart factory. *IEEE Access* 6: 55419–55430.

**59** Javaid, M. and Haleem, A. (2019). Industry 4.0 applications in medical field: a brief review. *Current Medicine Research and Practice* 9 (3): 102–109.

**60** Chawla, M.N. (2020). AI IoT and wearable technology for smart healthcare – a review. *International Journal of Green Energy* 7 (1): 9–13.

**61** Pawar, U., O'Shea, D., Rea, S., and O'Reilly, R. (2020). Explainable ai in healthcare. *Proceedings of International Conference on Cyber Situational Awareness, Data Analytics and Assessment (CyberSA)*, 1–2.

**62** Krupitzer, C. et al. (2020). A survey on human machine interaction in Industry 4.0.

**63** Meske, C. and Bunde, E. (2020). Transparency and trust in human-AI-interaction: The role of model-agnostic explanations in computer vision-based decision support. *Proceedings of International Conference, AI-HCI*, 54–69

**64** Li, Z., Wang, Y., and Wang, K.-S. (2017). Intelligent predictive maintenance for fault diagnosis and prognosis in machine centers: Industry 4.0 scenario. *Advanced Manufacturing* 5 (4): 377–387.

**65** Paolanti, M., Romeo, L., Felicetti, A. (2018). Machine learning approach for predictive maintenance in Industry 4.0. *Proceedings of 14th IEEE/ASME International Conference on Mechtronic and Embedded Systems and Applications*, 1–6.

**66** Hrnjica, B. and Softic, S. (2020). Explainable AI in manufacturing: a predictive maintenance case study. *Proceedings of IFIP International Conference on Advances in Production Management Systems*, 66–73.

**67** Zajec, P., Rožanec, J.M., Novalija, I. et al. (2021). Towards active learning based smart assistant for manufacturing. *Advances in Production Management Systems. Artificial Intelligence for Sustainable and Resilient Production Systems*

**68** Nunes, M.L., Pereira, A., and Alves, A.C. (2017). Smart products development approaches for Industry 4.0. *Procedia Manufacturing* 13: 1215–1222.

**69** Frank, A.G., Dalenogare, L.S., and Ayala, N.F. (2019). Industry 4.0 technologies: implementation patterns in manufacturing companies. *International Journal of Production Economics* 210: 15–26.

**70** Tomiyama, T., Lutters, E., Stark, R., and Abramovici, M. (2019). Development capabilities for smart products. *CIRP Annals* 68 (2): 727–750.

**71** Malý, I., Sedláček, D. and Leitao, P. (2016). Augmented reality experiments with industrial robot in Industry 4.0 environment. *Proceedings of IEEE 14th International Conference on Industrial Informatics (INDIN)*, 176–181.

**72** Ervural, B.C. and Ervural, B. (2018). Overview of cyber security in the Industry 4.0 era. In: *Industry 4.0: Managing the Digital Transformation*, 267–284. Berlin, Germany: Springer.

**73** Darraj, E., Sample, C., and Justice, C. (2019). Artificial intelligence cybersecurity framework: Preparing for the here and now with AI. In: *Proceedings of 18th European Conference on Cyber Warfare and Security*, 132–141. Acad. Conf. Publishing Limited.

**74** Li, J.-H. (2018). Cyber security meets artificial intelligence: a survey. *Frontiers of Information Technology & Electronic Engineering* 19 (12): 1462–1474.

**75** Chang, M.-C. et al. (2019). AI city challenge 2019-city-scale video analytics for smart transportation. *Proceedings of CVPR Workshops*, 99–108.

**76** Zantalis, F., Koulouras, G., Karabetsos, S., and Kandris, D. (2019). A review of machine learning and IoT in smart transportation. *Future Internet* 11 (4): 94.

**77** Gupta, R., Tanwar, S., Kumar, N., and Tyagi, S. (2020). Blockchain-based security attack resilience schemes for autonomous vehicles in Industry 4.0: a systematic review. *Computers and Electrical Engineering* 86.

**78** Hill, J., Ford, W.R., and Farreras, I.G. (2015). Real conversations with artificial intelligence: a comparison between human-human online conversations and human-chatbot conversations. *Computers in Human Behavior* 49: 245–250.

**79** Qin, H. et al. (2019). Autonomous exploration and mapping system using heterogeneous UAVs and UGVs in GPS-denied environments. *IEEE Transactions on Vehicular Technology* 68 (2): 1339–1350.

**80** Benevolo, C., Dameri, R.P., and D'auria, B. (2016). *"Smart Mobility in Smart City" in Empowering Organizations*, 13–28. Berlin, Germany: Springer.

**81** Zheng, P. et al. (2018). Smart manufacturing systems for Industry 4.0: conceptual framework scenarios and future perspectives. *Frontiers of Mechanical Engineering* 13 (2): 137–150.

**82** Roblek, V., Meško, M., and Krapež, A. (2016). A complex view of Industry 4.0. *SAGE Open* 6 (2).

**83** Xu, W. (2019). Toward human-centered AI: a perspective from human-computer interaction. *Interactions* 26 (4): 42–46.

**84** Ahmed, I., Ahmad, M., Jeon, G., and Piccialli, F. (2021). A framework for pandemic prediction using big data analytics. *Big Data Research* 25.

# 3

# Industry 4.0: Design Principles, Challenges, and Applications

*K.K. Girish, Sunil Kumar, and Biswajit R. Bhowmik*

BRICS Laboratory, Department of Computer Science and Engineering, National Institute of Technology Karnataka, Mangalore, Karnataka, India

## 3.1 Introduction

Throughout history, the world has witnessed significant revolutions with immense potential to transform and restructure the entire human race. These transformations have consistently led to the establishment of a new global order with massive changes not only in various sectors but also in the mindset of the people. The industrial revolutions, which began in the late eighteenth century and continued through the nineteenth and twentieth centuries, were a series of profound changes in how goods were produced, marking a significant shift from manual labor to machine-based manufacturing. These revolutions were substantial because they brought about unprecedented technological advancements, transportation, and communication, leading to substantial changes in society, culture, and the economy. The First Industrial Revolution [1], originating in late eighteenth-century Britain, brought forth the implementation of novel machinery and production techniques, resulting in a remarkable surge in productivity and the establishment of fresh industries. This transformation significantly influenced society, fueling the expansion of urban areas, the development of the middle class, and the rise of contemporary capitalism.

The Second Industrial Revolution [2], which started in the late nineteenth century, witnessed the initiation of novel technologies such as electricity, the internal combustion engine, the telephone, and so on. This revolution had made a significant imprint on several advancements in transportation, communication, and manufacturing, making it possible to manufacture goods on a larger scale at a lower cost.

The term "digital revolution" is frequently used to describe the Third Industrial Revolution [3], which started in the middle of the twentieth century, and remains strong today. The Internet, social media, and artificial intelligence (AI) are just a few of the new technologies that have emerged due to the rapid development of computer technology. Consequently, there have been significant changes in several industries, including business, healthcare [4–8], and education. This has changed the way individuals connect with the environment around them.

All industrial revolutions shared similar goals: to stimulate economic growth, enhance communication and transportation, improve living standards, and promote innovation and creativity in the production and manufacturing industries [1–3, 9]. The industrial revolution's social, cultural, and economic impacts were profound, and their repercussions were wide-ranging. Increased productivity, urbanization, the rise of the middle class, the advent of modern capitalism, breakthroughs in technology, and changes in social and cultural norms are just a few of the primary effects of the industrial revolution. The industrial revolutions were notable for their significant impact on manufacturing goods, resulting in remarkable technological advancements and tremendous human transformations. These revolutions established the foundation for the technological advancements and innovations that we benefit from today and have left a lasting impact on our daily lives, work, and social interactions.

*Topics in Artificial Intelligence Applied to Industry 4.0*, First Edition. Edited by Mahmoud Ragab AL-Refaey, Amit Kumar Tyagi, Abdullah Saad AL-Malaise AL-Ghamdi, and Swetta Kukreja.
© 2024 John Wiley & Sons Ltd. Published 2024 by John Wiley & Sons Ltd.

## 3.2 Organization of Chapter

The rest of the chapter is organized as follows. Section 3.3 discusses a general overview of industrial revolutions. Section 3.4 explores the evolution of industrial revolutions. Section 3.5 discusses the transition to Industry 4.0. Section 3.6 explores various characteristics of Industry 4.0. Section 3.7 discusses the technologies under Industry 4.0. Section 3.8 discusses various design principles of Industry 4.0. Section 3.9 discusses multiple applications of Industry 4.0. Section 3.10 explores the recent trends in Industry 4.0. Section 3.11 discusses challenges in Industry 4.0. Section 3.12 discusses the related works. Section 3.13 discusses Industry 5.0. Section 3.14 discusses the future research dimension. Finally, Section 3.15 provides the conclusions.

## 3.3 Industrial Revolutions

The term "industrial revolution" describes a time when significant changes in how commodities and services were produced and provided to the public. Before the industrial revolution, agriculture was the mainstay of economies worldwide, with most people residing in rural areas and working on farms or in small-scale cottage industries that produced things by hand [10–12]. Living on a subsistence level and experiencing little economic growth was the norm. Production processes used a lot of manual labor and minimal machinery and technology.

The preindustrial era, which included the Middle Ages to the eighteenth century, is frequently used in Europe to refer to the time before the industrial revolution. Most manufacturing at this time was done by trained artisans utilizing hand tools and basic machinery. It was common for products to be made in small numbers in homes, workshops, or the cottage industry. Transportation could have been more active and affordable, hampering trade and the flow of people and products. Feudal systems dominated social and economic structures, with the nobles and the church having the most sway and influence [13].

First Industrial Revolution (1770–1850s): This era witnessed a massive transition toward technological breakthroughs in the manufacturing sector, including the development of new machinery, the introduction of steam power, and the growth of the factory system.

Second Industrial Revolution (1870s–1914): This period was characterized by the widespread use of electricity, the development of the internal combustion engine, and the growth of mass production techniques.

Third Industrial Revolution (1960s–present): This period has been marked by the rise of computers, automation, and the Internet, leading to significant changes in how people work and communicate.

Some historians argue that we are currently in the midst of a Fourth Industrial Revolution characterized by the rise of technologies such as cyber-physical systems (CPS), AI, the Internet of Things (IoT), and blockchain. Figure 3.1 shows a holistic view of industrial revolutions.

## 3.4 Generations of Industrial Revolutions

The industrial revolution is frequently divided into four categories based on significant technological developments and modifications to manufacturing techniques. The four industrial revolution generations are listed in the next section.

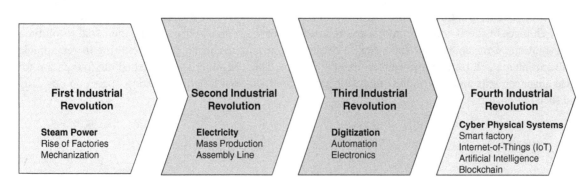

**Figure 3.1** Industrial revolutions.

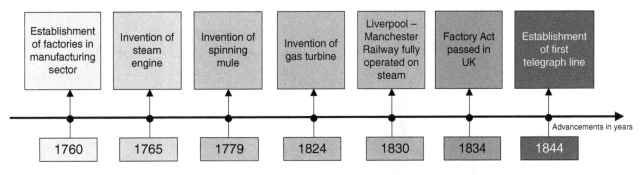

**Figure 3.2** Advancements during Industry 1.0.

### 3.4.1 First Industrial Revolution

The First Industrial Revolution, also known as "the industrial revolution" or "Industry 1.0," occurred from the mid-eighteenth century to the mid-nineteenth century, primarily in Europe and North America. It was a period of significant technological advancements in the manufacturing industry and saw the emergence of the factory system. Significant advances during this period are shown in Figure 3.2.

The main aim of the First Industrial Revolution was to increase productivity and efficiency in the manufacturing industry by replacing manual labor with machines. Several vital developments during the First Industrial Revolution transformed how goods were produced. The invention of new machinery, such as the spinning jenny and the power loom, revolutionized the textile industry and led to increased production and lower costs. The development of the steam engine, which was used to power machines, led to the growth of factories and the mass production of goods [14, 15]. Transportation also underwent significant changes during this time. The development of the steam-powered locomotive and the expansion of railway networks allowed for faster and more efficient transport of goods and people. This helped to connect markets and led to increased trade and economic growth.

The First Industrial Revolution also had significant social and economic impacts. The growth of factories and the urbanization that accompanied it led to the formation of a working class. In contrast, the development of trade and commerce created new opportunities for entrepreneurship and economic advancement. However, the new industrial system also led to significant social and economic inequalities and challenges, such as harsh working conditions, environmental degradation, and urban poverty [16].

### 3.4.2 Second Industrial Revolution

The Second Industrial Revolution, or the technological revolution or Industry 2.0, occurred from the late nineteenth century to the early twentieth century, primarily in Europe and North America. It was a period of significant technological advancements in the manufacturing industry and saw the emergence of new forms of power, such as electricity and the internal combustion engine [17]. Significant upgrades during this period are shown in Figure 3.3.

The Second Industrial Revolution aimed to further increase productivity, efficiency, and innovation in the manufacturing industry by building upon the achievements of the First Industrial Revolution. The invention of new types of machinery, such as the Bessemer converter for steel production and the assembly line for mass production, revolutionized the manufacturing industry and led to increased productivity and lower costs. The development of new forms of power, such as electricity and the internal combustion engine, also significantly impacted industry and transportation. The widespread adoption of electricity allowed for the growth of new initiatives, such as electronics, while the internal combustion engine led to the development of the automobile industry and the expansion of global transportation networks [2].

The Second Industrial Revolution profoundly impacted society, transforming how goods were produced and consumed and leading to significant social and economic changes. It also saw the rise of large corporations, the expansion of global trade and imperialism, and the emergence of new social and political movements. The Second Industrial Revolution faced significant challenges, which are still relevant today, including social inequality, labor exploitation, environmental degradation, resource depletion, political instability, and so on. However, it also paved the way for further technological innovation and economic growth, which has continued to shape the modern world.

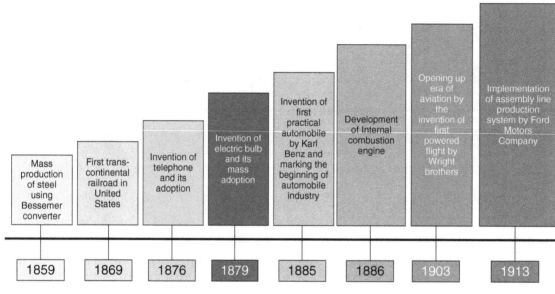

**Figure 3.3** Advancements during Industry 2.0.

### 3.4.3 Third Industrial Revolution

The Third Industrial Revolution, also known as "the digital revolution" or "Industry 3.0," is a period of rapid technological change that began in the late twentieth century and continues to the present day. It is characterized by the widespread adoption of digital technologies and the emergence of the Internet and other forms of digital communication [3]. Significant advancements during this period are shown in Figure 3.4.

The Third Industrial Revolution builds upon the technological advancements of the Second Industrial Revolution but differs in several key ways. Rather than focusing primarily on physical production, the Third Industrial Revolution is characterized by digitizing information and developing new forms of communication and collaboration. One of the key motivations behind the Third Industrial Revolution was the need to address the challenges posed by the Second Industrial Revolution, such as environmental degradation, social inequality, and resource depletion. Digital technologies and automation can reduce waste, improve efficiency, and create a more sustainable and equitable economic system [18].

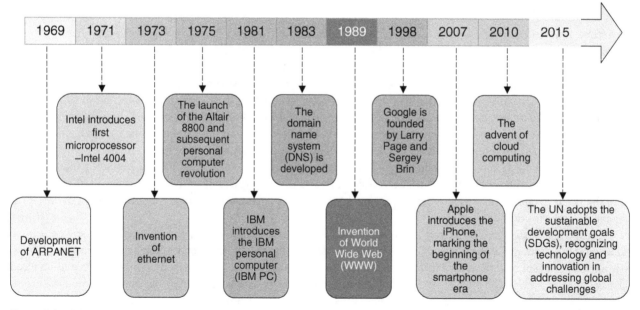

**Figure 3.4** Advancements during Industry 3.0.

Some key features of the Third Industrial Revolution include the following:

- The Boom of Information Technology: The Third Industrial Revolution witnessed the widespread adoption of information technology, including personal computers, the Internet, and mobile devices. These innovations have transformed how people interact, work, and access information.
- The Emergence of E-commerce: The Third Industrial Revolution has led to the growth of e-commerce and online retail as consumers increasingly purchase goods and services through digital channels.
- The Development of Automation: The Third Industrial Revolution has also led to the development of automation technologies, such as robotics and AI, which have transformed how goods are produced and services are delivered.
- The Growth of the Gig Economy: The Third Industrial Revolution has led to the emergence of the gig economy, as more people work as freelancers or contractors and use digital platforms to find work.

## 3.5 Transformation to Industry 4.0

The core of every industrial revolution is to increase productivity. Starting from steam power, electricity, and digitization, all contributed to this common goal. However, as the industries flourished beyond a saturation point, it became easier to efficiently handle large volumes of big data with the help of intelligent analytical tools [19]. It has become time to integrate embedded intelligence and predictive technologies into mass production to achieve expected efficiency within the stipulated time. Subsequently, this has pushed the world into a new technological revolution.

The Fourth Industrial Revolution, popularly known as "Industry 4.0," refers to the current trend of automation and data exchange in manufacturing technologies, including CPSs, the IoT, cloud computing, and cognitive computing [20–22]. Industry 4.0 originated from a national strategic initiative by the German government in 2011 [23]. The transformation to Industry 4.0 is an ongoing process involving integrating these technologies into companies' manufacturing processes across different sectors. This involves a significant shift in manufacturing processes from traditional mass production methods to highly automated and data-driven systems.

The transformation to Industry 4.0 from previous industrial revolutions involves eliminating human factors from the production line. While the last industrial processes exhibited this to some extent, the technologies under the umbrella of Industry 4.0 allow for the total elimination of human error, thereby ensuring complete automation of the production [24]. The first step in transforming to Industry 4.0 is the digitization of processes. This involves using sensors and other digital devices to collect data on various aspects of the manufacturing process, from production output to equipment performance. The next step is integrating this data into a centralized system, such as a cloud-based platform. This allows for real-time manufacturing process monitoring and provides insights into areas that need improvement. Once the data is integrated, the next step is to automate specific processes using CPSs and robotics. This allows for faster and more efficient production and greater precision and accuracy in manufacturing.

Another critical aspect of Industry 4.0 is the implementation of cognitive technologies, such as AI and machine learning. These technologies can help optimize manufacturing processes and improve decision-making by analyzing large amounts of data and providing insights into patterns and trends. Finally, the transformation to Industry 4.0 involves the development of new business models that leverage these technologies to create new products and services. This can include the development of new software platforms, the creation of new value chains, and using predictive analytics to anticipate customer needs and preferences.

## 3.6 Characteristics of Industry 4.0

Industry 4.0 is characterized by integrating advanced digital technologies into manufacturing and other industries. As shown in Figure 3.5, the key characteristics of Industry 4.0 include the following:

- Interconnectivity: This refers to the ability of machinery, systems, and humans to communicate effortlessly with each other in real time through the IoT and other digital technologies.
- Automation: Industry 4.0 is characterized by automating processes using robotics, IoT, AI, and machine learning algorithms. This allows companies to maximize their production with greater efficiency and less human intervention.

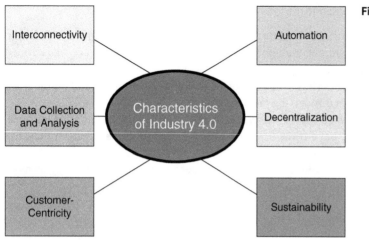

**Figure 3.5** Characteristics of Industry 4.0.

- Data Collection and Analysis: Industry 4.0 involves gathering massive amounts of data from diverse sources through sensors, machinery, and other devices. This data is employed in enhancing decision-making and optimizing processes.
- Decentralization: Industry 4.0 is characterized by decentralizing decision-making, with greater autonomy given to machines and systems. This allows for greater flexibility and responsiveness to changing conditions.
- Customer-Centricity: Industry 4.0 focuses on meeting customer needs and preferences through personalized and customized products and services.
- Sustainability: Industry 4.0 aims to reduce the environmental impact of manufacturing and other sectors through sustainable and green technologies.

## 3.7 Technologies Under Industry 4.0

Over the last decade, the global industrial landscape has significantly changed due to technological advancements and innovations. Industry 4.0 is characterized by a wide range of diverse digital technologies that have the potential to transform manufacturing and other related sectors substantially. Some of the critical technologies under Industry 4.0 include CPSs, IoT, big data and analytics, cloud technology, AI, blockchain, simulation and modeling, visualization technology (augmented and virtual reality), automation and industrial robots, additive manufacturing, and so on [25, 26]. Figure 3.6 shows technologies under Industry 4.0.

### 3.7.1 Cyber-Physical Systems

CPSs play a pivotal role in Industry 4.0. The term "cyber-physical system" was first introduced by Helen Gill in 2006 at the National Science Foundation, United States [27]. A CPS is an association of interconnected, complex heterogeneous, and networked systems whose operations are controlled, monitored, and coordinated by computing, communication, and components interacting with the physical environment [28]. It combines the physical and virtual worlds, creating intelligent factories for real-time monitoring, control, and optimization of biological processes in the manufacturing sector. Figure 3.7 shows a conceptual diagram of a CPS.

In Industry 4.0, CPSs enable a seamless interaction between the physical and digital components, creating intelligent and interconnected systems. Here are some key aspects and benefits of CPSs in Industry 4.0:

- Real-Time Monitoring: CPS collects real-time data through sensors and other data collection devices attached to physical objects. Then the collected data is transferred to a processing system for analysis, enabling real-time monitoring and control of the physical processes. This capability facilitates enhancement in production, downtime reduction, and efficiency improvement.
- Communication and Connectivity: CPS promotes flawless communication and data interchange between physical components, machines, and systems. CPS in smart factories enables communication and connectivity among smart objects

**Figure 3.6** Technologies under Industry 4.0.

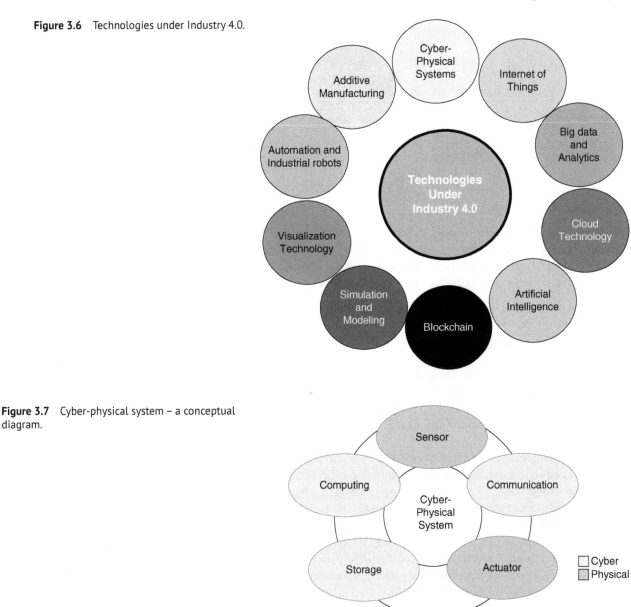

**Figure 3.7** Cyber-physical system – a conceptual diagram.

and facilitates among themselves reconfiguring the necessary system resources based on particular system dynamics [29].

- Autonomous Decision-Making: CPS exhibits the capability of autonomous decision-making based on the combination of data collected through smart objects and AI technologies. This enables smart objects in the production line to make efficient decisions and predictions without human intervention, resulting in improved productivity and responsiveness.
- Predictive Maintenance: CPS facilitates predictive maintenance by regular and periodic monitoring and condition checking of the machines, equipment, and tools used in smart factories. With proper data pattern analysis, self-learning, self-diagnosis, and self-coordination [29], CPS can predict faults, costly breakdowns, upcoming maintenance, and so on.
- Adaptability and Flexibility: CPS transforms production line easily reconfigurable and flexible. With the help of efficient digital interfaces and control systems, CPS allows manufacturing companies to respond rapidly to dynamic market demands, corresponding resource supply, current trends, challenges, or last-minute changes in customer orders [29].
- Improved Security: CPS encompasses novel and advanced measures to protect interconnected components from cyber threats, thereby ensuring the safety and security of the production systems. It also aids in the protection of critical infrastructure, sensitive data, and intellectual property from cyber threats.

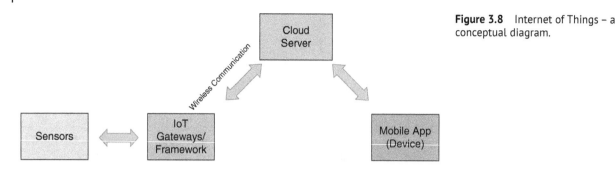

### 3.7.2 Internet of Things

The IoT is an essential component of Industry 4.0, which works to create flexible connections and communication of physical objects and systems through the Internet. The CPS enhances smart factory productivity using IoT technology by seamlessly integrating devices, sensors, and machines [30]. Figure 3.8 shows a conceptual diagram of IoT.

Here are the key aspects and benefits of IoT in Industry 4.0:

- Connectivity and Data Exchange: IoT facilitates interaction between a wide range of physical objects, from sensors and machinery to products and automobiles, creating an interconnection network where data can be collected and exchanged in real time. This makes not only the entire production process but also the delivery systems highly interactive and intelligent.
- Real-Time Monitoring: IoT equipment gathers data from the surrounding physical environments through intelligent devices and sensors and transmits it to processing systems for analysis. This real-time monitoring and analytics provide valuable insights into manufacturing processes, equipment performance [31], energy consumption, and other relevant dynamics. Organizations can use this data to optimize operations, identify deficiencies, and make appropriate, timely decisions.
- Predictive Maintenance: IoT enables predictive maintenance by regularly monitoring the condition and performance of machines and equipment. While continuously analyzing various data patterns and detecting deficiencies, IoT can predict maintenance needs and identify potential failures before they happen. This proactive approach helps minimize downtime, reduce maintenance costs, and optimize equipment lifespan.
- Supply Chain Optimization: IoT enhances every aspect of the supply chain, from logistics to warehouse management, mitigating all inefficiencies. With IoT-enabled RFID tracking and GPS monitoring, companies can obtain real-time information about inventory levels, product location, and delivery status, ensuring on-time delivery and minimizing losses. This transparency enables better inventory management and asset tracking and prevents counterfeiting and substitution, improving supply and demand planning [32].
- Enhanced Product and Service: IoT facilitates the development of intelligent and connected products that have the potential to exhibit advanced functionalities and services. IoT helps companies collect usage data, provide remote support, and offer customized experiences by embedding sensors and other connectivity devices in the manufactured product. This connectivity also allows for remote updates and upgrades, warranty servicing, extending the product life cycle, and improving customer satisfaction.
- Energy Efficiency and Sustainability: IoT plays a significant role in promoting energy efficiency and sustainability. With regular and continuous monitoring of industrial processes based on energy consumption data, companies can point out areas for improvement, reduce waste, and minimize environmental impact [33].
- Safety and Security: IoT-enabled solutions also exhibit efficient security measures to ensure data integrity and safety of the connected components. IoT security becomes paramount to mitigate cybersecurity risks and protect sensitive information as many devices are connected.

### 3.7.3 Big Data and Analytics

Big data and analytics play a critical role in Industry 4.0 by equipping the capability of data to discover valuable correlations, enhancing processes, and enabling data-driven decision-making. Here are the key aspects and benefits of big data and analytics in Industry 4.0:

- Data Collection and Integration: Industry 4.0 demands the gathering of large amounts of data from different sources, including sensors, smart objects, machines, equipment, systems, and other components in the production line. Big data technologies facilitate the collection, storage, and amalgamation of these diverse and vast data from structured and unstructured sources [34].

- Real-Time Monitoring: The capability of big data analytics to process data in real-time allows real-time monitoring and analysis of manufacturing processes, equipment performance [31], machinery efficiency, and various other operational benefits. This enables companies to identify deficiencies, mishaps, and faults as they occur, facilitating proactive decision-making and timely interventions.

- Predictive and Prescriptive Analytics: Big data analytics uses AI technologies and other statistical models to analyze vast volumes of historical and real-time data, enabling predictive and prescriptive analytics. Predictive analytics forecasts future events, such as equipment failures, demand patterns, or maintenance needs with the help of historical data. Prescriptive analytics takes it further by providing recommendations, suggestions, and optimal actions to address predicted outcomes.

- Process Optimization and Efficiency: Big data analytics aid in the identification of deficiencies, performance bottle-necks [35], and optimization opportunities in manufacturing processes. Companies can gain insights into operational performance, energy consumption, quality control, and supply chain dynamics by analyzing large datasets. This information allows for process optimization, waste reduction, and improved efficiency.

- Quality Control and Defect Detection: Big data analytics helps provide real-time quality control by continuously monitoring production data and pointing out deficiencies, deviations, faults, or anomalies. Advanced analytics methods can enable companies to detect correlations and hidden patterns that help identify the root causes, predict quality issues, and promptly initiate corrective actions.

- Supply Chain Optimization: Big data analytics creates better decisions for all supply chain-related operations by combining data and statistical methodologies. Big data analytics helps in supplier relationship management, assists the manufacturers in product design and development, demands planning, and optimizes inventory levels and logistics management. Companies can enhance supply chain efficiency, reduce costs, and improve customer satisfaction by analyzing data from various sources, such as sales data, supplier data, and market trends [36].

- Product and Service Innovation: Big data analytics promote innovations in products and services by unwrapping customer insights, demands, preferences, and market trends. By adequately analyzing customer data, user feedback, and current market trends, companies can develop and customize products and services that can satisfy their customer's needs well.

- Decision Support and Strategic Planning: Big data analytics exhibits data-driven recommendations and suggestions for decision-makers so that productive steps are taken during all the phases of manufacturing till the delivery. It also helps managers make informed decisions, eliminate risks, and optimize resource allocation. As big data analytics enables scenario modeling and predictive modeling for long-term planning, it benefits strategic planning very well.

### 3.7.4 Cloud Technology

Cloud technology is a fundamental enabler of Industry 4.0, facilitating a flexible, interactive, and scalable platform for storing, processing, and accessing data and applications. Figure 3.9 shows an architecture of cloud technology. Cloud technology offers several benefits in the context of Industry 4.0:

- Data Storage and Management: Cloud technology provides extensive centralized storage infrastructure to companies for storing huge volumes of data securely. This mitigates the requirement for vast on-premises data storage and facilitates a scalable solution to accommodate the ever-growing data needs of Industry 4.0 applications [37].

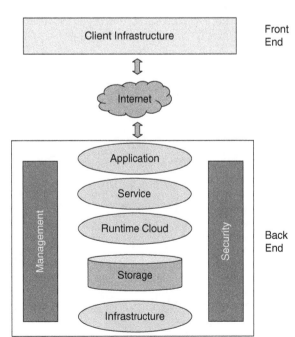

**Figure 3.9** Cloud technology architecture.

- Data Processing and Analytics: Cloud platforms exhibit powerful computing resources and data-processing capabilities. Companies can use cloud-based analytics tools and services to perform complex data analysis, machine learning algorithms, and real-time processing of sensor data, enabling advanced analytics and actionable insights.
- Collaboration and Connectivity: Cloud technology facilitates collaboration and connectivity between various stakeholders across different geographical locations. Cloud-based collaboration tools enable real-time data, documents, and workflow sharing, fostering seamless collaboration and promoting efficient remote work.
- Flexibility and Scalability: Cloud-based solutions provide a flexible and scalable platform for companies to adjust their computing resources and storage capacity based on their needs. This flexibility is beneficial in the dynamic and evolving landscape of Industry 4.0, where data volumes and computing requirements can vary significantly.
- Cost Efficiency: The shift from capital expenditure to operational expenditure by cloud technology helps companies optimize production costs. Rather than investing in expensive infrastructure and maintenance, companies can take advantage of cloud services on a pay-per-use basis, minimizing initial outlay and achieving cost streamlining.
- Rapid Deployment and Innovation: Cloud platforms support the speedy rollout of applications, minimizing the time to market for new services and solutions. Industries can harness cloud infrastructure to investigate, prototype, and rapidly scale innovative ideas without significant upfront investments [38].
- Reliability and Security: Cloud service providers offer decisive security measures, including data encryption, access controls, and regular backups. They also extend high availability and redundancy, ensuring critical data and applications are protected and accessible even during hardware failures or disasters.

### 3.7.5 Artificial Intelligence

AI plays a groundbreaking role in Industry 4.0, transforming heterogeneous factors of manufacturing and associated industries. AI is defined as machines' ability to imitate human intelligence, enabling them to learn, reason, and make decisions [39]. Figure 3.10 shows various AI technologies in Industry 4.0. Here are vital aspects and applications of AI in Industry 4.0:

- Machine Learning: Machine learning is a subset of AI that concentrates on algorithms and models that encourage machines to learn from data and improve their performance over time [35, 40, 41]. In Industry 4.0, machine learning algorithms can analyze large datasets to identify hidden patterns, correlations, and anomalies [39]. This facilitates predictive maintenance, quality control, demand forecasting, and optimization of industrial production processes.
- Autonomous Systems: AI supports the development of autonomous systems that can perform tasks without human intervention. For instance, autonomous robots can handle repetitive or hazardous tasks on the factory floor, improving productivity and worker safety. Autonomous vehicles and drones are used to improve efficiency and minimize costs in logistics and transportation.
- Cognitive Computing: AI techniques such as natural language processing, speech recognition, and computer vision help cognitive computing systems to exhibit human-like cognitive capabilities. These systems promote advanced human-machine interactions, such as voice commands, natural language interfaces, and visual recognition. They are mainly employed in customer service, product design, and quality control [42].
- Predictive Analytics: AI-powered predictive analytics utilizes historical and real-time data to forecast future events and trends. In Industry 4.0, predictive analytics can optimize maintenance schedules, pinpoint potential equipment failures, and foresee demand instabilities. This proactive approach helps minimize downtime, optimize resource allocation, and enhance operational efficiency.
- Intelligent Virtual Assistants: AI-powered intelligent virtual assistants provide conversational interfaces and support tasks such as information retrieval, scheduling, and data analysis. These assistants, such as chatbots or voice-based assistants, improve productivity, deliver real-time information, and enhance customer support.

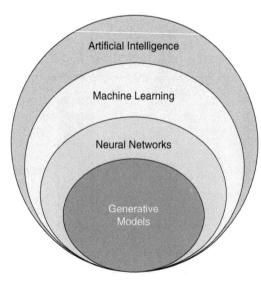

**Figure 3.10** Artificial intelligence technologies.

- Computer Vision: Computer vision technology uses AI techniques to enable machines to interpret and analyze visual data, such as images or videos. Computer vision can be used in Industry 4.0 for object recognition, quality inspection, and gesture control. It improves accuracy, speed, and efficiency in tasks that require visual perception.
- Optimization and Decision Support: Complex systems are optimized using AI algorithms by analyzing multiple variables, constraints, and objectives. These algorithms can optimize production schedules, resource allocation, and supply chain logistics, improving efficiency and cost reduction. AI-powered decision support systems assist managers by providing data-driven insights and recommendations for strategic decision-making.
- Intelligent Data Analysis: AI techniques support intelligent data analysis, allowing organizations to extract valuable correlations from vast volumes of data. AI algorithms can analyze unstructured data sources, such as text documents, social media, and sensor data, to uncover patterns, sentiments, and trends. This information can be used for market analysis, customer profiling, and product innovation.

### 3.7.6  Blockchain

Blockchain technology can benefit Industry 4.0 by enhancing trust, security, and transparency in various processes. As shown in Figure 3.11, vital aspects and applications of Blockchain in Industry 4.0 include the following:

- Supply Chain Management: Blockchain helps enhance supply chain visibility by facilitating a decentralized and immutable ledger that records transactions and the mobility of goods. It enables end-to-end traceability, ensuring transparency and authenticity of products. Blockchain also facilitates the verification of certifications, regulatory compliance, and provenance throughout the supply chain.
- Smart Contracts: Blockchain allows the execution of smart contracts, self-executing agreements with predefined rules and conditions. Smart contracts can automate and streamline processes in Industry 4.0, such as payment settlements, asset transfers, and service-level agreements. They eliminate intermediaries, minimize administrative costs, and improve efficiency [43].
- Product Lifecycle Management: Product lifecycle management is the collection of managing activities of a company's product. Blockchain technology keeps track of a product throughout its entire lifecycle by securely recording and sharing its information. It helps manufacturers, dealers, suppliers, and customers to access and update product information, including design specifications, manufacturing history, maintenance records, and warranties. This transparency improves collaboration, minimizes errors, and ensures data integrity.
- Intellectual Property Protection: Blockchain provides a secure and tamper-proof platform for safeguarding intellectual property rights. It can be used to time-stamp and record the creation and ownership of digital assets, designs, patents, and copyrights. Blockchain-based solutions enable decentralized storage and management of intellectual property, preventing unauthorized modifications or breaches.

**Figure 3.11**  Blockchain applications.

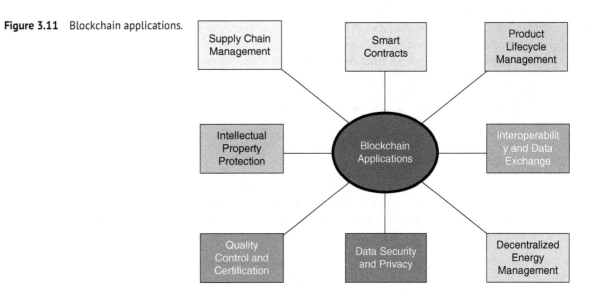

- Quality Control and Certification: Blockchain can confirm and ascertain product quality and certifications. It permits the safe storage of quality control data, inspection reports, and credentials on a distributed ledger. This allows easy access to validated information, minimizes counterfeit products, and improves consumer trust.
- Data Security and Privacy: Blockchain offers a decentralized and immutable ledger, ensuring data security and privacy in Industry 4.0. It supports cryptographic techniques to safeguard sensitive information, control access permissions, and stop unauthorized tampering. Blockchain can also help secure data sharing and identity management across multiple stakeholders [44].
- Decentralized Energy Management: Blockchain technology facilitates decentralized energy management systems where players securely record and share energy production and consumption data. This permits peer-to-peer energy trading, efficient grid management, and incentivizing renewable energy sources. Blockchain-based solutions can promote energy efficiency and sustainability.
- Interoperability and Data Exchange: Blockchain can foster secure and standardized data exchange between systems, devices, and organizations in Industry 4.0. It promotes interoperability by setting a common framework for data sharing and communication protocols. Blockchain-based data exchanges improve collaboration, interoperability, and trust among participants.

### 3.7.7 Simulation and Modeling

Simulation and modeling are essential components of Industry 4.0, enabling the virtual representation of physical systems and processes. Here are some of the critical applications and benefits of simulation and modeling in Industry 4.0:

- Design and Optimization: Simulation and modeling allow for creating and testing virtual models of products, machines, and processes. Engineers and designers can optimize designs, improve performance, and reduce costs before building physical prototypes. Virtual simulations also provide insights into the behavior of complex systems, allowing for the exploration of various scenarios and alternatives [45].
- Predictive Maintenance: Simulation and modeling can predict machines and equipment's potential failures and maintenance requirements. By analyzing sensor data and machine learning algorithms, simulations can predict the remaining useful life of components and identify optimal maintenance schedules. This minimizes downtime and maintenance costs and improves equipment's reliability and performance.
- Training and Education: Simulation and modeling provide an immersive and interactive environment for training and education. Virtual simulations permit for practicing skills and techniques in a safe and controlled environment, reducing risks and costs associated with real-world training. Virtual simulations enable remote learning and collaboration, providing access to training and education from anywhere.
- Process Optimization: Simulation and modeling can optimize manufacturing and production processes by identifying bottlenecks, optimizing workflows, and improving resource allocation. Simulating different scenarios can optimize process parameters to reduce waste, increase efficiency, and strengthen quality [46].
- Supply Chain Optimization: Simulation and modeling can optimize supply chain management by examining data and simulating diverse scenarios. This allows the optimization of inventory levels, transportation routes, and production schedules, enhancing the overall efficiency and stability of the supply chain.
- Safety and Risk Management: Simulation and modeling can simulate and analyze safety and risk scenarios, identifying potential hazards and developing appropriate safety measures. Virtual simulations evaluate different methods and interventions, improving systems' overall safety and reliability.

### 3.7.8 Visualization Technology

Visualization, characterized by integrating digital technologies into manufacturing and industrial processes, plays an essential role in Industry 4.0. Visualization enables companies to gain insights, make informed decisions, and optimize operations through data visualization, augmented reality (AR), virtual reality (VR), and other visual tools [47]. As shown in Figure 3.12, here are the key areas where visualization is applied in Industry 4.0:

- Data Visualization: With the proliferation of sensors and IoT devices, vast amounts of data are generated in manufacturing environments. Data visualization techniques like dashboards and real-time monitoring systems help transform raw data into actionable insights. It enables operators, managers, and decision-makers to understand complex data patterns, identify anomalies, and make data-driven decisions for improved efficiency and productivity.

- Augmented Reality and Virtual Reality: AR and VR technologies are widely used in Industry 4.0 to enhance manufacturing and maintenance processes. AR overlays digital information onto the physical world, providing workers real-time instructions, guidance, and visualizations. It can be used for tasks like assembly, quality control, and training, improving accuracy, reducing errors, and increasing efficiency [47]. Conversely, VR creates a simulated environment that allows users to experience virtual representations of real-world scenarios. It can be utilized for immersive training, simulation, and design reviews, leading to better understanding, collaboration, and innovation.

- Digital Twin Visualization: Digital twin technology creates virtual replicas of physical assets, machines, or entire production systems. Visualization plays a crucial role in digital twins by visually

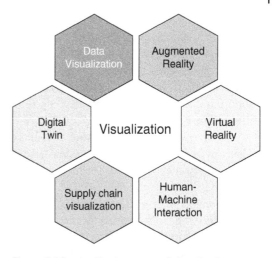

**Figure 3.12** Application areas of visualization technologies.

representing the physical entity and its real-time data. It helps manufacturers to monitor, analyze, and optimize performance, predict maintenance needs, and simulate scenarios for process improvement. By visualizing the digital twin, companies can comprehensively understand the physical system and make data-driven decisions for optimization [48].

- Supply Chain Visualization: Visualization techniques monitor and optimize supply chain processes in Industry 4.0. Using visual analytics, companies can track inventory levels, analyze demand patterns, identify bottlenecks, and optimize logistics operations. Supply chain visualization helps improve efficiency, reduce costs, and enhance overall performance by providing a clear view of the entire supply chain network and facilitating better decision-making.

- Human-Machine Interface: In Industry 4.0, human-machine interfaces (HMIs) have evolved to provide intuitive and interactive visualizations. Touchscreens, graphical user interfaces, and other visual tools allow operators to monitor and control complex manufacturing systems. Visualizing real-time data, alarms, and system status lets operators quickly respond to changes, diagnose issues, and take appropriate actions.

### 3.7.9 Automation and Industrial Robots

Industry 4.0 relies heavily on automation and industrial robots to revolutionize manufacturing processes, boost productivity, and enable greater flexibility and customization. As shown in Figure 3.13, aspects and uses of automation and industrial robots that are crucial to Industry 4.0 are listed here:

**Figure 3.13** Application areas of automation and industrial robots.

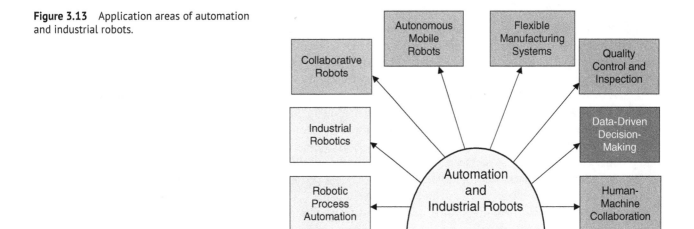

- Robotic Process Automation (RPA): RPA involves automating repetitive and rule-based processes typically completed by humans by employing software robots, sometimes called "bots." RPA can speed up administrative procedures, data entry, and data analysis, enhancing precision, efficacy, and economy [49].
- Industrial Robotics: Industrial robots are programmable devices capable of accurately, promptly, and reliably carrying out operations. They are employed in several production procedures, including material handling, welding, and assembling. Industrial robots enable tackling difficult or dangerous operations while increasing production and decreasing errors.
- Collaborative Robots: Collaborative robots (cobots) are machines that cooperate with people in the workplace. They can carry out activities like assembly or quality control that demand close human engagement. Cobots are equipped with safety measures like sensors and force-limiting systems to ensure that humans and robots work together safely [50].
- Autonomous Mobile Robots: Autonomous mobile robots (AMRs) are robots with mobility features that allow them to move independently inside a factory or warehouse. They can move things, supplies, or components, improving logistics and minimizing hand-holding. AMRs increase productivity, save labor costs, and improve material flow.
- Flexible Manufacturing Systems: The development of adaptable manufacturing systems that can change with changing production demands is made possible by automation and robotics. Robots can be quickly customized and reprogrammed to handle various activities or products. This adaptability enhances market response and allows mass customization.
- Quality Control and Inspection: Robotics and automation can improve quality control through thorough and reliable inspections. Robotic systems can use vision and sensors to inspect products for flaws, measure their dimensions, and check their quality. This lessens human mistakes, enhances quality control, and guarantees standard compliance.
- Data-Driven Decision-Making: Massive amounts of data are produced by automation and robotics, which can be analyzed to reveal new information and improve processes. Robot-generated data can be analyzed using data analytics and machine learning techniques to spot trends, optimize performance, and anticipate maintenance requirements. Decisions are made using data to increase operational effectiveness and decrease downtime [51].
- Human-Machine Collaboration: Robotics and automation enable collaborative human-machine work, in which both species cooperate to complete challenging tasks. Humans can better concentrate on higher-level decision-making, problem-solving, and creativity, while robots undertake laborious or repetitive jobs. This partnership improves output, effectiveness, and job satisfaction.

Industry 4.0 is primarily driven by automation and industrial robots, which alter manufacturing processes and allow for higher productivity levels, adaptability, and personalization. They increase productivity, quality assurance, and safety while creating new opportunities for competition and innovation in the digital age.

### 3.7.10 Additive Manufacturing

Industry 4.0 heavily relies on additive manufacturing, often known as "3D printing," a game-changing technology. It includes creating three-dimensional items layer by layer using computer models. As shown in Figure 3.14, the following are crucial elements and uses of additive manufacturing in Industry 4.0:

- Rapid Prototyping: Additive manufacturing makes producing prototypes quickly and affordably possible. Before committing to large-scale production, it helps designers and engineers to iterate designs, test functionality, and validate concepts rapidly. Time to market is shortened, and product development cycles are accelerated [52].
- Customization and Personalization: Fabricating highly customized and personalized products are made possible through additive manufacturing. Individual items can be quickly customized to meet unique consumer needs by utilizing digital design files, leading to higher customer satisfaction and market differentiation.
- Additive printing allows for design freedom, enabling the fabrication of complicated geometries and lightweight structures that are challenging or impossible to realize using conventional manufacturing techniques. This makes it possible to produce complex, optimized components with less material, which reduces weight and improves performance.
- Manufacturing on Demand: Additive manufacturing makes it possible to produce goods just when needed, lowering inventory costs and eliminating the need for large-scale production facilities. This makes it possible for a manufacturing model to be more flexible and responsive, accommodating changing demand and lowering the danger of having too much inventory.

**Figure 3.14** Application areas of additive manufacturing.

- Supply Chain Optimization: By decentralizing production, additive manufacturing can improve the supply chain. Products can be printed nearby or on-site rather than at a centralized manufacturing plant, which lowers lead times and transportation expenses. This provides a more resilient and sustainable supply chain and facilitates distributed manufacturing.
- Spare Parts Production and Obsolescence Management: On-demand creation of spare parts using additive manufacturing can cut inventory costs and lead times associated with conventional spare parts management. By allowing the creation of legacy parts that might no longer be accessible through traditional manufacturing methods, it also helps address issues with obsolescence.
- Tooling and Jig Production: Custom tooling, jigs, and fixtures for manufacturing processes can be made using additive manufacturing. These tools can be swiftly created and made to support particular manufacturing requirements, improving process efficiency and lowering costs related to manufacturing traditional tooling.
- Material Innovation and Lightweighting: Utilizing novel materials created especially for the process is made possible by additive manufacturing. This creates possibilities for material innovation, including creating specialized and high-performance materials. Additionally, additive manufacturing supports lightweight tactics by maximizing material consumption and lowering product weight overall [53].

Industry 4.0 manufacturing processes are being revolutionized by additive manufacturing, which has benefits including quick prototyping, customization, complicated geometries, and decentralized production. It enables businesses to be more imaginative, flexible, and creative, changing how things are developed, made, and delivered.

## 3.8 Design Principles of Industry 4.0

Design principles that underpin Industry 4.0 direct the incorporation of cutting-edge digital technology into manufacturing and other sectors. As shown in Figure 3.15, interoperability, virtualization, real-time capabilities, service orientation, modularity, information transparency, decentralization, innovative products, corporate social responsibility (CSR), technical support, resource efficiency, and so on [54], are some of the design principles mentioned in the next section.

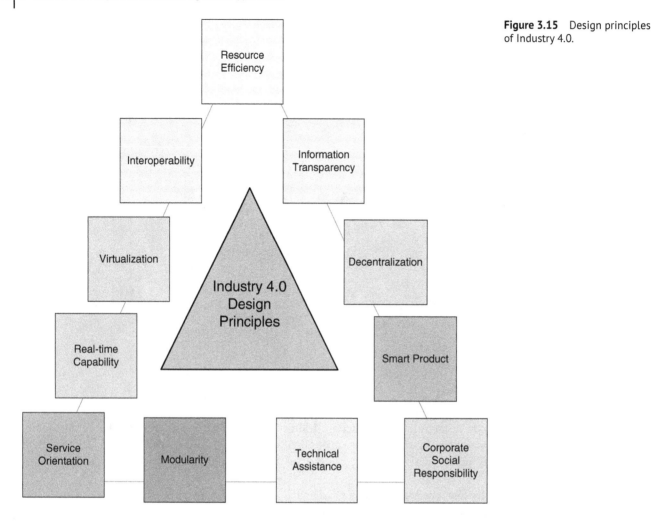

**Figure 3.15** Design principles of Industry 4.0.

### 3.8.1 Interoperability

The ability of machines, devices, and systems to converse and exchange data is known as "interoperability" and is a crucial element of Industry 4.0. Interoperability in Industry 4.0 is made possible by open standards and protocols that allow various devices and processes to communicate and share data [55].

For several reasons, interoperability is significant in Industry 4.0. The first benefit is that it makes it easier for various platforms and systems to collaborate and integrate. For instance, machines from many manufacturers can coexist peacefully, enabling more effective and adaptable production processes.

Second, increased data sharing and analysis are made possible by interoperability. Businesses can gather and analyze data from various sources, enabling more precise and well-informed decision-making by integrating equipment and systems.

Third, interoperability helps businesses to create production systems that are more adaptable and scalable. Companies can quickly add new equipment and techniques to their current infrastructure by leveraging open standards and protocols, which enable them to react to changing business needs and consumer expectations.

### 3.8.2 Virtualization

Industry 4.0's key technology, virtualization, makes creating digital replicas of real-world products and operations possible [56]. Virtualization is employed in Industry 4.0 in several ways, which include the following:

- Virtual Product Design and Testing: Businesses can test and improve items before they are manufactured physically, cutting down on the time and expense of development.

- Virtual Efficiency: Virtualization enables businesses to model and optimize production processes, reducing errors and downtime while increasing productivity.
- Virtual Commissioning: To reduce risk and downtime during installation, virtualization can also be utilized to commission and test production systems before physical deployment.
- Virtual Training and Education: Industry 4.0 also uses virtualization for training and education, allowing employees to learn and hone their abilities in a secure setting.
- Virtual Maintenance and Repair: In addition to providing predictive maintenance and lowering downtime, virtualization may be used to monitor and diagnose issues with devices and systems.

Virtualization, in general, is a critical technology in Industry 4.0, providing increased flexibility, efficiency, and cost savings by allowing businesses to build digital twins of natural things and processes.

### 3.8.3 Real-Time Capability

A key component of Industry 4.0 is real-time capability, which enables machines and systems to function and communicate in real time with little delay or latency. Real-time functionality is achieved in Industry 4.0 by utilizing cutting-edge digital technologies, including the IoT, cloud computing, and edge computing [57].

For various reasons, Industry 4.0 places a premium on real-time functionality. It increases productivity and lowers downtime by allowing machines and systems to react promptly to changes in production processes. To minimize breakdowns and lower repair costs, sensors, for instance, can spot equipment irregularities and initiate real-time maintenance processes.

Real-time capability makes more excellent connectivity and cooperation between machines and systems possible. For instance, real-time data exchange enables devices to coordinate operations and optimize manufacturing processes to increase efficiency and quality.

Companies with real-time capabilities may gather and analyze data in real time, allowing for more precise and well-informed decision-making. Real-time data analytics, for instance, can assist businesses in locating and eliminating production bottlenecks, which will decrease waste and boost throughput.

### 3.8.4 Service Orientation

Service orientation, which refers to using digital services to improve production procedures and develop new business models, is a significant concept in Industry 4.0. Service orientation is accomplished in Industry 4.0 by utilizing cutting-edge digital technologies, including the IoT, cloud computing, and big data analytics [58].

Industry 4.0 places a lot of emphasis on service orientation. It enables businesses to develop new revenue streams and business models by providing digital services like asset tracking, remote monitoring, and predictive maintenance.

Service orientation makes greater personalization and customization of products and services possible. For instance, businesses can customize their products and services to match their client's unique needs and tastes by gathering and analyzing data from sensors and other sources.

Service orientation makes greater collaboration and integration between various businesses and industries possible. For instance, companies can work with other businesses along the value chain to optimize production processes and boost efficiency by providing digital services.

### 3.8.5 Modularity

Modularity, which refers to a machine or system's ability to be quickly reconfigured and integrated into various manufacturing processes, is a crucial concept in Industry 4.0. Utilizing standardized interfaces and parts and cutting-edge digital innovations like IoT and cloud computing, Industry 4.0 achieves modularity [59].

Industry 4.0 places a high priority on modularity for several reasons. It enables businesses to create production systems that are more flexible and responsive. Companies can readily change machines and designs by employing standardized components and interfaces to satisfy varying business needs and consumer expectations.

Greater scalability and cost reductions are made possible through modular design. Companies can save the cost of production, development, maintenance, and repair by adopting standardized components.

Modularity makes greater collaboration and integration between various businesses and industries possible. Companies can quickly combine equipment and systems from many manufacturers by adopting standardized interfaces, enabling more effective and adaptable production procedures.

### 3.8.6 Information Transparency

The capacity of machines, systems, and people to access and share information in real time across organizational and geographical borders is a critical component of Industry 4.0. Information transparency is accomplished in Industry 4.0 by utilizing cutting-edge digital technologies, including the IoT, big data analytics, and cloud computing [60].

Industry 4.0 places a lot of emphasis on information transparency. Greater visibility and control over production processes are made possible by this, allowing businesses to spot issues as they arise and take immediate action to fix them. This decreases downtime and boosts productivity.

Information transparency makes greater collaboration and integration between diverse businesses and industries possible. Organizational and regional barriers can be overcome so that companies can collaborate more successfully to streamline manufacturing procedures and raise quality.

Information transparency facilitates improved decision-making by providing real-time data and analytics to support strategic and operational decision-making. Companies can find trends, patterns, and insights that can guide corporate strategy and enhance performance by analyzing data from sensors and other sources.

### 3.8.7 Decentralization

Decentralization, which refers to the spread of decision-making and control across several levels of a production system, is a significant concept in Industry 4.0. Decentralization in Industry 4.0 is made possible by applying cutting-edge digital technologies like the IoT, edge computing, and AI [56].

For several reasons, decentralization is significant in Industry 4.0. It allows for increased adaptability and reactivity in production processes, enabling machines and other systems to decide based on analytics and real-time data.

Greater scalability and adaptability of production systems are made possible by decentralization. Companies can add or remove equipment and procedures to satisfy changing company needs and consumer requirements by distributing decision-making and control across several levels.

By lowering reliance on centralized control systems, decentralization reduces the risk of system failures and downtime. Companies can lower the risk of single points of failure by dispersing decision-making and control across various levels, enhancing system reliability and resilience.

### 3.8.8 Smart Product

An intelligent product has been improved with cutting-edge digital technologies like AI, connection, and sensors. As a result, it can gather and analyze data, interact with other products and systems, and make decisions on its own [61]. Industry 4.0 comprises many intelligent products, allowing businesses to develop new business strategies, increase consumer value, and boost productivity.

Industry 4.0 places a lot of importance on innovative products for various reasons. They enable businesses to provide cutting-edge digital services and business models like pay-per-use, remote monitoring, and predictive maintenance.

Intelligent products make greater customization and personalization of goods and services possible. Companies can customize their products and services to match their customers' unique needs and tastes by gathering and analyzing data from sensors and other sources.

Greater collaboration and integration between many businesses and industries are made possible by smart products. Smart products can optimize production processes and increase efficiency throughout the value chain by exchanging data and connecting with other products and systems.

### 3.8.9 Corporate Social Responsibility

A company's commitment to conducting business in a socially responsible manner, considering how its operations affect the environment, society, and other stakeholders, is known as "CSR." CSR is gaining importance in Industry 4.0 as businesses strive to balance their economic goals and social and environmental responsibilities [62].

Industry 4.0 places a lot of emphasis on CSR for various reasons. Building credibility and trust with stakeholders, such as clients, staff, and investors, can benefit businesses. Companies can improve their reputation and brand value by committing to social and environmental responsibility.

By implementing sustainable practices and technologies, CSR may assist businesses in lowering costs and increasing efficiency. For instance, companies can save money and boost their bottom line by reducing energy use and waste.

By coordinating their corporate aims with social and environmental objectives, CSR can assist firms in generating shared value. For instance, businesses can open new markets and revenue streams while promoting a more sustainable future by creating goods and services that solve social and environmental issues.

As businesses strive to balance their economic goals and social and environmental responsibilities, CSR is crucial in Industry 4.0. Companies can improve their reputation, cut expenses, and contribute to a more sustainable future by embracing sustainable practices and technology and creating shared value.

### 3.8.10 Technical Assistance

Industry 4.0 emphasizes technical support as businesses look to adopt cutting-edge digital technology and reshape their business models. Technical support, which includes training, advising, and maintenance services, is given to companies to help them deploy and use new technology.

Industry 4.0 places a premium on technical support for several reasons. It can assist businesses in overcoming the difficulties associated with deploying new technologies and procedures. Technical support providers can aid companies in acquiring the skills and knowledge necessary to successfully implement new technologies by offering training and consulting services.

Companies can increase their productivity and operations efficiency with the help of technical support. Technical assistance providers can aid businesses in ensuring that their systems and equipment are working optimally, reducing downtime, and increasing productivity by offering maintenance and support services.

Companies might benefit from technical assistance to stay current with emerging technologies and market trends. Technical support providers can assist businesses in maintaining their competitiveness and innovation by sharing their knowledge and advice on new technologies and best practices [63].

### 3.8.11 Resource Efficiency

Industry 4.0 strongly emphasizes resource efficiency as businesses look to minimize their adverse environmental effects while also increasing their bottom line. The efficient use of energy, materials, and other resources in manufacturing processes is referred to as resource efficiency.

There are various ways to increase resource efficiency with Industry 4.0. For instance, businesses may track real-time resource and energy use and spot optimization opportunities using sensors and data analytics. As a result, there may be cost savings, environmental advantages from waste reduction, and improved energy efficiency.

Another way Industry 4.0 may increase resource efficiency is by utilizing intelligent systems and goods. Companies can decrease material waste, increase production efficiency, and improve operational sustainability by integrating interests and procedures and optimizing their performance. Along the entire value chain, Industry 4.0 also makes it easier to collaborate and integrate, improving resource efficiency and reducing waste. Companies can find opportunities to cut waste and boost resource efficiency throughout the value chain by sharing data and working with suppliers and consumers [64].

Industry 4.0 prioritizes resource efficiency as businesses aim to increase profits while maximizing resource utilization, cutting waste, and environmental performance. Employing sustainable practices and utilizing new digital technologies can help enterprises save money, positively influence the environment, and be more sustainable overall.

## 3.9 Applications of Industry 4.0

Industry 4.0, called the Fourth Industrial Revolution, refers to incorporating modern digital technology into industrial processes to develop smart factories and promote transformative changes in manufacturing and other industries. It expands on earlier industrial revolutions by combining CPSs, IoT, and data analytics to enable automation, connectivity, and intelligent decision-making [65–67]. Some instances of Industry 4.0 applications in various industries are shown in Figure 3.16.

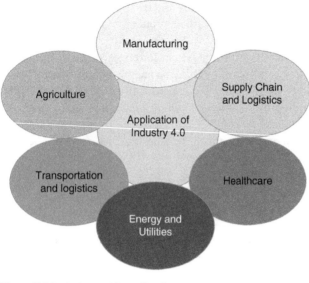

**Figure 3.16** Industry 4.0 application.

- Manufacturing: Industry 4.0 technologies enable smart factories with networked systems, real-time data analytics, and automation. This enables optimized production operations, predictive maintenance, and increased quality control. Robotics, IoT, additive manufacturing (3D printing), and AI are essential in modernizing industrial operations.
- Supply Chain and Logistics: The digitalization and automation of supply chain and logistical operations are enabled by Industry 4.0. This includes real-time cargo tracking, inventory management, predictive demand forecasting, and autonomous vehicles. These technologies increase efficiency, save costs, and improve supply chain transparency and traceability.
- Healthcare: Industry 4.0 makes building smart hospitals and healthcare systems easier. IoT devices, wearables, and remote patient monitoring allow real-time health data collection, analysis, and personalization. AI and machine learning algorithms aid in the diagnosis of diseases, the prediction of outcomes, and the improvement of patient care [4–8].
- Energy and Utilities: Smart grid systems produced by Industry 4.0 technology provide intelligent energy monitoring, management, and optimization. It allows for more efficient energy use, grid stability, and incorporation of renewable energy sources. IoT devices and sensors offer real-time energy consumption monitoring and predictive infrastructure maintenance [33].
- Transportation and Logistics: It has substantial benefits from Industry 4.0 technologies. Intelligent transportation networks, autonomous vehicles, and networked fleets enhance efficiency, safety, and sustainability. By utilizing real-time data analytics and optimization algorithms, routes can be optimized, fuel consumption can be reduced, and overall logistical operations can be improved.
- Agriculture: Precision agriculture is made accessible by Industry 4.0 owing to the usage of IoT gadgets, drones, and analytics powered by AI. Farmers may monitor and improve crop health, irrigation, and fertilizer consumption based on real-time data. Robotic harvesters are one kind of automated farming equipment that can boost output while requiring less labor [10, 11, 26].

## 3.10 Trends in Industry 4.0

Industry 4.0 integrates advanced technologies into industrial processes to create a more connected, automated, and efficient system. Manufacturers are incorporating new technology into their manufacturing facilities and techniques, such as IoT, AI, machine learning, cloud computing, and analytics [26, 68, 69]. According to SNS Insider Research [70], the Industry 4.0 market was estimated to be worth US$ 78.44 billion in 2022 and is expected to rise at a high compound annual growth rate (CAGR) of 18.5% from 2023 to 2030, reaching US$ 305.01 billion.

Here are some recent technological trends that have been shaping Industry 4.0.

- Edge Computing: Edge computing has gained significant momentum in Industry 4.0. Bringing computing power and data storage closer to edge devices and machines enables real-time data processing, reduced latency, and improved response times. This trend allows for more efficient and autonomous decision-making at the network's edge (Figure 3.17).
- CPSs Security: With the growing connectivity and integration of CPSs, ensuring the security of these systems has become a critical concern. Recent trends focus on developing robust cybersecurity measures, advanced encryption techniques, and secure communication protocols to protect industrial systems from cyber threats and maintain data integrity.
- Digital Twin Technology: Digital twin technology continues to evolve and find applications in various industries. It involves creating virtual replicas of physical assets, processes, or systems, enabling real-time monitoring, simulation, and analysis. Digital twins facilitate predictive maintenance, optimization, and innovation by providing a deeper understanding of the physical world and enabling virtual experimentation.

**Figure 3.17** Industry 4.0 market size 2023–2030 (US$ in billions).

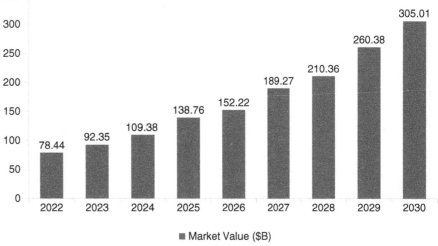

- AI Advancements: AI continues advancing and gaining widespread adoption in Industry 4.0. Machine learning algorithms, deep learning techniques, and neural networks are being applied to analyze vast amounts of data, improve predictive maintenance, optimize production processes, and enhance decision-making capabilities.
- 5G Connectivity: The rollout of 5G networks provides high-speed, low-latency connectivity, which is crucial for implementing Industry 4.0 technologies. 5G enables real-time communication and reliable connectivity for IoT devices and supports the massive data transmission required in industrial applications, paving the way for more advanced and interconnected systems.
- Additive Manufacturing Innovations: Additive manufacturing, or 3D printing, continues to evolve with advancements in materials, printing techniques, and production scale. Industries are exploring new applications, such as printing complex components, customized products, and 3D printing with multiple materials or functional properties.
- Collaborative Robotics: Collaborative robots, or cobots, are gaining traction in Industry 4.0. These robots are designed to work alongside humans safely, assisting with repetitive tasks, enhancing productivity, and improving workplace ergonomics. Recent cobot trends include sensor technology advancements, human-robot collaboration techniques, and intuitive programming interfaces.
- Blockchain in Supply Chain: Blockchain technology is being increasingly explored for supply chain management in Industry 4.0. Its decentralized and immutable nature allows for transparent and secure tracking of goods, verification of transactions, and traceability of products across the supply chain. Blockchain can enhance supply chain efficiency, reduce fraud, and enable trustworthy stakeholder collaboration.
- Human-Centric Design: A growing trend in Industry 4.0 is the focus on human-centered design principles. Recognizing the importance of user experience, ergonomics, and worker well-being, industries are designing intuitive interfaces and user-friendly systems and optimizing human-technology interactions to enhance productivity, safety, and job satisfaction.
- Sustainability and Green Manufacturing: Industry 4.0 increasingly focuses on sustainability and green manufacturing practices. Technologies such as energy-efficient systems, smart grids, and resource optimization solutions are being leveraged to reduce environmental impact and promote sustainable production.

## 3.11 Challenges of Industry 4.0

While Industry 4.0 presents numerous opportunities and advancements, it also brings forth certain challenges that must be addressed. As shown in Figure 3.18, the key challenges associated with Industry 4.0 include the following:

- Workforce Adaptation: Implementing Industry 4.0 technologies requires a highly skilled workforce to operate, maintain, and troubleshoot advanced systems. However, the rapid adoption of new technologies may result in a skills gap where the existing workforce needs more expertise. Upskilling and reskilling initiatives are necessary to ensure that the workforce can adapt to the changing demands of Industry 4.0 [71, 72].

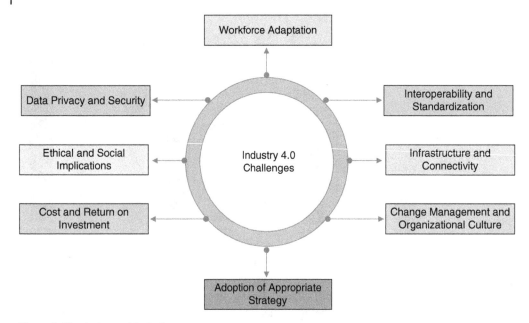

**Figure 3.18** Industry 4.0 challenges.

- Data Privacy and Security: The interconnected nature of Industry 4.0 systems and the vast amount of data generated pose significant concerns regarding privacy and security. Protecting sensitive data from cyber threats, ensuring secure communication between devices, and complying with data protection regulations are essential challenges that must be addressed to maintain trust in the digital ecosystem [26, 73].
- Interoperability and Standardization: Industry 4.0 integrates various technologies, devices, and systems from different vendors. However, ensuring seamless interoperability among these components can be challenging due to differing protocols, communication standards, and data formats. Establishing common standards and protocols is crucial to enable effective communication and interoperability across the industrial ecosystem.
- Ethical and Social Implications: The increasing automation and digitization in Industry 4.0 raise ethical and social concerns. Job displacement due to automation, the impact on workers' well-being, and the ethical implications of AI and robotics in decision-making must be carefully considered. Developing policies, regulations, and frameworks that address these moral and social implications is crucial for the responsible and sustainable implementation of Industry 4.0.
- Infrastructure and Connectivity: Industry 4.0 relies heavily on robust and reliable connectivity, such as high-speed Internet access and stable network infrastructure. However, in some regions or industries, the existing infrastructure may need to be revised to adequately support the requirements of Industry 4.0. Expanding connectivity and upgrading infrastructure, including broadband networks and communication technologies, are necessary to leverage the benefits of Industry 4.0 fully.
- Cost and Return on Investment: Implementing Industry 4.0 technologies often requires significant equipment, software, and training investments. The initial costs associated with adopting new technologies can be a barrier, especially for small and medium-sized enterprises (SMEs). Additionally, quantifying the return on investment (ROI) and demonstrating the long-term benefits of Industry 4.0 solutions can be challenging, requiring careful planning and assessment.
- Change Management and Organizational Culture: Industry 4.0 often requires a transformational change in the organizational culture and processes. Embracing new technologies, adapting workflows, and fostering a culture of innovation and continuous learning can be challenging for established organizations. Change management strategies, leadership commitment, and employee engagement are essential to navigate this transition successfully [74].
- Adoption of Appropriate Strategy: The difficulties and factors that organizations must consider while creating and implementing strategies to fully utilize Industry 4.0 technology and practices are referred to as "strategic challenges" in this context. Aligning an Industry 4.0 strategy with the organization's overarching business goals and objectives is necessary for its development. For the system to be implemented effectively, it must support the organization's long-term vision and mission [75].

## 3.12   Related Works

The difficulties of Industry 4.0 have been extensively covered in works and studies. Researchers, academics, and business professionals know the challenges and potential complexity that organizations may have when adopting and implementing Industry 4.0. These difficulties cut across several domains, such as technology, labor, management, and organizational elements.

Ozkan and Kazancoglu [76] identify and examine 13 distinct workforce development concerns. The research shows that the lack of IT/digital skills substantially impacts other issues as well as workforce development. This lack of abilities affects several areas, including more time dedicated to learning and specialized training, the inability to think analytically and deal with complexity, and the need for transdisciplinary thinking and action. The report also needs to include the problems with decentralized decision-making and the need for workers with the necessary skill sets in the labor market. These difficulties have more significant macro-effects on other challenges in workforce development. The research's findings also imply that most problems are linked. This indicates the interconnected and complex character of these difficulties in the context of Industry 4.0. It shows that the challenges in workforce development are not isolated but have links.

Avdibasic et al. [77] highlight data security concerns by looking at various fields, including healthcare, smart manufacturing, the IoT, and CPSs. The study identifies factors that include ignorance, a lack of experts, companies' unpreparedness, a broad attack surface caused by numerous entry points, vulnerable devices connected to networks, weak supply chains, unsafe data exchange, theft of sensitive information for personal gain, blackmailing, DoS attacks, endangering people's safety, financial harm, default passwords, risky updates, and interruptions in service delivery caused by connection loss as challenges to data security and privacy. Zeid et al. [78] highlight the difficulties with interoperability and standardization in intelligent manufacturing. These difficulties may make it easier for data and information to flow between systems without interruption. Various variables, including variations in data transfer between similar or dissimilar systems, compatibility problems between software versions, misunderstandings of terminology, nonstandardized documentation, and insufficient testing of conformant applications, cause the challenges of achieving interoperability and standardization.

Peckham [79] emphasizes the varied moral ramifications of new technology and societal transformations. Cognitive acuity (the impact of technology on cognitive abilities), interpersonal relationships (the effect of technology on interpersonal relationships), freedom and privacy (the balance between technological advancements and individual privacy rights), moral agency (the responsibility and accountability in decision-making), loss of work (the potential impact of automation on employment), and perception of reality (challenges posed by VR or AR) are just a few examples of factors to consider.

Different viewpoints on the infrastructure difficulties related to the use of digital twins for smart infrastructure are illustrated by Broo et al. [80]. The systemic perspective emphasizes the importance of stakeholder engagement in creating the digital twin's system architecture. Sharing knowledge about the system's parts, talking about the needs and expectations of stakeholders, and figuring out the KPIs that matter to them are all parts of this process. This viewpoint aims to build adaptable and reliable data models to deploy the digital twin successfully. The informational approach emphasizes essential factors, including adaptability, transparency, modularization, and integration. The ability of the digital twin to adapt to changing requirements, facilitate interoperability, and successfully connect with other systems and data sources depends critically on these elements. The organizational perspective covers the non-technological factors that influence the creation and use of digital twins in smart infrastructure systems. Governance, stakeholder involvement, corporate culture, and change management are a few examples of these elements. Certain factors are essential for digital twins to implement and use in smart infrastructure projects successfully.

Rossini et al. [81] stress the importance of large expenditures on cutting-edge technology. Investing in new technology is frequently essential for organizations to remain competitive, promote innovation, increase efficiency, and react to shifting market needs. These investments involve buying and installing cutting-edge equipment, software, automation, tools, and other digital solutions. The problem of unpredictable investment returns is brought up by Singh et al. [75]. Uncertain returns can have inconsistent or variable results or benefits that can be obtained through investments. Numerous factors, including market instability, technical advancements, competitive dynamics, governmental regulations, and economic situations, might contribute to this uncertainty. Whysall et al. [82] highlight the difficulties with change management in organizations. It explicitly outlines numerous elements, such as talent acquisition, new core competencies, crucial talent positions, and transformation of talent management, that present difficulty for change management.

**Table 3.1** Summary of Industry 4.0 challenges.

| Author(s) | Challenge group | Challenge |
|---|---|---|
| Ozkan and Kazancoglu [76] | Workforce adaptation challenge | • Lack of technological skills<br>• Lack of Industry 4.0 skilled workers |
| Avdibasic et al. [77] | Data privacy and security challenge | • Inability to extract knowledge from data<br>• Lack of expertise in dealing with cyber threats<br>• Need for data protection |
| Zeid et al. [78] | Interoperability and standardization challenge | • Inability in establishing uniform data interchange standards<br>• Misunderstanding of terminologies<br>• Insufficient testing |
| Peckham [79] | Ethical and social implications challenge | • Reduced cognitive acuity<br>• Lack of morality<br>• Loss of work due to increased automation |
| Broo et al. [80] | Infrastructure and connectivity challenge | • Adaptability to changing infrastructure<br>• Lack of transparency<br>• Lack of better connectivity |
| Rossini et al. [81], Singh et al. [75] | Cost and return on investment challenge | • Need for large investment in novel technologies<br>• Uncertainty over returns of investments |
| Whysall et al. [82] | Change management and organizational culture challenge | • Lack of proper talent acquisition<br>• Equipping acquired talent to face changing scenarios |
| Moktadir et al. [83] | Adoption of appropriate strategy challenge | • Absence of a dynamic strategic plan |

The absence of a dynamic strategic plan to facilitate the implementation of Industry 4.0 in the industrial sector, with a particular focus on the leather industry in Bangladesh, is highlighted by Moktadir et al. [83]. The best-worst method (BWM), a revolutionary multi-criteria decision-making technique, is used by researchers to analyze strategic problems and formulate solutions. The results indicate that a critical barrier to the successful adoption of Industry 4.0 in the leather industry is the need for a dynamic strategic plan. A summary of past literature in this regard is shown in Table 3.1. Besides the aforementioned works, Karnik et al., Chen et al., and Choi et al. [84–86] highlight various technology-specific challenges in Industry 4.0.

## 3.13 Paradigm Shift Toward Industry 5.0

The paradigm shift from Industry 4.0 to Industry 5.0 represents the next stage of industrial transformation, focusing on integrating advanced technologies with human capabilities to create more collaborative and personalized manufacturing systems. While Industry 4.0 emphasized automation and connectivity, Industry 5.0 took it a step further by emphasizing the role of humans in the production process [63, 87, 88]. As shown in Figure 3.19, here are the critical aspects of the transition from Industry 4.0 to Industry 5.0:

- Human-Centric Approach: Industry 5.0 recognizes the importance of human skills, creativity, and problem-solving capabilities in manufacturing. It seeks to create a harmonious collaboration between humans and machines, where humans contribute their cognitive abilities, intuition, and adaptability to complement the capabilities of automated systems.
- Customization and Personalization: Industry 5.0 strongly emphasizes customer-centric manufacturing and the ability to deliver personalized products and experiences. Advanced technologies like AI, machine learning, and robotics are leveraged to enable flexible and efficient customization, allowing for mass production with individualization.
- Decentralized Production: Industry 5.0 promotes decentralized production, where manufacturing processes are distributed across various locations and closer to the end consumers. This approach aims to reduce transportation costs, shorten supply chains, and enhance local production capabilities.

- Sustainability and Circular Economy: Industry 5.0 strongly emphasizes sustainability and circular economy principles. It aims to reduce waste, optimize resource utilization, and implement eco-friendly manufacturing practices. Technologies like additive manufacturing and advanced recycling systems contribute to sustainable production and reduced environmental impact.
- Edge Computing and Real-Time Processing: With the increasing complexity and volume of data generated in Industry 5.0, edge computing plays a crucial role. Edge computing involves processing data near the source, enabling real-time decision-making and reducing latency. This allows for faster response times, improved operational efficiency, and better utilization of resources.
- Human-Machine Interfaces: Industry 5.0 introduces advanced HMIs, such as AR and VR, to enhance collaboration and communication between humans and machines. These interfaces provide intuitive ways to interact with devices, visualize data, and assist in training, maintenance, and troubleshooting processes.
- Ethical and Social Considerations: Industry 5.0 acknowledges advanced technologies' ethical and social implications. It seeks to ensure responsible and inclusive deployment, addressing concerns related to job displacement, privacy, fairness, and transparency. Ethical frameworks, regulations, and guidelines are developed to guide the ethical adoption of Industry 5.0 technologies.

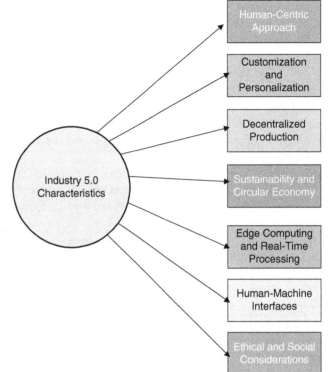

**Figure 3.19**    Industry 5.0 characteristics.

## 3.14  Future Challenges and Research

Industry 4.0 continually evolves as new technologies emerge and existing ones mature, driving the transformation of industries and reshaping the way businesses operate [89–91]. It is expected to face several challenges in the future. Here are some potential challenges that may arise.

- Data Management and Privacy: The increasing volume of data generated by interconnected devices and systems poses data storage, processing, and privacy challenges. Managing and analyzing large datasets while ensuring data security and privacy will remain a significant challenge for organizations.
- Skills Gap and Workforce Transformation: Industry 4.0 requires a highly skilled workforce that can adapt to new technologies and work collaboratively with automation and AI systems. Closing the skills gap, upskilling employees, and ensuring a smooth workforce transformation will be crucial for successful implementation.
- Ethical and Social Implications: The advancements in Industry 4.0 raise ethical concerns related to job displacement, privacy, fairness, and transparency. Addressing these concerns and establishing ethical frameworks, guidelines, and regulations will be essential to ensure responsible and inclusive deployment.
- Interoperability and Standardization: As more technologies and systems are integrated into Industry 4.0, ensuring interoperability and standardization among different devices, platforms, and protocols will become increasingly complex. Establishing common standards and protocols will enable seamless communication and integration.
- Security and Cyber Threats: With the increased connectivity and reliance on digital systems, Industry 4.0 faces heightened security risks. Protecting critical infrastructure, intellectual property, and sensitive data from cyber threats will require robust cybersecurity measures, advanced encryption techniques, and constant monitoring.
- Infrastructure and Connectivity: The widespread adoption of Industry 4.0 requires reliable and high-speed connectivity, which may pose challenges in regions with limited infrastructure. Expanding broadband networks, improving connectivity in remote areas, and ensuring seamless communication will be vital for the future of Industry 4.0.

- Change Management and Organizational Culture: Embracing Industry 4.0 requires organizations to significantly change processes, workflows, and organizational culture. Managing the cultural shift, overcoming resistance to change, and fostering a culture of innovation and continuous learning will be vital for successful implementation.
- Environmental Impact and Sustainability: While Industry 4.0 offers opportunities for optimizing resource utilization and reducing waste, it must also address its environmental impact. Ensuring sustainable practices, minimizing energy consumption, and promoting circular economy principles will be necessary for the long-term sustainability of Industry 4.0.
- Cost and ROI: Implementing Industry 4.0 technologies often involves significant upfront investments, particularly for SMEs. Demonstrating the ROI and quantifying the long-term benefits of adopting these technologies will be crucial to encourage wider adoption.
- Regulatory and Legal Frameworks: The rapid advancement of Industry 4.0 may outpace existing regulations and legal frameworks. Governments and regulatory bodies must adapt and establish appropriate frameworks to address data privacy, intellectual property rights, liability, and safety issues in emerging technologies.

## 3.15   Conclusion

This chapter set out on a voyage through the intriguing world of industrial revolutions, tracing the development from the beginnings of mechanization to the emergence of Industry 4.0 and beyond. It specifically examined the idea of industrial revolutions and revealed the several generations that have influenced the development of society. The dawn of Industry 4.0, a transformational era characterized by the fusion of cutting-edge technology and the digitization of industries, came next into focus. The technologies underlying Industry 4.0, highlighting the fantastic developments in areas like robotics, AI, the IoT, and big data analytics, are explored in greater detail and also reveal the design guiding principles for Industry 4.0 solutions, highlighting the value of interoperability, decentralization, real-time data, and modularization. Nevertheless, it also looked at the difficulties of the shift to Industry 4.0 and the chances and achievements. The related works also looked at relevant publications and research that clarified challenges faced and lessons learned from global Industry 4.0 initiatives. A brief insight into the paradigm shift toward Industry 5.0, a new idea that combines human-centered methods with cutting-edge technology, is also provided. The road ahead is paved with promising potential and tremendous obstacles that need coordinated efforts, creative problem-solving, and sound judgment.

## References

1 Mohajan, H. (2019). The First Industrial Revolution: Creation of a New Global Human Era. *Journal of Social Sciences and Humanities* 5 (4): 377–387.

2 Haradhan, M. (2020). The Second Industrial Revolution Has Brought Modern Social and Economic Developments. *Journal of Social Sciences and Humanities* 6 (1): 1–14.

3 Mohajan, H. (2021, 2021). Third Industrial Revolution Brings Global Development. *Journal of Social Sciences and Humanities* 7 (4): 239–251.

4 Hindu, A. and Bhowmik, B. (2022). An iot-enabled stress detection scheme using facial expression. *2022 IEEE 19th India Council International Conference (INDICON)*, 1–6, Kochi, India (24–26 Nov 2022). Kerala, India: IEEE.

5 Bhowmik, B., Varna, S. A., Kumar, A., and Kumar, R. (2021). Reducing false prediction on covid-19 detection using deep learning. *2021 IEEE International Midwest Symposium on Circuits and Systems (MWSCAS)*, 404–407, Lansing, MI, USA (09–11 Aug 2021). East Lansing, Michigan USA: IEEE.

6 Guria, M. and Bhowmik, B. (2022). Iot-enabled driver drowsiness detection using machine learning. *2022 Seventh International Conference on Parallel, Distributed and Grid Computing (PDGC)*, 519–524, Himachal Pradesh, India (25–27 Nov 2022). Solan, Himachal Pradesh, India: IEEE.

7 Bhowmik, B.R., Varna, S.A., Kumar, A., and Kumar, R. (2021). Deep neural networks in healthcare systems. In: *Machine Learning and Deep Learning in Efficacy Improvement of Healthcare Systems*, 195–226. CRC Press, Taylor and Francis.

8 Gupta, S., Cherukuri, A.K., Subramanian, C.M., and Ahmad, A. (2022). *Comparison, Analysis and Analogy of Biological and Computer Viruses*, 3–34. Singapore: Springer Singapore https://doi.org/10.1007/978-981-16-6542-4_1.

9 Ross, P. and Maynard, K. (2021). Towards a 4th industrial revolution. *Intelligent Buildings International* 13 (3): 159–161.

**10** Amit Kumar, K. and Bhowmik, B. (2023). Rice cultivation and its disease classification in precision agriculture. *2023 International Conference on Artificial Intelligence and Smart Communication (AISC)*, 1–6, Greater Noida, India (27–29 Jan 2023). Uttar Pradesh, India: IEEE.

**11** Verma, A. and Bhowmik, B. (2022). Automated detection of maize leaf diseases in agricultural cyber-physical systems. *2022 30th Mediterranean Conference on Control and Automation (MED)*, 841–846, Vouliagmeni, Greece (28 June - 01 July 2022). Athens, Greece: IEEE.

**12** Gragnolati, U.M., Moschella, D., and Pugliese, E. (2014). The spinning jenny and the guillotine: technology diffusion at the time of revolutions. *Cliometrica* 8: 5–26.

**13** Mokyr, J. (2001). The industrial revolution and the economic history of technology: lessons from the British experience, 1760–1850. *The Quarterly Review of Economics and Finance* 41 (3): 295–295.

**14** Tann, J. (2015). Borrowing brilliance: technology transfer across sectors in the early industrial revolution. *The International Journal for the History of Engineering and Technology* 85 (1): 94–114.

**15** Nurdiana, N. and Pandin, M.G.R. (2021). Industrial revolution: a history of industrial revolution and its influence in manufacturing companies. *Historia Madania: Jurnal Ilmu Sejarah* 5 (2): 137–151.

**16** Coleman, D. (1962). Growth and decay during the industrial revolution: the case of east anglia. *Scandinavian Economic History Review* 10 (2): 115–127.

**17** Mokyr, J. and Strotz, R.H. (1998). The second industrial revolution, 1870–1914. *Storia dell'economia Mondiale* 21945 (1): 1–16.

**18** Rifkin, J., Carvalho, M., Consoli, A., and Bonifacio, M. (2008). Leading the way to the third industrial revolution. *European Energy Review* 1: 4–18.

**19** Lee, J., Kao, H.-A., and Yang, S. (2014). Service innovation and smart analytics for Industry 4.0 and big data environment. *Procedia CIRP* 16: 3–8.

**20** Behrens, V. and Viete, S. (2020). A note on Germany's role in the fourth industrial revolution. *Arbeitspapier, Tech. Rep. Report No.: 09/2020*.

**21** Prathyusha, M. R. and Bhowmik, B.2023). Iot evolution and recent advancements. *2023 IEEE 9th International Conference on Advanced Computing and Communication Systems (ICACCS)*, 1–6, Coimbatore, India (17–18 March 2023). Coimbatore, India: IEEE.

**22** Prathyusha, M. and Bhowmik, B. (2023). Iot-enabled smart applications and challenges. *2023 IEEE 8th International Conference on Communication and Electronics Systems (ICCES)*, 1–6, Coimbatore, India (01–03 June 2023). Coimbatore, India: IEEE.

**23** Aceto, G., Persico, V., and Pescapé, A. (2020). Industry 4.0 and health: internet of things, big data, and cloud computing for healthcare 4.0. *Journal of Industrial Information Integration* 18: 100129.

**24** Popkova, E.G., Ragulina, Y.V., and Bogoviz, A.V. (2019). *Industry 4.0: Industrial Revolution of the 21st Century*, vol. 169. Springer.

**25** Tay, S.I., Lee, T., Hamid, N., and Ahmad, A.N.A. (2018). An overview of Industry 4.0: definition, components, and government initiatives. *Journal of Advanced Research in Dynamical and Control Systems* 10 (14): 1379–1387.

**26** Tyagi, A.K., Dananjayan, S., Agarwal, D., and Thariq Ahmed, H.F. (2023). Blockchain—internet of things applications: opportunities and challenges for Industry 4.0 and society 5.0. *Sensors* 23 (2): 1–30. https://www.mdpi.com/1424-8220/23/2/947.

**27** Baheti, R. and Gill, H. (2011). Cyber-physical systems. *The Impact of Control Technology* 12 (1): 161–166.

**28** Rajkumar, R., Lee, I., Sha, L., and Stankovic, J. (2010). Cyber-physical systems: the next computing revolution. *Proceedings of the 47th design automation conference*, 731–736, Anaheim, California, USA (13–18 June 2010). New York, United States: Association for Computing Machinery.

**29** Sinha, D. and Roy, R. (2020). Reviewing cyber-physical system as a part of smart factory in Industry 4.0. *IEEE Engineering Management Review* 48 (2): 103–117.

**30** Saravanan, G., Parkhe, S.S., Thakar, C.M. et al. (2022). Implementation of iot in production and manufacturing: an Industry 4.0 approach. *Materials Today Proceedings* 51: 2427–2430. International Conference on Advances in Materials Science https://www.sciencedirect.com/science/article/pii/S2214785321076264.

**31** Bhowmik, B. (2022). Ann-based performance prediction in mocs. In: *International Symposium on Artificial Intelligence*, 133–144. Springer.

**32** Mastos, T.D., Nizamis, A., Vafeiadis, T. et al. (2020). Industry 4.0 sustainable supply chains: an application of an iot enabled scrap metal management solution. *Journal of Cleaner Production* 269: 122377.

**33** Sajja, K. and Bhowmik, B. (2023). Iot systems and battery-based energy sources. *2023 International Conference on Artificial Intelligence and Smart Communication (AISC)*, 1–6, Greater Noida, India (27-29 Jan 2023). Uttar Pradesh, India: IEEE.

**34** Greco, L., Maresca, P., and Caja, J. (2019). Big data and advanced analytics in Industry 4.0: a comparative analysis across the european union. *Procedia Manufacturing* 41: 383–390.

**35** Bhowmik, B.R. (2022, pp. 99–128). Ai technology in networks-on-chip. In: *Industrial Transformation: Implementation and Essential Components and Processes of Digital Systems* (ed. O.P. Jena, S.S. Patra, M. Panda, et al.). CRC Press.

**36** Patil, A., Dwivedi, A., Moktadir, M.A. et al. (2023). Big data-Industry 4.0 readiness fac-tors for sustainable supply chain management: towards circularity. *Computers & Industrial Engineering* 178: 109109.

**37** Mourtzis, D. (2022). Introduction to cloud technology and Industry 4.0. In: *Design and Operation of Production Networks for Mass Personalization in the Era of Cloud Technology*, 1–12. Elsevier.

**38** Ivanov, D., Dolgui, A., and Sokolov, B. (2022). Cloud supply chain: integrating Industry 4.0 and digital platforms in the "supply chain-as-a-service". *Transportation Research Part E: Logistics and Transportation Review* 160: 102676.

**39** Murugesan, U., Subramanian, P., Srivastava, S., and Dwivedi, A. (2023). A study of artificial intelligence impacts on human resource digitalization in Industry 4.0. *Decision Analytics Journal* 7: 100249.

**40** Kale, P., Hazarika, P., Jain, S., and Bhowmik, B. (2022). Performance evaluation in 2d nocs using ann. In: *Advanced Information Networking and Applications*, 360–369. Cham: Springer International Publishing.

**41** Bhowmik, B., Hazarika, P., Kale, P., and Jain, S. (2021). Ai technology for noc performance evaluation. *IEEE Transactions on Circuits and Systems II: Express Briefs* 68 (12): 3483–3487.

**42** Ahmad, T., Zhu, H., Zhang, D. et al. (2022). Energetics systems and artificial intelligence: applications of Industry 4.0. *Energy Reports* 8: 334–361.

**43** Nuttah, M.M., Roma, P., Nigro, G.L., and Perrone, G. (2023). Understanding blockchain applications in Industry 4.0: from information technology to manufacturing and operations management. *Journal of Industrial Information Integration* 33: 100456.

**44** Patil, B. S., Sharma, M., Soubhari, T. et al. (2023). Quantitative assessment of blockchain applications for Industry 4.0 in manufacturing sector. *Materials Today: Proceedings* 28 April 2023.

**45** Cimino, A., Gnoni, M.G., Longo, F. et al. (2023). Modeling & simulation as Industry 4.0 enabling technology to support manufacturing process design: a real industrial application. *Procedia Computer Science* 217: 1877–1886.

**46** Tao, F., Qi, Q., Wang, L., and Nee, A. (2019). Digital twins and cyber–physical systems toward smart manufacturing and Industry 4.0: correlation and comparison. *Engineering* 5 (4): 653–661.

**47** Allen, L., Atkinson, J., Jayasundara, D. et al. (2021). Data visualization for Industry 4.0: a stepping-stone toward a digital future, bridging the gap between academia and industry. *Patterns* 2 (5): 100266.

**48** Zhu, Z., Liu, C., and Xu, X. (2019). Visualisation of the digital twin data in manufacturing by using augmented reality. *Procedia CIRP* 81: 898–903.

**49** Gradim, B. and Teixeira, L. (2022). Robotic process automation as an enabler of Industry 4.0 to eliminate the eighth waste: a study on better usage of human talent. *Procedia Computer Science* 204: 643–651.

**50** Ikumapayi, O.M., Afolalu, S.A., Ogedengbe, T.S. et al. (2023). Human-robot co-working improvement via revolutionary automation and robotic technologies–an overview. *Procedia Computer Science* 217: 1345–1353.

**51** Papulová, Z., Gažová, A., and Šufliarský, L. (2022). Implementation of automation technologies of Industry 4.0 in automotive manufacturing companies. *Procedia Computer Science* 200: 1488–1497.

**52** Elhazmiri, B., Naveed, N., Anwar, M.N., and Haq, M.I.U. (2022). The role of additive manufacturing in Industry 4.0: an exploration of different business models. *Sustainable Operations and Computers* 3: 317–329.

**53** Ashima, R., Haleem, A., Bahl, S. et al. (2021). Automation and manufacturing of smart materials in additive manufacturing technologies using internet of things towards the adoption of Industry 4.0. *Materials Today Proceedings* 45: 5081–5088.

**54** Dikhanbayeva, D., Shaikholla, S., Suleiman, Z., and Turkyilmaz, A. (2020). Assessment of Industry 4.0 maturity models by design principles. *Sustainability* 12 (23): 9927.

**55** Lelli, F. (2019). Interoperability of the time of Industry 4.0 and the internet of things. *Future Internet* 11 (2): 36.

**56** Brettel, M., Friederichsen, N., Keller, M., and Rosenberg, M. (2017). How virtualization, decentralization and network building change the manufacturing landscape: an Industry 4.0 perspective. *FormaMente* 12: 37–44.

**57** Ghobakhloo, M. (2020). Industry 4.0, digitization, and opportunities for sustainability. *Journal of Cleaner Production* 252: 119869.

**58** Reis, J. Z. and Gonçalves, R. F. (2018). The role of internet of services (ios) on Industry 4.0 through the service oriented architecture (soa). *Advances in Production Management Systems. Smart Manufacturing for Industry 4.0: IFIP WG 5.7 International Conference, APMS 2018* (August 26–30, 2018, Seoul, Korea; *Proceedings, Part II*. Springer, 2018, 20–26.

**59** Gupta, P. (2019). Modularity enablers: a tool for Industry 4.0. *Life Cycle Reliability and Safety Engineering* 8 (2): 157–163.

**60** Nolting, L., Priesmann, J., Kockel, C. et al. (2019). Generating transparency in the worldwide use of the terminology Industry 4.0. *Applied Sciences* 9 (21): 4659.

**61** Zawadzki, P. and Żywicki, K. (2016). Smart product design and production control for effective mass customization in the Industry 4.0 concept. *Management and Production Engineering Review* 7: 105–112.

**62** Adamik, A. and Nowicki, M. (2019). Pathologies and paradoxes of co-creation: a contribution to the discussion about corporate social responsibility in building a competitive advantage in the age of Industry 4.0. *Sustainability* 11 (18): 4954.

**63** Xu, X., Lu, Y., Vogel-Heuser, B., and Wang, L. (2021). Industry 4.0 and Industry 5.0 – inception, conception and perception. *Journal of Manufacturing Systems* 61: 530–535.

**64** Lasi, H., Fettke, P., Kemper, H.-G. et al. (2014). Industry 4.0. *Business & Information Systems Engineering* 6: 239–242.

**65** Vaidya, S., Ambad, P., and Bhosle, S. (2018). Industry 4.0–a glimpse. *Procedia Manufacturing* 20: 233–238.

**66** Kumar, R., Singh, R.K., and Dwivedi, Y.K. (2020). Application of Industry 4.0 technologies in smes for ethical and sustainable operations: analysis of challenges. *Journal of Cleaner Production* 275: 124063.

**67** Zheng, T., Ardolino, M., Bacchetti, A., and Perona, M. (2021). The applications of Industry 4.0 technologies in manufacturing context: a systematic literature review. *International Journal of Production Research* 59 (6): 1922–1954.

**68** Xu, L.D., Xu, E.L., and Li, L. (2018). Industry 4.0: state of the art and future trends. *International Journal of Production Research* 56 (8): 2941–2962.

**69** Sharma, A.K., Bhandari, R., Pinca-Bretotean, C. et al. (2021). A study of trends and industrial prospects of Industry 4.0. *Materials Today Proceedings* 47: 2364–2369.

**70** Yugandhara, Y. R. (2023). Industry 4.0 market size, growth and trends report 2023. *Report No: SNS/ICT/1226*.

**71** Schröder, C. (2016). *The Challenges of Industry 4.0 for Small and Medium-sized Enterprises*. Friedrich-Ebert-Stiftung: Bonn, Germany.

**72** Mohamed, M. (2018). Challenges and benefits of Industry 4.0: an overview. *International Journal of Supply and Operations Management* 5 (3): 256–265.

**73** Culot, G., Fattori, F., Podrecca, M., and Sartor, M. (2019). Addressing Industry 4.0 cybersecurity challenges. *IEEE Engineering Management Review* 47 (3): 79–86.

**74** Luthra, S. and Mangla, S.K. (2018). Evaluating challenges to Industry 4.0 initiatives for supply chain sustainability in emerging economies. *Process Safety and Environmental Protection* 117: 168–179.

**75** Singh, S., Mahanty, B., and Tiwari, M. (2018). Framework and modelling of inclusive manufacturing system. *International Journal of Computer Integrated Manufacturing* 32: 1–19.

**76** Ozkan-Ozen, Y.D. and Kazancoglu, Y. (2022). Analysing workforce development challenges in the Industry 4.0. *International Journal of Manpower* 43 (2): 310–333.

**77** Avdibasic, E., Toksanovna, A.S., and Durakovic, B. (2022). Cybersecurity challenges in Industry 4.0: a state of the art review. *Defense and Security Studies* 3: 32–49.

**78** Zeid, A., Sundaram, S., Moghaddam, M. et al. (2019). Interoperability in smart manufacturing: Research challenges. *Machines* 7 (2): 1–17. https://www.mdpi.com/2075-1702/7/2/21.

**79** Peckham, J.B. (2021). The ethical implications of 4ir. *Journal of Ethics in Entrepreneurship and Technology* 1 (1): 30–42.

**80** Gürdür Broo, D., Bravo-Haro, M., and Schooling, J. (2022). Design and implementation of a smart infrastructure digital twin. *Automation in Construction* 136: 104171. https://www.sciencedirect.com/science/article/pii/S0926580522000449.

**81** Rossini, M., Costa, F., Tortorella, G.L. et al. (2022). Lean production and Industry 4.0 integration: how lean automation is emerging in manufacturing Industry. *International Journal of Production Research* 60 (21): 6430–6450.

**82** Whysall, Z., Owtram, M., and Brittain, S. (2019). The new talent management challenges of Industry 4.0. *Journal of Management Development* 38: 118–129.

**83** Moktadir, M.A., Ali, S.M., Kusi-Sarpong, S., and Shaikh, M.A.A. (2018). Assessing challenges for implementing Industry 4.0: implications for process safety and environmental protection. *Process Safety and Environmental Protection* 117: 730–741. https://www.sciencedirect.com/science/article/pii/S0957582018301344.

**84** Karnik, N., Bora, U., Bhadri, K. et al. (2022). A comprehensive study on current and future trends towards the characteristics and enablers of Industry 4.0. *Journal of Industrial Information Integration* 27: 100294. https://www.sciencedirect.com/science/article/pii/S2452414X21000911.

**85** Chen, Y., Lu, Y., Bulysheva, L., and Kataev, M.Y. (2022). Applications of blockchain in Industry 4.0: a review. *Information Systems Frontiers* 24: 1–15.

**86** Choi, T.-M., Kumar, S., Yue, X., and Chan, H.-L. (2022). Disruptive technologies and operations management in the Industry 4.0 era and beyond. *Production and Operations Management* 31 (1): 9–31.

**87** Aslam, F., Aimin, W., Li, M., and Ur Rehman, K. (2020). Innovation in the era of IoT and Industry 5.0: absolute innovation management (aim) framework. *Information* 11 (2): 124.

**88** Jafari, N., Azarian, M., and Yu, H. (2022). Moving from Industry 4.0 to Industry 5.0: what are the implications for smart logistics? *Logistics* 6 (2): 26.

**89** Zhou, K., Liu, T., and Zhou, L. (2015). Industry 4.0: towards future industrial opportunities and challenges. In: *12th International Conference on Fuzzy Systems and Knowledge Discovery (FSKD)*, 2147–2152. IEEE.

**90** Gröger, C. (2018). Building an Industry 4.0 analytics platform: practical challenges, approaches and future research directions. *Datenbank-Spektrum* 18 (1): 5–14.

**91** Derigent, W., Cardin, O., and Trentesaux, D. (2021). Industry 4.0: contributions of holonic manufacturing control architectures and future challenges. *Journal of Intelligent Manufacturing* 32 (7): 1797–1818.

# 4

# Detection from Chest X-Ray Images Based on Modified Deep Learning Approach

*Jyoti Dabass[1], Manju Dabass[2], and Ananda K. Behera[3]*

[1] *DBT Centre of Excellence Biopharmaceutical Technology, IIT, New Delhi, India*
[2] *EECE Department, The Northcap University, Gurugram, Haryana, India*
[3] *Artificial Intelligence and Machine Learning Programme, Liverpool John Moores University, Liverpool, UK*

## 4.1 Introduction

A lung condition known as "tuberculosis" (TB) is caused by the *Mycobacterium tuberculosis* bacteria. After COVID-19, it is the second most contagious illness that spreads through airborne droplets from a TB patient's coughing and sneezing. The World Health Organization (WHO) guesstimates that 10 million people would be infected by TB by 2020, and 1.5 million will die from it. TB is a curable disease and can be treated through four antimicrobial drugs taken over six months [1]. To avoid the spread of the infection and treatment of TB-affected patients, timely diagnosis and detection of TB are necessary. Currently used septum smear microscopy and WHO-recommended rapid molecular assay tests to detect TB are both costly and time taking [2]. The expense of these tests restricts the detection of the disease because the majority of TB-afflicted persons live in low- to middle-income South Asian nations. Expert radiologists may additionally detect TB with the use of chest X-ray (CXR) images. However, the procedure of diagnosing TB in such regions is hampered by the shortage of skilled radiologists. Deep learning and convolutional neural network (CNN) models for applications based on image recognition have grown rapidly in recent years. The CNN models can extract additional information from the images that are challenging for humans to see without the need for prior domain expertise. This encouraged the researchers to utilize these models for computer-aided diagnosis (CAD) for the early finding of cancer, TB, pneumonia, and other illnesses.

From the year 2010, the annual ImageNet Large Scale Visual Recognition Challenge (ILSVRC) for the classification of object categories from millions of images has led to the development of many powerful CNN models in the last decade [3]. In recent years, the development of CNN models like VGGNet, GoogLeNet, and ResNet improved the image classification accuracy with a top-five error rate of 7.3%, 6.7%, and 3.6%, respectively [4]. As CNNs developed, researchers began using them for various disease-detection tasks involving medical image classification. With a deep neural network-based approach, Cireşan et al. [5] outperformed all the previous mitosis detection methods and obtained an F1-score of 78.2%. Rao et al. [6] used a CNN model with only two convolution layers and achieved 76% accuracy for tumor detection in the thoracic CT scans. Hooda et al. [7] used a CNN model with seven convolution layers and three fully connected layers to achieve a validation accuracy of 82% on the Montgomery County and Shenzhen (MCS) X-ray datasets for TB detection. Later, transfer learning was used for disease prediction tasks using pretrained image classification models. The medical image dataset was used to refine these pretrained models, which significantly improved prediction accuracy while requiring less training time. For an anatomy object detection task, a fine-tuned AlexNet and a modified AlexNet model got 74% and 81% accuracy, respectively [8]. Ho et al. [9] experimented with pretrained ResNet-152, Inception-ResNet, and DenseNet-121 with CXR14 dataset for TB detection and obtained an area under the curve (AUC) of 0.98 with DenseNet-121.

A significant amount of data is necessary for CNNs and other deep learning models to train for the prediction task. The most widely used TB dataset like Montgomery County (MC) dataset and Shenzhen dataset contains 662 X-rays and 138 X-rays, respectively [10]. There are only 152 and 150 X-rays in the DA and DB collection, respectively. The generalizability of the trained models is constrained by the dearth of easily accessible CXRs. In this regard, Liu et al. [11] collected 11,200

X-ray images classified into healthy, sick and non-TB and TB X-rays. Expert radiologists from renowned hospitals annotated the affected region of the lung in the TB images. In the field of TB diagnostics, the report suggested two next possibilities. One is to identify TB from X-ray images as healthy, sick but non-TB, and TB. Classifying the patient's TB infection as latent TB, active TB, or class-agnostic TB is another goal. Additionally, the majority of past studies focused on enhancing just one aspect of CNN models, such as image segmentation, image enhancement, or ensemble learning, in order to achieve improved performance.

In our proposed work, we intend to create a more generalized TB classification model with better performance. We will be using TBX11K dataset for training and various public datasets like MC, Shenzhen, and DA and DB datasets for testing to create a more generalized model. For improved learning of our model, segmented lung images paired with various image enhancements for different channels will be used. The performance and reliability of our model will be improved by an ensemble learning approach based on lightweight CNN networks. The following are the primary contributions of our work:

- To let the model learn more from the diseased region of the lungs, the U-Net-based lung segmentation model will be used.
- In order to fine-tune the pretrained network, it shall be fed with improved images. For improving the images, contrast limited adaptive histogram equalization (CLAHE) filter, high-emphasis filter (HEF), and bilateral low-pass filter (BF) will be used.
- To use safe-level SMOTE in order to deal with class imbalance.
- To use data augmentation techniques comprising shearing, rotation, horizontal flip, and scaling in order to reduce overfitting.
- To increase the reliability and to reduce the information loss, ensemble models will be used.
- To use models like EfficientNet-B0, ChexNet, and SqueezeNet for creating a voting-based ensemble model.
- To use a supplementary layer comprising of activation, SoftMax activation, and batch normalization above the concatenated features with ChexNet, EfficientNet-B0, and SqueezeNet as feature extractors.

## 4.2 Related Works

Kute et al. [12, 13] have discussed the challenges and realities in e-healthcare applications and emerging technologies. This section deals with current studies that aimed to enhance CXR image-based medical diagnosis. In a study using VGG16, VGG19, ResNet152, and DenseNet-121 models for the TB classification on the Shenzhen dataset, Cao et al. [14] observed that the DenseNet-121 outperformed other models with an accuracy of 0.90. Rajpurkar et al. [15] fine-tuned DenseNet-121 with more than 100,000 CXRs to detect 14 lung-related diseases from the CXR 14 dataset, and it has performed better than the existing models. It reached an accuracy of 0.92, 0.92, and 0.93 in cardiomegaly, hernia, and emphysema diseases, respectively. Oloko-Oba and Viriri [16] fine-tuned EfficientNet network models on the MCS dataset and obtained a classification accuracy of 93.3% with EfficientNet-B0 and 97% with EfficientNet-B7. In comparison to ResNet-50 and DenseNet-169, which used 26 M and 14 M parameters, respectively, EfficientNet-B0 was able to achieve greater accuracy with only 5.4 M parameters. EfficientNet-B7 used 66 M parameters for training. Liu et al. [11] evaluated the efficacy of different models with four publicly available datasets. With segmented lung image, ChexNet and SqueezeNet models obtained an accuracy of 98.14% and 96.6%, respectively. Lakhani and Sundaram [17] used GoogLeNet and AlexNet networks to evaluate the performance of pretrained models, CLAHE image enhancement, and ensemble-based learning compared to the untrained models. In the study, it was found that image augmentation and enhancement increased the untrained model's AUC from 0.90 to 0.95. With augmentation and ensemble learning, pretrained models increased their AUC from 0.98 to 0.99. The study was ineffective in inspecting the impact of lung segmentation on the performance of the classification task.

Image enhancement improves the performance of the CNN models by highlighting the smaller features, which enable the models to learn from them [18]. In an experiment to detect COVID-19-induced pneumonia detection, Heidari et al. [19] used image enhancement techniques of histogram equalization (HE) filter and bilateral low-pass filter (BF) and the original image as the three-channel input for fine-tuning a VGG16 model. The model's accuracy to detect pneumonia with the COVID-19 infection was 98.1%. By using unsharp masking (UM) for image enhancement with a fine-tuned EfficientNet-B4 network, Munadi et al. [20] achieved a 94.8% AUC for the TB classification on the Shenzhen dataset. An ensemble of VGG16 and InceptionV3 models with the CLAHE enhancement technique by Dasanayaka and Dissanayake [21] obtained an average accuracy of 97.1%. In a COVID-19 classification task, the ChexNet model with HE filter got an average accuracy of 94.3% [22]. CLAHE, when compared with gamma correction and HE, performed better on a bone segmentation task conducted by Ikhsan et al. [23].

The CXR images are used to train the CNN classifiers for diagnosing TB because it is a lung disease. In addition to the lung region, the X-ray image also shows the diaphragm, airways, additional thoracic tissues, and so on. The models are unable to learn more from the lung region due to these extra image attributes. Recent developments in medical image segmentation, like SegNet, U-Net, and so forth, allowed research to segment the lung region from the X-ray image and use it for the training of the classifiers to improve the accuracy. Rahman et al. [24] performed lung segmentation to improve the accuracy of the DenseNet-201 model from 95.1% to 98.6% for the TB classification task. All the other models, including ResNet, ChexNet, VGG19, Inception V3, SqueezeNet, and MobileNet, were also able to improve their accuracy by more than 2% through image segmentation. Dasanayaka and Dissanayake [21] were able to achieve a classification accuracy of 97.1% and a sensitivity of 97.9% on a CXR dataset.

By introducing some bias to each learner, ensemble-based models lower the variance in prediction errors while learning from a variety of base classifiers. The resultant model thus performs better than the individual contributing models [25]. The flexibility and adaptability of the ensemble models encouraged their application in bioinformatics [26]. Rashid et al. [27] used ResNet, Inception-ResNet, and DenseNet as feature extractors and SVM as a classifier to obtain an accuracy of 90.5% on the Shenzhen X-ray dataset. A stacked generalization (SG) ensemble with features extracted from VGGNet, EfficientNet, and DenseNet models was able to get the highest accuracy of 95% on a TB dataset without using any image enhancement technique like CLAHE, HEF, or thresholding [28]. A classifier trained on deep concatenated features with its features extracted from a proposed CNN, GoogLeNet, and ResNet-18 obtained an accuracy of 98.9% and recall of 98.5% on a COVID-19 dataset [29]. For a pneumonia-detection task, an ensemble of feature maps generated from optimized ChexNet and VGG19 with random forest classifier achieved 98.93% accuracy. The VGG19 and ChexNet model separately obtained an accuracy of 89.26% and 92.63%, respectively [30]. In our study, we suggested a novel model based on current experiments to develop a reliable, accurate, and generalized model on the TBX11K CXR dataset for TB identification.

### 4.2.1 Significance of the Study

Since TB is an infectious disease, it is important to have a quick diagnosis in order to stop its spread and lower the cost of treatment. An accurate and reliable CAD can aid in this effort for low-income countries that lack experienced radiologists for TB detections. Despite the fact that there are numerous deep learning-based CADs on the market right now, their adoption is modest because the models were only built using a few datasets. Our research intends to create models using the largest TB dataset to date, the TBX11K dataset, which contains more than 11,000 CXR pictures. To ensure that our models are widely accepted, we test their effectiveness with that of other state-of-the-art techniques on the most popular TB datasets, such as the Shenzhen, MC, DA and DB datasets, and so on. When used in practical settings, our model will help doctors identify TB from CXRs more precisely and reliably. By lessening the burden of TB sickness worldwide, it will aid society.

## 4.3 Research Methodology

For binary classification tasks of TB and non-TB, the majority of prior studies on TB detection employed the MCS datasets. With the rise in COVID-19 and other lung diseases, we found it appropriate to experiment with a multi-class classifier, which can identify not only TB and healthy images but also other sick lung images. Liu et al. [11] provided an opportunity to study a new dataset called "TBX11K" with three classes of images, namely TB, sick but non-TB, and healthy. Along with other binary classification datasets, we utilized this dataset to create our model and generalize it.

The earlier experiments employed the same grayscale X-ray image in all channels, whereas fine-tuned CNN models needed a three-channel input (a colored image). We decided to use three-filtered images for our investigation, one for each input channel. Each channel of the three-channel image will have a separate enhanced characteristic because the same image has multiple filters applied to it. This shall help our network to extract more data from the input image. We have separated our area of interest by lung segmentation, followed by concentrating the network's learning on it. We used the popular U-Net for segmentation because it excels in situations with little training data.

We chose three pretrained models – ChexNet, EfficientNet-B0, and SqueezeNet – when choosing CNN models for the TB prediction task. Since the ChexNet model has previously been optimized across more than 100,000 CXR images, we decided to employ it. From the recent benchmarking of models, we found that the SqueezeNet model has used its parameters most efficiently compared to other CNN models. Furthermore, with few training settings, EfficientNet-B0 has demonstrated efficiency and accuracy in image recognition tasks. Although individual models can achieve performance on a particular dataset, an ensemble model will provide us with a better-generalized prediction for the unseen datasets. Therefore, for our three-class classification experiment, we have worked on two most popular ensemble models, SG and the sum of probabilities (SOP) ensemble model.

### 4.3.1 Dataset Description

We have selected the TBX11K dataset for our experiment, along with the MC, Shenzhen, DA, and DB datasets. Due to the accessibility of their segmentation masks, which will aid in the validation of our segmentation model, we used the MCS datasets for the U-Net lung segmentation.

We used the Kaggle TBX11 repository, which includes images from the TBX11K dataset, the MC dataset, the Shenzhen dataset, and the DA and DB datasets, among other sources. These images are accessible in $512 \times 512$ dimensions. Table 4.1 provides the number of TB and non-TB images used.

#### 4.3.1.1 TB X 11K Dataset

The 11,200 de-identified CXR images in the TBX11K collection have a resolution of $512 \times 512$. These CXR images have all been annotated with bounding boxes by skilled radiologists from some of the best hospitals. There are 1,200 TB images and 10,000 non-TB images in the CXR images. Five categories – healthy, sick, non-TB, active TB, latent TB, and uncertain TB – are used to categorize the images. The test dataset contains all of the uncertain TB cases needed to carry out a class-based TB classification. The dataset is split into 6,600, 1,800, and 2,800 for training, validation, and testing sets, respectively [11].

#### 4.3.1.2 Montgomery County Chest X-Ray Dataset

The MC dataset comprises 138 images, 80 of which are healthy, and 58 of which are TB related. We will do our experiment using a $512 \times 512$ resolution dataset provided by Kaggle. The age and sex of each X-ray are also provided for a more comprehensive analysis [10]. The segmentation masks for each of these images are also included. This enables the dataset to be used for both training our U-Net segmentation model and testing the classification model.

#### 4.3.1.3 Shenzhen Chest X-Ray Dataset

Included in this dataset's 652 CXR images are 326 normal cases and 326 TB cases. While the original dataset is of the resolution $3,000 \times 3,000$ pixels, the dataset made available by the Kaggle TB11X dataset has a resolution of $512 \times 512$ pixels and is used in our experiment [10].

#### 4.3.1.4 DA and DB Chest X-Ray Dataset

The National Institute of Tuberculosis and Respiratory Disease in New Delhi has two datasets called "TDA" and "DB," which have been collected from two separate X-ray equipment. DA dataset comprises 78 non-TB images and 78 TB images, while DB dataset has 75 TB and 75 non-TB images. We have used the training and testing ratio of 2:1 for both datasets. All the images are of the resolution $1,024 \times 1,024$ pixels [31].

**Table 4.1** Non-TB and TB images in a different dataset.

| Data source | TB images | Non-TB images | Total images |
|---|---|---|---|
| TBX11K | 1,200 | 10,000 | 11,200 |
| DA and DB | 75 | 75 | 150 |
| Montgomery County | 58 | 80 | 138 |
| Shenzhen | 326 | 326 | 652 |

### 4.3.2 Data Preprocessing

Data preprocessing helps in enhancing the image's feature set, creating an additional image for training and regularization, and enhancing the model learning process's focus on the targeted area. During this phase, we used data augmentation, lung segmentation, and image enhancement to accomplish the aforementioned tasks.

#### 4.3.2.1 Lung Image Segmentation

The CNN models perform better by focusing more on the diseased region of the lungs. It is because lung image segmentation eradicates the nonessential part of the X-ray images, which help in better learning.

Figure 4.1 displays the U-Net model. The U-Net model has been a popular medical image segmentation technique due to its ability to fuse both low-level and high-level features by combining low-resolution feature maps with high-resolution feature maps through a skip connection [32]. Skourt et al. [33], using a U-Net model, obtained a 95% Dice coefficient index for a lung CT image segmentation task. A U-Net and a bidirectional convolution LSTM (BiConv-LSTM)-based U-Net were compared for lung image segmentation tasks by Rahman et al. [24]. In the experiment, U-Net performed better than BiConv-LSTM and obtained a 96.19% Dice coefficient index with Adam optimizer. Islam and Zhang [34] used a modified U-Net model to get a Dice coefficient of 98.6% on the MC dataset.

For our experimental purpose, we have chosen the U-Net model for lung segmentation due to its high effectiveness in the smaller dataset [35]. The network is trained using datasets of CXRs taken from Shenzhen and MC. The resolution of all the images and masks has been reduced to $256 \times 256$.

The U-Net model is a U-shaped CNN, which comprises an encoder network that contracts and a decoder network that expands. A rectified linear unit (ReLU), a $2 \times 2$ filter for maximum pooling, and two $3 \times 3$ convolutions are all that make up the contracting route of a down-sampling network. The expanding path at each step is made up of two $3 \times 3$ convolutions, two $2 \times 2$ convolutions, a concatenation of feature maps from the corresponding contracting path, and an up-sampling step.

The final layer contains a $1 \times 1$ filter for pixel-wise convolution [35].

#### 4.3.2.2 Contrast-Limited Adaptive Histogram Equalization

The original image is separated into smaller areas known as "Tiles for the CLAHE algorithm." These areas are all separately contrast-boosted. Bilinear interpolation is then used to combine the regions. This helps redistribute the lightness value of the image to produce a better-quality image [36].

#### 4.3.2.3 High-Frequency Emphasis Filter

The high-frequency emphasis filter enables the sharpening and highlighting of the image's edges. Fourier transformation, high-pass Gaussian filter, and an inverse Fourier transformation applied on the original image provide the filtered image. The HE technique is applied to this filtered image to get the final enhanced image [20]. Figure 4.2 displays the flow of HEF image creation.

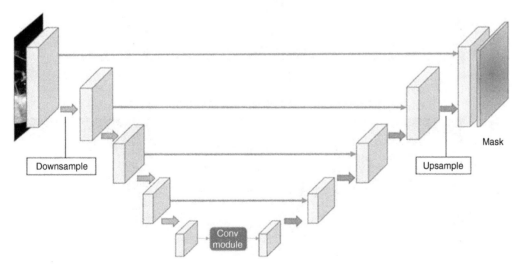

**Figure 4.1** U-Net model.

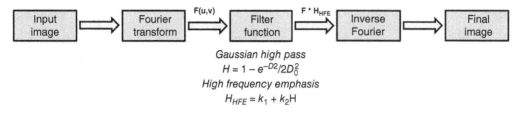

$$\text{Gaussian high pass}$$
$$H = 1 - e^{-D2/2D_0^2}$$
$$\text{High frequency emphasis}$$
$$H_{HFE} = k_1 + k_2 H$$

**Figure 4.2** High-emphasis filter image creation flow. *Source:* Image enhancement using histogram equalization and high-frequency emphasis filtering in Python | easy machine learn (2022).

#### 4.3.2.4 Bilateral Filter

BF, a nonlinear filter, helps in eradicating the artifact noise in an X-ray image. It smoothens the images by replacing individual pixels with the weighted average of the pixels in the neighborhood [37]. Sample filter images obtained using filters discussed earlier are provided in Figure 4.3.

#### 4.3.2.5 Data Augmentation

Shearing, scaling, rotation, and flipping are used for data augmentation, which helps in improving the generalizability of the model. Figure 4.4 shows the augmented dataset.

### 4.3.3 Transfer Learning

As there is a scarcity of CXRs for CNN model training, we have used the pretrained CNN models such as EfficientNet, ChexNet, and SqueezeNet and fine-tuned these models on the available CXR images through the method of transfer learning. Figure 4.5 provides steps to fine-tune a CNN model.

#### 4.3.3.1 Fine-Tuning a CNN Model

Following are the phases to fine-tune a CNN model.

In the proposed work, the last layers of all pretrained models are replaced with three-class SoftMax activation layers. Dropout layer of 0.4 is added before SoftMax layer for the generalizability of the network.

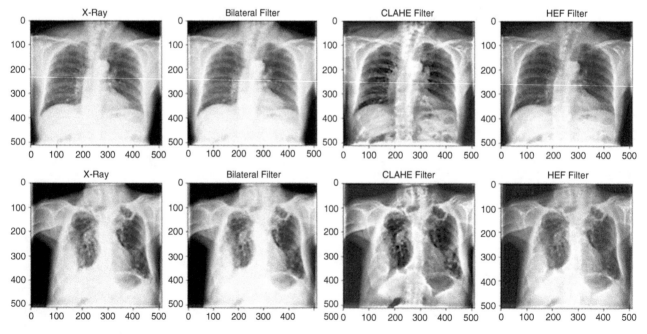

**Figure 4.3** Bilateral filter, CLAHE, HEF of the original image.

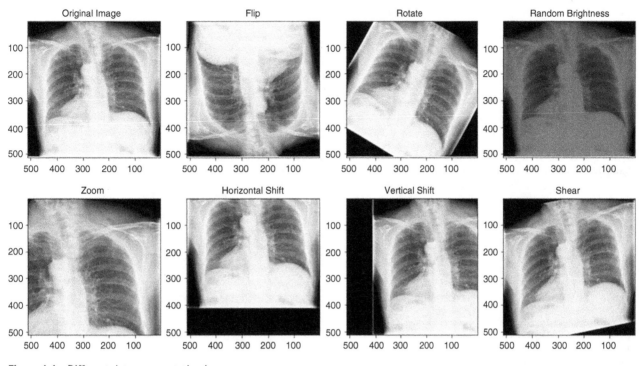

**Figure 4.4** Different data augmentation images.

**Figure 4.5** Phases to fine-tune a CNN model.

```
1. Load the pre-trained model with its weights (base model)
2. Remove the last Dense Layer (SoftMax/Sigmoid Activation layer)
   from the model.
3. Use GlobalAveragePooling to flatten output from the last layer
   after Step 2.
4. Add BatchNormalization and/or Dropout layer for regularization of
   the network (Optional).
5. Add a SoftMax/Sigmoid activation with the number of neurons based
   on the classification task.
6. Freeze the layers of the base model and train only the final layer
   of the network.
7. Once the final layer is trained, unfreeze all the layers in the
   base model. And, retrain the network with a lower learning rate.
```

***ChexNet Model*** A 121-layer DenseNet model known as "ChexNet" has been fine-tuned over more than 100,000 CXR images. Originally, it was built to predict pneumonia from the CXR14 dataset [15]. The cost of training will be meaningfully lower since the network will combine more quickly on the CXRs because it has previously been trained on a large number of X-ray pictures. Figure 4.6 provides parameters for the ChexNet model.

### 4.3.3.2 EfficientNet-B0 Model

EfficientNet is a CNN model that utilizes the compound scaling technique to scale the network's breadth, dimensions, and resolution uniformly with a set ratio. The EfficientNet network family's baseline network is called "EfficientNet-B0." In order to capture fine-grained patterns as input resolution rises, feature channels must be expanded, and network depth must be increased for larger receptive fields. This is accomplished using compound scaling, which uniformly scales each network dimension. EfficientNet-B0 has obtained 93.3% accuracy on the ILSVRC challenge with only 5.3 M parameters and 0.39 B floating-point operations, better than what ResNet-50 and DenseNet-169 have achieved using 26 and 14 M parameters, respectively [38]. Figure 4.7 provides EfficientNet-B0 training parameters.

The EfficientNet-B0 model imported from TensorFlow Keras has around 4 M training parameters.

```
global_average_pooling2d (Glob    (None, 1024)         0         ['relu[0][0]']
alAveragePooling2D)

dropout (Dropout)                 (None, 1024)         0         ['global_average_pooling2d[0][0]'
                                                                  ]

dense_1 (Dense)                   (None, 3)            3075      ['dropout[0][0]']
================================================================================================
Total params: 7,040,579
Trainable params: 6,956,931
Non-trainable params: 83,648
```

**Figure 4.6** Parameters for the ChexNet model.

```
global_average_pooling2d (Glob    (None, 1280)         0         ['top_activation[0][0]']
alAveragePooling2D)

dropout (Dropout)                 (None, 1280)         0         ['global_average_pooling2d[0][0]'
                                                                  ]

dense (Dense)                     (None, 3)            3843      ['dropout[0][0]']
================================================================================================
Total params: 4,053,414
Trainable params: 3,969,375
Non-trainable params: 84,039
```

**Figure 4.7** Training parameters of EfficientNet-B0.

### 4.3.3.3 SqueezeNet Model

SqueezeNet is a CNN architecture that, when applied to an ImageNet dataset with 50 times fewer parameters, achieves an accuracy level comparable to that of the AlexNet network. It employs fire modules with two layers – the expansion layer and the squeeze layer. A $1 \times 1$ convolution filter is used in the squeezing layer, while a combination of $1 \times 1$ and $3 \times 3$ convolution filters is used in the expansion layer. The network decreases the number of parameters by nine times by employing a squeeze layer. In a recent benchmark study done, SqueezeNet performed the best on accuracy density, which measured how efficiently a network used its parameters. For the purpose of creating models, we used the official SqueezeNet repository (rcmalli/keras – squeezenet: SqueezeNet implementation with Keras Framework).

Figure 4.8 depicts the architectural layout of a SqueezeNet model's squeeze and excite layers that we created.

Only 1.2 M parameters make up the SqueezeNet model, significantly fewer than ChexNet and EfficientNet-B0. Figure 4.9 provides SqueezeNet parameters.

### 4.3.4 Experimental Approaches

We have experimented with two ensemble models: the SG ensemble and the SOP ensemble in an effort to create an appropriate ensemble model. ChexNet, EfficientNet-B0, and SqueezeNet models that have already been refined are used to generate these ensemble models. We freeze the training of the base models while creating these ensemble models. The feature concatenation layer, dense layer, SoftMax layer, average layer, dropout, and batch normalization layers are just a few of the new layers that can be added to the initial model to accommodate any further learning.

### 4.3.4.1 Approach1: Sum of Probabilities Ensemble

The probability score of each fine-tuned model for a class was combined and used in the SOP ensemble. By averaging the aggregate score's weighted components, the final probability score for the class is determined. This method enhances network performance without the need for extra training. However, because we are taking into account the average score of the underlying models, the score of the weakest classifiers may have an impact on the final forecast. The SOP ensemble model is shown in Figure 4.10.

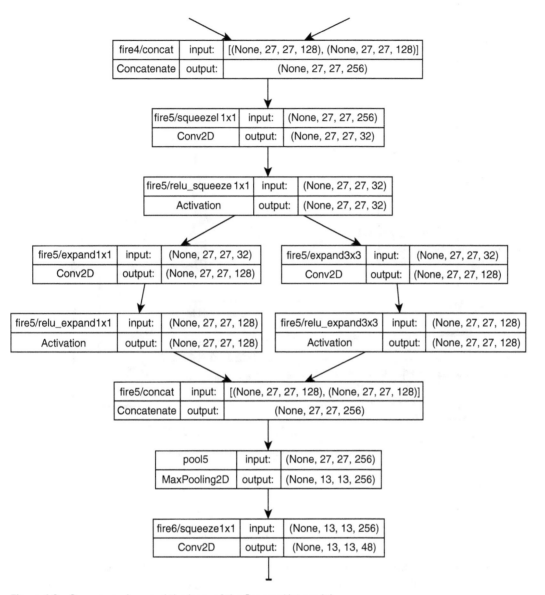

**Figure 4.8** Squeeze and expand the layer of the SqueezeNet model.

```
relu_conv10 (Activation)          (None, 13, 13, 1000   0           ['conv10[0][0]']
                                  )

global_average_pooling2d_1 (Gl    (None, 1000)          0           ['relu_conv10[0][0]']
obalAveragePooling2D)

dropout (Dropout)                 (None, 1000)          0           ['global_average_pooling2d_1[0][
                                                                    0]']

dense (Dense)                     (None, 3)             3003        ['dropout[0][0]']
==============================================================================================
Total params: 1,238,499
Trainable params: 1,238,499
Non-trainable params: 0
```

**Figure 4.9** SqueezeNet params.

### 4.3.4.2 Approach 2: Stacked Generalization Ensemble

The refined models serve as a feature extractor in the SG ensemble technique. To get feature channels with identical dimensions, we deleted the final few layers from each model. The global average pooling layer is then used to concatenate and flatten these features. The output is connected to two extra dense layers with dropout and ReLU activations

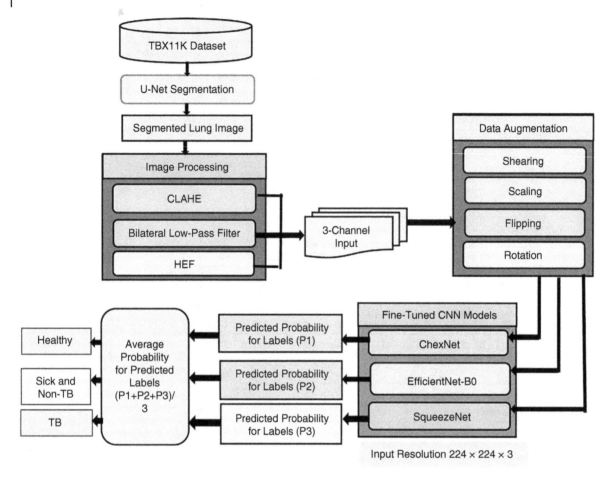

**Figure 4.10** Sum of probabilities ensemble.

using batch normalizations. The final classification output is created by passing the resultant output through a three-class SoftMax activation layer.

After deleting the final few layers from ChexNet, EfficientNet-B0, and SqueezeNet, respectively, we were able to get (7, 7, 1024), (7, 7, 1280), and (7, 7, 1000) features for our experiment. For our ensemble learning, we combined these characteristics to create (7, 7, 2304) features. After the feature layers, we add two additional fully connected layers and a three-class SoftMax layer to build our SG ensemble model. The SG ensemble model is shown in Figure 4.11.

### 4.3.5 Evaluation Metrics

An analysis of the model's performance on unobserved data yields a confusion matrix. A number of attributes that can be used to compare the efficacy of several models in a certain classification task can be created using this matrix. A confusion matrix can be found in Table 4.2.

The attributes that come from the performance matrix are listed here.

$$\text{Accuracy} = \left(\text{TP} + \text{TN}\right) / \left(\text{TP} + \text{TN} + \text{FP} + \text{FN}\right) \tag{4.1}$$

$$\text{Precision or True Positive Rate}\left(\text{TPR}\right) = \text{TP} / \left(\text{TP} + \text{FP}\right) \tag{4.2}$$

$$\text{Sensitivity} = \text{TP} / \left(\text{TP} + \text{FN}\right) \tag{4.3}$$

$$\text{Recall} = \text{TP} / \left(\text{TP} + \text{FN}\right) \tag{4.4}$$

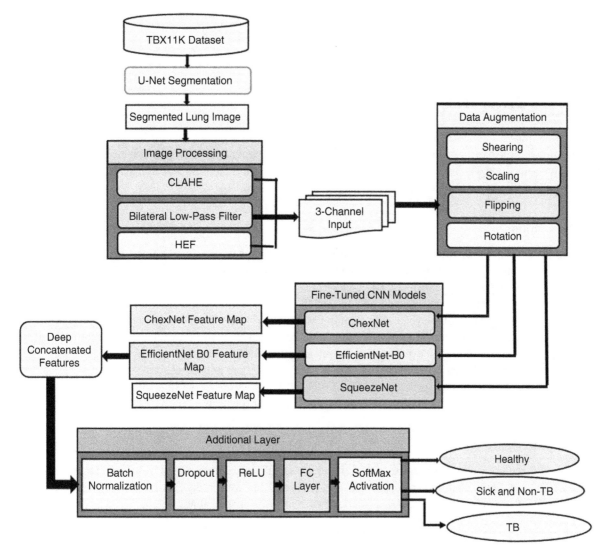

**Figure 4.11** Stacked generalization ensemble.

**Table 4.2** Confusion matrix.

|  | Actual negative (0) | Actual positive (1) |
| --- | --- | --- |
| Predicted negative (0) | True negative (FN) | False negative (TN) |
| Predicted positive (1) | False positive (FP) | True positive (TP) |

$$\text{Dice Coefficient}\left(F1-\text{score}\right)=\left(2^{*}\text{TP}\right)\big/\left(2^{*}\text{TP}+\text{FN}+\text{FP}\right) \tag{4.5}$$

$$\text{Intersection over Union}\left(\text{IoU}\right)\text{or Jaccard Index}=\text{TP}\big/\left(\text{TN}+\text{FN}+\text{FP}\right) \tag{4.6}$$

For the lung image segmentation task, we will be using the standard performance metrics of accuracy, IoU and Dice coefficient index. We will compare our model with existing models using accuracy, precision, sensitivity, and recall in order to evaluate the performance of our model. A quick, accurate, and affordable method of TB detection has become essential for its eradication. We aim to automate the TB detection process using ensemble learning and deep learning techniques to support such a procedure. To generate the model described in our study, we used a strategy that included data preprocessing, image processing, lung area extraction, pretrained model fine-tuning on our dataset, and ensemble model generation. We implemented the procedures using TensorFlow, Keras, and OpenCV.

## 4.4 Results and Discussions

The data preparation; image augmentation; image segmentation by U-Net; image enhancements by HEF, CLAHE, and BF filters; and fine-tuning of the base classifiers like EfficientNet-B0, ChexNet, and SqueezeNet processes are covered in more detail in the following sections. The process of developing ensemble models utilizing the SOP and SG to categorize CXRs as sick, healthy, or TB is also covered in detail.

### 4.4.1 Data Preparation

We have gathered the TBX11K, Shenzhen, MC, and DA and DB datasets for the training, validation, and evaluation of our models. The healthy, sick, and TB class of data are included in the TBX11K dataset. Only healthy and TB X-rays are included in other databases. We employed 3,000, 3,000, and 600 CXRs from the TBX11K dataset, categorized into the healthy, sick, and TB classes, respectively, for the training of our proposed models. The 800, 800, and 200 CXRs for the validation process came from the same source. We also employed CXRs from the Montgomery, Shenzhen, and DA and DB datasets for the generalization of our model. We selected 60% and 20% of the CXRs from each of these datasets for the training and validation procedures, respectively. We have tried oversampling as well as assigning class weights inversely related to their image counts while computing the loss function in order to address the problem of class imbalance.

These CXR images are kept in the train/val/test – health|sick|TB folder structure so that we can use TensorFlow's ImageDataGenerator class's flow_from_directory function to import data for the training phase.

### 4.4.2 Image Enhancements

Image augmentation is required to highlight some aspects that the trained model would otherwise miss and to increase the quality of the raw images. Three well-known filters, CLAHE, HEF, and BF, were used to enhance the images. For our model training, a three-channel image is produced by merging each of these image enhancements.

1) The first step in the CLAHE enhancement is to convert the input image to grayscale.
2) Produce a CLAHE object with a clip limit of 40 and $8 \times 8$ tile dimensions.
3) Use the HE approach to enhance the image's overall contrast.
4) Apply CLAHE to the Step 2 equalized image.

#### 4.4.2.1 Applying Bilateral Filter to the Image

How to use the HEF:

1) Read the input image and convert it to a grayscale image.
2) Apply the image to the fast Fourier transform (FFT).
3) Position the FFT image in the middle.
4) Put a Gaussian high-pass filter in place and then calculate the filter value.
5) To create the HEF image, apply the "high_filter" to the original image and then use the inverse Fourier transform.

We used the OpenCV merge function to combine all of the enhanced images into a three-channel image after they had been formed. Our model training will now use these three-channel images as its input. Figure 4.12 shows sample images for image enhancements.

### 4.4.3 Image Augmentation

To expand the amount of training data and to add regularization to the developed model, image augmentation is required. We may produce enhanced images while the model is being trained using the ImageDataGenerator in the Keras package.

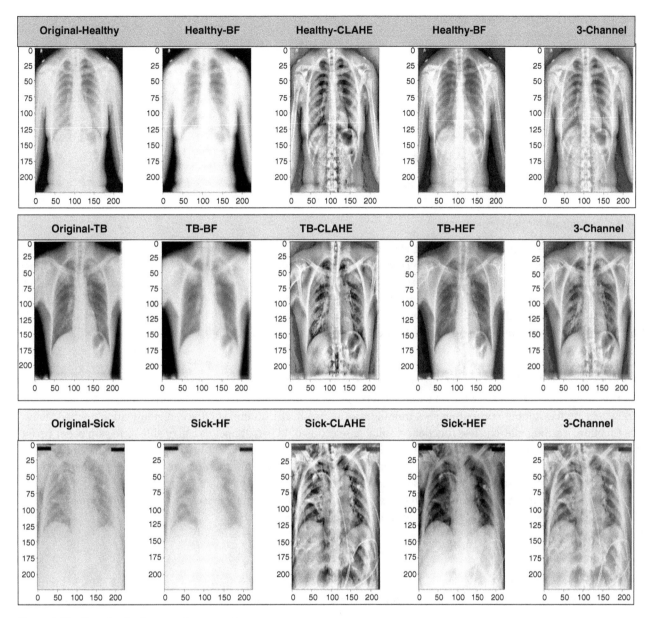

**Figure 4.12** Samples for image enhancements.

As a result, it helps us in avoiding the need for extra processing logic for its development as well as the storage space needed for the augmented images.

### 4.4.3.1 Lung Segmentation and Region of Interest Extraction

For the purpose of separating the area of interest (ROI) from the provided X-rays, a U-Net lung segmentation model is created. For training of the model, we have used the MCS datasets [10], which also contain the lung masks of corresponding X-ray images annotated by expert radiologists. U-Net model hyperparameters are given in Table 4.3. The procedures we took to construct the U-Net model are listed here.

Step 1: Define the U-Net model with a 256×256 input shape.

Step 2: To create training and validation images for the model training, define the train and validation generator. While we preserved the original data for validation, we used augmentation for training images.

The TBX11K dataset's masks were predicted using the U-Net model once it was developed.

**Table 4.3** U-Net model hyperparameters.

| Parameter | Value |
|---|---|
| Input shape | $256 \times 256$ |
| Output shape | $256 \times 256$ |
| Epochs | 500 |
| Optimizer | Adam |
| Loss function | Dice coefficient loss |
| Learning rate | 1e-4 |

### 4.4.4 Pretrained Model Fine-tuning

In order to accurately complete a variety of classification tasks, CNN models require a large amount of training data. Transfer learning helped us to overcome the lack of large datasets in the medical domain as we can now use pretrained models from other similar domains for our disease classification tasks. On our own dataset of CXR images, we have refined three models that were trained on more than 10 million real-world images. These algorithms will be able to accurately predict TB, sick, and healthy images with the aid of transfer learning. The selection of the hyperparameters and the callbacks required for effectively training our model are discussed in the section that follows.

### 4.4.5 Model Hyperparameter Selection

The fixed parameters that a CNN model uses to determine its performance are called "hyperparameters." The following parameters are fixed: input size, output size, training epochs, batch size, optimizer choice, regularization method, activation function, learning rate, and loss function. A suitable selection of hyperparameters can enable us to gain better performance for our chosen models. We have only conducted experiments to select the best optimizers and regularization techniques from among all the available hyperparameters. Based on past information and resource parameters, other parameters were chosen. Table 4.4 provides a first model evaluation for hyperparameter selection.

The following hyperparameter settings will be used during the model training phase based on the preliminary findings of the model training. Hyperparameters for model training are shown in Table 4.5.

### 4.4.6 Model Callbacks

When a model reaches a particular, planned stage of training, a series of instructions known as "model callbacks" must be followed. These callbacks include model checkpoint (storing weights), Reduce LR On Plateau (decreasing learning rate), and early stopping (quitting training early when the model can no longer learn).

**Table 4.4** Initial model assessment for hyperparameter selection.

| Base classifiers | Regularization method | Optimizer | Validation accuracy | Validation loss |
|---|---|---|---|---|
| EfficientNet-B0 | Batch normalization | Adam | 96.36 | 0.1161 |
| EfficientNet-B0 | Dropout (0.4) | Adam | 96.65 | 0.1075 |
| EfficientNet-B0 | Batch normalization | SGD | 92.32 | 0.2107 |
| SqueezeNet | Batch normalization | Adam | 94.05 | 0.2069 |
| SqueezeNet | Dropout (0.3) | Adam | 92.18 | 0.2013 |
| SqueezeNet | Batch normalization | SGD | 88.58 | 0.3050 |
| ChexNet | Batch normalization | Adam | 89.07 | 0.3253 |
| ChexNet | Dropout (0.4) | Adam | 96.56 | 0.0982 |
| ChexNet | Batch normalization | SGD | 87.11 | 0.4393 |

**Table 4.5** Hyperparameters for model training.

| Parameter | Value |
| --- | --- |
| Input shape | $(224 \times 224 \times 3)$ |
| Batch size | 32 |
| Output shape | $(1,3)$ |

### 4.4.7 Model Training

After completing the procedures outlined earlier (procedures to fine-tune a CNN model), we have adjusted our classifiers. By importing the pretrained models, we first built our model. ChexNet has already been trained on X-rays, so we decided to train every layer from the very beginning. For SqueezeNet and EfficientNet-B0, we decided to train the final dense layer first.

Step 3: After building the models, we assembled them and ran the training model. We used the callbacks defined in Figure 4.4 (callbacks for the model training).

Step 4: Except for the batch normalization layer, all other layers of the EfficientNet-B0 and SqueezeNet model were unfrozen, following the training of the final layer. The model was created and then retrained using a slower learning rate.

### 4.4.8 Ensemble Model Development

We used our classifiers after fine-tuning to produce SOP and SG classifiers. The result from the individual model is transmitted to the average layer, which generates the SOP ensemble as the final output from the ensemble model. We removed the final four layers from each model before extracting the features for the SG ensemble. As shown in the subsequent section, we combined these features and added further layers.

### 4.4.9 Model Evaluation

Without any further augmentation, we utilized TensorFlow's ImageDataGenerator to build a test generator in order to assess the model. Following that, the generator is utilized to predict batch data. The model's precision, recall, and F1-score were then obtained using the classification_report function of Sklearn. In the part that follows, we went into detail about the outcomes of applying our segmentation model, fine-tuning models, and built ensemble models. The results' importance has been explored, and they have been compared with earlier research.

### 4.4.10 U-Net Model Performance

We created the U-Net model using both the original and augmented CLAHE images. We have compared the obtained model to other study results in Table 4.6. The U-Net model's performance is also shown in Table 4.6.

We were able to increase both the classification accuracy and the Dice coefficient loss by enhancing the images using CLAHE. Several examples of predicted masks produced by our developed model are shown in the next section, along with a comparison to the real mask. Figure 4.13 provides the U-Net model's predicted masks.

**Table 4.6** U-Net model performance.

| Image type | Binary accuracy | Dice coefficient loss |
| --- | --- | --- |
| Original image | 97.06 | 96.12 |
| CLAHE image | 98.45 | 98.04 |

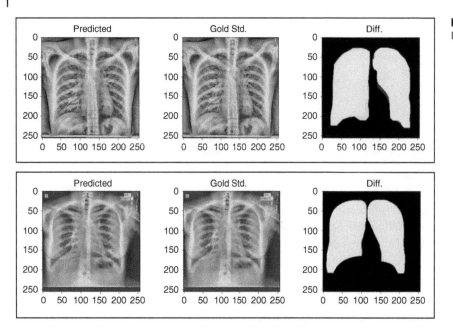

**Figure 4.13** Predicted masks of the U-Net model.

### 4.4.11 Performance of Fine-Tuned Model

On the following sorts of images, the fundamental classifiers EfficientNet-B0, ChexNet, and SqueezeNet are improved.

- Original images
- Enhanced images with lung segmentation (referred to as SegmentedBCH for further discussion)
- Enhanced images with CLAHE, HEF, and BF filter
- The fusion of HEF, CLAHE, and BF into a single image (referred to as BCH in the discussion that follows)

We have decided to preserve the weights for the epoch with the lowest validation loss because the reliability and precision of the model prediction rely on the loss function value.

### 4.4.12 EfficientNet-B0 Model Performance

As a result of the experiment, we discovered that HEF enhancement boosted the EfficientNet-B0 model's performance, whereas other enhancements provided slight decrement in cross-entropy loss. The appropriate settings for each enhancement still need to be explored because they are all produced using default arguments. The optimal weights for this model were found using the three-channel filter image fine-tuning. The performance summary of EfficientNet-B0 is provided in Table 4.7.

**Table 4.7** Performance analysis of EfficientNet-B0.

| Image type | Precision | Recall | Categorical cross-entropy loss | Accuracy |
| --- | --- | --- | --- | --- |
| Original | 96.70 | 96.65 | 0.1075 | 96.65 |
| CLAHE | 96.50 | 96.36 | 0.1109 | 96.41 |
| BF | 96.90 | 96.80 | 0.1097 | 96.85 |
| HEF | 96.80 | 96.70 | 0.0991 | 96.70 |
| **BCH** | **97.09** | **96.97** | **0.0872** | **97.03** |
| Segmented BCH | 95.54 | 95.29 | 0.1257 | 95.36 |

### 4.4.13 ChexNet Model Performance

A DenseNet-121 model called "ChexNet" has already been refined using more than 100,000 X-ray images. In addition to the BF, which reduces image noise, CLAHE and HEF images had a considerable decrease in cross-entropy loss value. The performance summary of ChexNet is given in Table 4.8.

### 4.4.14 SqueezeNet Model Performance

The model's insufficient ability to learn from the source image is caused by the fact that it only has 73 layers, as opposed to ChexNet's 121 layers and EfficientNet-B0's 428 layers. As more features were emphasized for model learning, the performance of the SqueezeNet model significantly improved as a result of the BF and CLAHE augmentation strategies. SqueezeNet worked best when the image noise was eliminated by a BF. Table 4.9 provides a performance summary of the SqueezeNet classifier.

### 4.4.15 Performances of Ensemble Models

We developed ensemble models for each form of image after the models were optimized on several sorts of images. Each SOP ensemble model outperformed at least one of the constituent models in terms of performance.

### 4.4.16 Performance Analysis of Sum of Probabilities Ensemble Model

We were successful in enhancing the lowest individual model's performance through the SOP ensemble. Among the various image types that we utilized to train our model, the BF, CLAHE, HEF (BCH) image yielded the best results. Table 4.10 provides the performance of the SOP ensemble model.

**Table 4.8** Performance analysis of ChexNet.

| Image type | Precision | Recall | Categorical cross-entropy loss | Accuracy |
|---|---|---|---|---|
| Original | 96.70 | 96.56 | 0.0982 | 96.56 |
| CLAHE | 93.43 | 93.11 | 0.2070 | 93.26 |
| BF | 96.30 | 96.11 | 0.1057 | 96.21 |
| HEF | 93.29 | 93.01 | 0.2198 | 93.06 |
| **BCH** | **97.16** | **96.97** | **0.0918** | **97.03** |
| Segmented BCH | 96.57 | 96.45 | 0.1114 | 96.51 |

**Table 4.9** Performance summary of SqueezeNet classifier.

| Image type | Precision | Recall | Loss | Accuracy |
|---|---|---|---|---|
| Original | 93.36 | 91.24 | 0.2013 | 92.18 |
| **BF** | **95.26** | **94.88** | **0.1292** | **95.13** |
| CLAHE | 94.02 | 93.65 | 0.1595 | 93.75 |
| HEF | 88.00 | 87.70 | 0.4501 | 87.80 |
| BCH | 93.93 | 93.75 | 0.1576 | 93.87 |
| Segmented BCH | 91.57 | 90.31 | 0.2423 | 90.99 |

**Table 4.10** Sum of probabilities ensemble.

| Image type | Precision | Recall | Categorical cross-entropy loss | Accuracy |
|---|---|---|---|---|
| Original | 97.41 | 96.26 | 0.1153 | 96.85 |
| BF | 96.80 | 96.75 | 0.0915 | 96.80 |
| CLAHE | 97.18 | 96.80 | 0.1062 | 96.95 |
| HEF | 96.23 | 95.47 | 0.1340 | 95.72 |
| **BCH** | **98.06** | **97.81** | **0.0782** | **97.94** |
| Segmented BCH | 97.38 | 96.90 | 0.0978 | 97.15 |

### 4.4.17  Performance Analysis of Stacked Generalization Model

Each of these models for the SG models outperformed the constituent models. Deep feature concatenation, which gives the models more features to train from, is responsible for this improvement in performance. Additionally, the two extra dense layers with ReLU activation enhanced the learning of the SG model. Table 4.11 provides performance of SG ensemble models.

Table 4.11 demonstrates that SG models were better able to learn from the input images. In comparison to other image kinds, BCH and BF images gave us the best results.

### 4.4.18  Overall Results of Model Building

Table 4.12 shows the comparison of performance for different methodologies used. According to the findings of the individual model fine-tuning, image enhancement can improve performance without incurring additional training expenditures. We were able to outperform models trained on the original image in all cases using three-channel BCH images. In

**Table 4.11** Performance of stacked generalization ensemble models.

| Image type | Precision | Recall | Categorical cross-entropy loss | Accuracy |
|---|---|---|---|---|
| Original | 96.70 | 96.75 | 0.0915 | 96.65 |
| BF | 97.38 | 97.19 | 0.0861 | 97.24 |
| CLAHE | 96.69 | 96.41 | 0.1096 | 96.50 |
| HEF | 94.91 | 94.48 | 0.1699 | 94.73 |
| **BCH** | **97.41** | **97.10** | **0.0793** | **97.23** |
| SegmentedBCH | 97.06 | 97.03 | 0.1047 | 97.03 |

**Table 4.12** Performance comparison for the methodology components analysis.

| Methodology | Precision | Accuracy | Recall |
|---|---|---|---|
| Original image (ChexNet) | 96.70 | 96.56 | 96.56 |
| BCH image (EfficientNet-B0) | 97.09 | 97.03 | 96.97 |
| SOP ensemble + SegmentedBCH | 97.38 | 97.15 | 96.90 |
| SOP ensemble + BCH | 98.06 | 97.94 | 97.81 |

the original image, ChexNet was revealed to be the best-performing individual model. This is due to the fact that it had previously undergone more than 100,000 X-ray adjustments. Our models of ChexNet and EfficientNet provided close to 97% accuracy, precision, and recall on the TBX11K dataset. EfficientNet-B0 reduced loss to the lowest level, improving the accuracy of its classification of BCH images. We achieved the best accuracy of 95.13% on the three-class classification tasks using SqueezeNet, which only has 1.2 M parameters.

All original images, including BCH, BF, CLAHE, and HEF, have been used in our experiments. The lung-segmented BCH image was also used. Although previous works have been done on the TB diagnosis [24], Jaeger et al. [10] suggest that the use of lung segmentation led to improvement in the performance of the model, for our three-class classification experiment, the lung segmentation image in our experiment, led to a reduction in the performances. This may have been brought on by the presence of abnormal images (original X-ray, segmented and dilated lung X-rays) whose important features may have been outside the lung region. Without these characteristics, the difference between sick and normal images might disappear entirely. After the lung segmentation experiment, there may have been more misclassifications as a result of this. In order to capture these important features that are outside the lung region and remove the features that are not required for our experiment, we have tried numerous iterations with mask dilation of kernel size 12×12. Even so, we were unable to achieve performance levels above those of the original BCH image. In comparison to our best-performing model, we were able to achieve accuracy, precision, and recall with lung segmentation and BCH images of 97.03%, 97.06%, and 97.03%, respectively. Therefore, we made the decision to skip the lung segmentation step of creating the final model. It assisted us in lowering the cost of the training without sacrificing performance. In Figure 4.14, the original, segmented, and dilated lung X-rays are displayed.

**Figure 4.14** Original X-ray, segmented, and dilated lung X-rays.

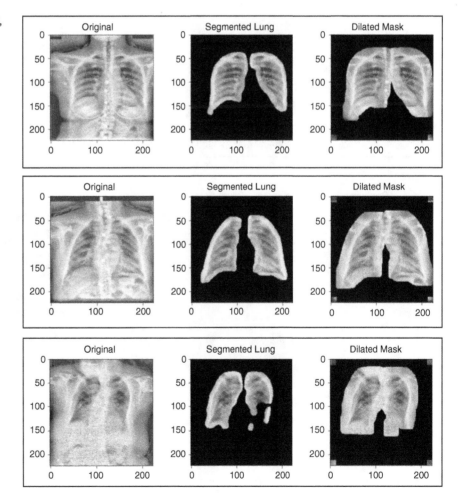

#### 4.4.19 Selection of Best Model

After examining the outcomes from various models across six distinct image types, we discovered that the SOP ensemble, trained on the BCH image, got the lowermost cross-entropy loss. The SOP ensemble is the most credible model since it has the lowest loss. On the TBX11K dataset, we achieved 98.06% precision, 97.94% accuracy, and 97.81 recall. For the TB class, our model's recall, F1-score, and precision were 93.57%, 94.91%, and 96.28%, respectively. Figure 4.15 provides the confusion matrix for the SOP ensemble model on the TBX11K dataset.

Additionally, our model performed well on datasets from Shenzhen and MC. Our accuracy, precision, recall, and F1-score for the Shenzhen dataset were 94.78%, 94.87%, 94.78%, and 94.78%, respectively.

Our accuracy, precision, recall, and F1-score for the MC dataset were 89.29%, 89.38%, 89.29%, and 89.22%, respectively. These results outperform the majority of models that were specially trained for two-class classification tasks using these datasets. These outcomes support the model's generalizability. Figure 4.16 provides the confusion matrix for the MC and Shenzhen datasets.

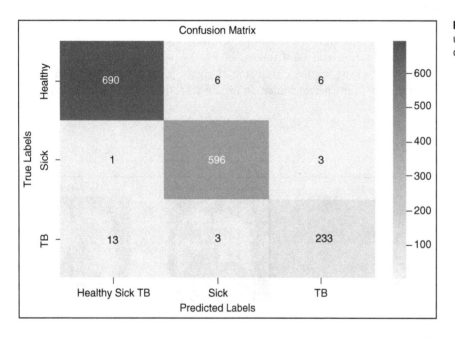

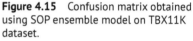

**Figure 4.15** Confusion matrix obtained using SOP ensemble model on TBX11K dataset.

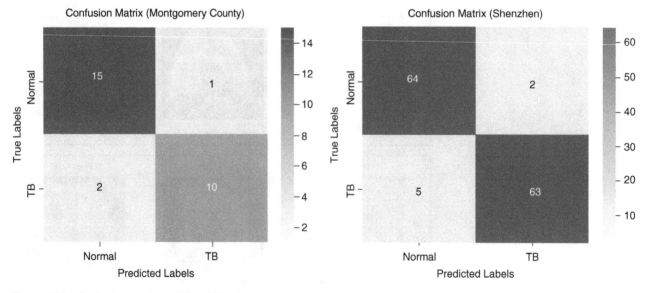

**Figure 4.16** Confusion matrix for MC and Shenzhen dataset.

**Table 4.13** Comparison of performance on the TBX11K dataset.

| Authored work | Method | Recall | Accuracy | Precision |
|---|---|---|---|---|
| Liu et al. [11] | **Faster R-CNN** | **90.5** | **89.7** | **87.7** |
| | SSD | 83.8 | 84.7 | 82.1 |
| | FCOS | 89.2 | 88.9 | 86.6 |
| | RetinaNet | 83.8 | 84.7 | 82.1 |
| | ChexNet | 82.5 | 95.9 | 80.9 |
| | VGG-19 | 83.5 | 95.7 | 79.1 |
| | **EfficientNet-B5-FPA** | **88.5** | **96.8** | **83.9** |
| Our proposed work | EfficientNet-B0 | 96.97 | 97.03 | 97.06 |
| | SqueezeNet | 94.88 | 95.13 | 95.26 |
| | ChexNet | 96.97 | 97.03 | 97.16 |
| | BCH + Lung Segmentation + SG | 97.03 | 97.03 | 97.06 |
| | **SOP-BCH Ensemble** | **97.81** | **97.94** | **98.06** |
| | **SG-BCH Ensemble** | **97.10** | **97.23** | **97.41** |

### 4.4.20 Comparison of Research Findings on TBX11K Dataset

Only two studies have examined the TBX11K dataset since it became available in the year 2020, according to our search. These studies concentrate on classifying TB as well as detecting regions. Our research work mainly focuses on TB classification problems. Our proposed experiment outperformed previous research work significantly on the classification tasks. Table 4.13 provides a summary of the comparison.

**Findings:** Compared to whose proposed model required 30 M parameters to train on, our best-performing SG model needed only 12 M parameters. We have enhanced classification accuracy and improved almost 10% on precision and recall metrics compared to the previous works.

### 4.4.21 Comparison of Model Performance with Other Binary Classification Models

Our model has been compared to previous binary TB classification models. On the Shenzhen dataset, our model fared better than the majority of the other models.

For the Montgomery dataset, our model has a comparative result with the similar work done by Tao Hwa et al. [39] albeit with far fewer training parameters. A comparison of performance with other TB classification models is given in Table 4.14. Tao Hwa et al. [39] used an ensemble of a heavier network that included VGG16 and InceptionV3 (158 M parameter training), 90% of total data for training, and 2,000 epochs of training to obtain accuracy and recall of 93.5%, and 92.31% respectively. Our model, which used only a 12.2 M training parameter, was able to outperform with accuracy and recall of 94.78% and 94.78%, respectively.

## 4.5 Conclusions

We have used segmentation and classification to accomplish our goal of building an accurate, dependable, and generalized TB detection system that can distinguish between sick, TB, and normal images from a CXR. We were successful in achieving our goal of creating a U-Net model for an improved segmentation model. With binary classification accuracy and Dice coefficient of 98.48% and 98.04%, respectively, the segmentation model performed remarkably well. Three image filters, that is, BF, CLAHE, and HEF, have been used to achieve the goal of image enhancement. To enhance images for model training and regularization, we used TensorFlow's ImageDataGenerator. In order to stabilize the model, we also included the dropout layer before the SoftMax layer. For the selection of the hyperparameters, we have performed preliminary experiments. We also employed the best-performing hyperparameters to further enhance our model. The ChexNet,

**Table 4.14** Performance comparison with other TB classification models.

| Authored work | Dataset | Method | Recall | F1-score | Acc. | Prec. | Training params |
|---|---|---|---|---|---|---|---|
| Rashid et al. [27] | Shenzhen | Ensemble of ResNet-152, Inception-ResNet-v2 and DenseNet-169 | 89.4 | | 90.50 | – | 138 M |
| Munadi et al., [20] | Shenzhen | EfficientNet-B4 | – | – | 89.92 | – | 19.34 M |
| Tao Hwa et al. [39] | Shenzhen | Ensemble of VGG16 and Inception V3 | – | 92.31 | 93.5 | – | 158.5 M |
| | Montgomery County | | | | 78.3 | | |
| Cao et al., [14] | Shenzhen | DenseNet-121 | 90.53 | 90.36 | 90.38 | 90.33 | 8 M |
| Oloko-Oba and Viriri [16] | Shenzhen | Ensemble of EfficientNet B2, B3, B4 | – | – | 97.44 | – | 40.2 M |
| | Montgomery County | | – | – | 95.82 | – | |
| **Our proposed work** | **Shenzhen** | **Ensemble of EfficientNet-B0, SqueezeNet, ChexNet** | **94.78** | **94.78** | **94.78** | **94.87** | **12.2 M** |
| | **Montgomery County** | | **89.29** | **89.22** | **89.29** | **89.38** | |

EfficientNet-B0, and SqueezeNet models were trained using a set of hyperparameters and enhanced images. SOP ensemble model and an SG ensemble model are then developed using these trained models. We evaluated our model's performance with other works. Our model outperformed the other few models created for the TBX11K dataset. Also, our model performed better than the majority of the current models in the Shenzhen and MC datasets. We have accomplished our primary goal of developing an ensemble model that can accurately categorize the TBX11K dataset into healthy, sick, and TB by the successful execution of the stated objective. On the TBX11K dataset, our original SOP ensemble model outperformed all others for the three-class classification task.

For model training, our investigation was restricted to BF, CLAHE, and HEF-enhanced images. In future, the same approach can be tested with other common enhancement techniques like UM and adaptive HE. Additionally, using datasets such as CXRs 14 and 18, our technique can be extended to develop models for the multi-class lung disease categorization.

# References

**1** WHO, (2021) Tuberculosis – WHO | World Health Organization. [online] World Health Organization. https://www.who.int/news-room/fact-sheets/detail/tuberculosis (accessed 23 April 2022).

**2** de Figueredo, L.J.A., de Miranda, S.S., dos Santos, L.B. et al. (2020). Cost analysis of smear microscopy and the Xpert assay for tuberculosis diagnosis: average turnaround time. *Revista da Sociedade Brasileira de Medicina Tropical* 53, 1–6.

**3** Russakovsky, O., Deng, J., Su, H., et al., (2014) Image Net Large Scale Visual Recognition Challenge. https://arxiv.org/abs/1409.0575 (accessed 23 April 2022).

**4** Pak, M. and Kim, S., (2017) A review of deep learning in image recognition. In *2017 4th International Conference on Computer Applications and Information Processing Technology (CAIPT)*. pp. 1–3.

**5** Cireşan, D.C., Giusti, A., Gambardella, L.M., and Schmidhuber, J. (2013). Mitosis detection in breast Cancer histology images with deep neural networks. In: *Medical Image Computing and Computer-Assisted Intervention – MICCAI 2013* (ed. K. Mori, I. Sakuma, Y. Sato, et al.), 411–418. Berlin, Heidelberg: Springer Berlin Heidelberg.

**6** Rao, P., Pereira, N.A. and Srinivasan, R., (2016) Convolutional neural networks for lung cancer screening in computed tomography (CT) scans. In: *2016 2nd International Conference on Contemporary Computing and Informatics (IC3I)*. pp. 489–493.

**7** Hooda, R., Sofat, S., Kaur, S., Mittal, A. and Meriaudeau, F., (2017) Deep-learning: A potential method for tuberculosis detection using chest radiography. In: *2017 IEEE International Conference on Signal and Image Processing Applications (ICSIPA)*. pp. 497–502.

**8** Khan, S. and Yong, S.-P., (2017) A deep learning architecture for classifying medical images of anatomy object. In: *2017 Asia-Pacific Signal and Information Processing Association Annual Summit and Conference (APSIPA ASC)*. pp. 1661–1668.

**9** Ho, T.K.K., Gwak, J., Prakash, O., Song, J.-I. and Park, C.M., (2019) Utilizing pretrained deep learning models for automated pulmonary tuberculosis detection using chest radiography. In: Asian conference on intelligent information and database systems. pp. 395–403.

**10** Jaeger, S., Candemir, S., Antani, S. et al. (2014). Two public chest X-ray datasets for computer-aided screening of pulmonary diseases. *Quantitative Imaging in Medicine and Surgery* 46: 475–477. https://pubmed.ncbi.nlm.nih.gov/25525580.

**11** Liu, Y., Wu, Y.-H., Ban, Y., Wang, H. and Cheng, M.-M., (2020) Rethinking computer-aided tuberculosis diagnosis. In: *Proceedings of the IEEE/CVF conference on computer vision and pattern recognition*. pp. 2646–2655.

**12** Deshmukh, A., Patil, D.S., Soni, G., and Tyagi, A.K. (2023). Cyber security: new realities for Industry 4.0 and Society 5.0. In: *Handbook of Research on Quantum Computing for Smart Environments* (ed. A. Tyagi), 299–325. IGI Global https://doi.org/10.4018/978-1-6684-6697-1.ch017.

**13** Kute, S.S., Tyagi, A.K., and Aswathy, S.U. (2022). Industry 4.0 challenges in e-healthcare applications and emerging technologies. In: *Intelligent Interactive Multimedia Systems for e-Healthcare Applications* (ed. A.K. Tyagi, A. Abraham, and A. Kaklauskas). Singapore: Springer https://doi.org/10.1007/978-981-16-6542-4_1.

**14** Cao, K., Zhang, J., Huang, M. and Deng, T., (2021) X-Ray classification of tuberculosis based on convolutional networks. In: *2021 IEEE International Conference on Artificial Intelligence and Industrial Design (AIID)*. pp. 125–129.

**15** Rajpurkar, P., Irvin, J., Zhu, K., Yang, B., Mehta, H., Duan, T., Ding, D., Bagul, A., Langlotz, C., Shpanskaya, K., Lungren, M.P. and Ng, A.Y., (2017) CheXNet: Radiologist-level pneumonia detection on chest x-rays with deep learning. https://arxiv.org/abs/1711.05225 (accessed 23 April 2022).

**16** Oloko-Oba, M. and Viriri, S. (2021). Ensemble of Efficient Nets for the diagnosis of tuberculosis. *Computational Intelligence and Neuroscience* 2021: 9790894. https://doi.org/10.1155/2021/9790894.

**17** Lakhani, P. and Sundaram, B. (2017). Deep learning at chest radiography: automated classification of pulmonary tuberculosis by using convolutional neural networks. *Radiology* 2842: 574–582.

**18** Koo, K.-M. and Cha, E.-Y. (2017). Image recognition performance enhancements using image normalization. *Human-centric Computing and Information Sciences* 71: 33. https://doi.org/10.1186/s13673-017-0114-5.

**19** Heidari, M., Mirniaharikandehei, S., Khuzani, A.Z. et al. (2020). Improving the performance of CNN to predict the likelihood of COVID-19 using chest X-ray images with preprocessing algorithms. *International Journal of Medical Informatics* 144: 104284.

**20** Munadi, K., Muchtar, K., Maulina, N., and Pradhan, B. (2020). Image enhancement for tuberculosis detection using deep learning. *IEEE Access* 8: 217897–217907.

**21** Dasanayaka, C. and Dissanayake, M.B. (2021). Deep learning methods for screening pulmonary tuberculosis using chest X-rays. *Computer Methods in Biomechanics and Biomedical Engineering: Imaging and Visualization* 91: 39–49. https://doi.org/10.1080/21681163.2020.1808532.

**22** Rahman, T., Khandakar, A., Qiblawey, Y. et al. (2021). Exploring the effect of image enhancement techniques on COVID-19 detection using chest X-ray images. *Computers in Biology and Medicine* 132: 104319.

**23** Ikhsan, I.A.M., Hussain, A., Zulkifley, M.A., Tahir, N.M. and Mustapha, A., (2014) An analysis of X-ray image enhancement methods for vertebral bone segmentation. In: *2014 IEEE 10th International Colloquium on Signal Processing and its Applications*. pp. 208–211.

**24** Rahman, T., Khandakar, A., Kadir, M.A. et al. (2020). Reliable tuberculosis detection using chest X-ray with deep learning, segmentation and visualization. *IEEE Access* 8: 191586–191601.

**25** Kundu, R., Basak, H., Singh, P.K. et al. (2021). Fuzzy rank-based fusion of CNN models using Gompertz function for screening COVID-19 CT-scans. *Scientific Reports* 111: 14133. https://doi.org/10.1038/s41598-021-93658-y.

**26** Cao, Y., Geddes, T.A., Yang, J.Y.H., and Yang, P. (2020). Ensemble deep learning in bioinformatics. *Nature Machine Intelligence* 29: 500–508. https://doi.org/10.1038/s42256-020-0217-y.

**27** Rashid, R., Khawaja, S.G., Akram, M.U. and Khan, A.M., (2018) Hybrid rid network for efficient diagnosis of tuberculosis from chest x-rays. In: *2018 9th Cairo International Biomedical Engineering Conference (CIBEC)*. pp. 167–170.

**28** Patel, M., Das, A., Pant, V.K. and Jayasurya M, (2021) Detection of tuberculosis in radiographs using deep learning-based ensemble methods. In: *2021 Smart Technologies, Communication and Robotics (STCR)*. pp. 1–7.

**29** Saad, W., Shalaby, W.A., Shokair, M. et al. (2022). COVID-19 classification using deep feature concatenation technique. *Journal of Ambient Intelligence and Humanized Computing* 134: 2025–2043. https://doi.org/10.1007/s12652-021-02967-7.

**30** Habib, N., Hasan, M.M., Reza, M.M., and Rahman, M.M. (2020). *Ensemble of CheXNet and VGG-19 Feature Extractor with Random Forest Classifier for Pediatric Pneumonia Detection*, vol. 16, 359. SN Computer Science https://doi.org/10.1007/s42979-020-00373-y.

**31** Chauhan, A., Chauhan, D., and Rout, C. (2014). Role of gist and PHOG features in computer-aided diagnosis of tuberculosis without segmentation. *PLoS One* 911: e112980–e112980. https://pubmed.ncbi.nlm.nih.gov/25390291.

**32** Lei, T., Wang, R., Wan, Y., Du, X., Meng, H. and Nandi, A.K., (2020) Medical Image Segmentation Using Deep Learning: A Survey, arXiv preprint arXiv:2009.13120, pp. 59.

**33** Ait Skourt, B., el Hassani, A., and Majda, A. (2018). Lung CT image segmentation using deep neural networks. *Procedia Computer Science* 127: 109–113.

**34** Islam, J. and Zhang, Y., (2018) Towards robust lung segmentation in chest radiographs with deep learning. https://arxiv.org/abs/1811.12638 (accessed 23 April 2022).

**35** Ronneberger, O., Fischer, P. and Brox, T., (2015) u-net: convolutional networks for biomedical image segmentation. https://arxiv.org/abs/1505.04597 (accessed 23 April 2022).

**36** Yadav, G., Maheshwari, S. and Agarwal, A., (2014) Contrast limited adaptive histogram equalization based enhancement for real time video system. In: *2014 International Conference on Advances in Computing, Communications and Informatics (ICACCI)*. pp. 2392–2397.

**37** Zhang, M., (2009) Bilateral filter in image processing. Thesis. https://digitalcommons.lsu.edu/gradschool_theses/1912 (accessed 23 April 2022).

**38** Tan, M. and Le, Q.v, (2019) Efficient net: rethinking model scaling for convolutional neural networks. [online] https://arxiv.org/abs/1905.11946 (accessed 23 April 2022).

**39** Tao Hwa, S.K., Bade, A., Hijazi, M.H.A. and Saffree Jeffree, M. (2020). Tuberculosis detection using deep learning and contrast-enhanced canny edge detected X-Ray images. *IAES International Journal of Artificial Intelligence (IJ-AI)* 9 (4): 713.

# 5

## Smart Technologies in Manufacturing Industries: A Useful Perspective

*V.M. Gobinath[1], A. Kathirvel[2], S.K. Rajesh Kanna[1], and K. Annamalai[3]*

[1] *Rajalakshmi Institute of Technology, Chennai, Tamil Nadu, India*
[2] *Panimalar Engineering College, Chennai, Tamil Nadu, India*
[3] *VIT University, Chennai, Tamil Nadu, India*

## 5.1  Introduction

The smart technology (ST) has a basic origin in information communication technology (ICT) and intelligent manufacturing. It involves many terminologies, including "time synchronization," "artificial intelligence" (AI), and "network communication" associated with accuracy in rapid and blistering work. Many manufacturing sectors have developed their own STs to transform their raw work into high-quality output with rapid and efficient production. The intention of STs is meteoric development and the adoption of vanguard technologies such as energy-saving efficiency, cloud manufacturing, cyber-physical production systems (CPPS), smart factories, intelligent manufacturing, and advanced manufacturing (refer Table 5.1). Smart manufacturing has attracted attention from industries, government organizations, and academia. Various consortia and discussion groups have been formed to develop architectures, roadmaps, standards, and research agendas. The overall concept of smart manufacturing systems in Figure 5.1 has to be translated into architectures that are quite specific. Efforts are underway to develop such architectures [1]. As the world is moving beyond the fourth generation, we are going toward an integrated and collaborative system of ICT; moreover, every field of science has its own tools and mechanisms to use it. In order to leverage automation control data, Industry 4.0 manufacturing systems require industrial devices to be connected to the network. Potentially, this could increase the risk of cyberattacks, which might compromise connected industrial devices to accumulate production data or gain control over the assembly process.

Globally, the extensive use of automated data and network operating systems is significant, though, in STs, the issue of cybersecurity arises. Many web search engines are utilized to identify potential cyberattacks, such as the Sentient Hyper-Optimized Data Access Network. Manufacturing operations can be disrupted by a cyberattack; consequently, organizations experience financial losses. The primary concern is safeguarding systems from cyberattacks, as these pose a serious risk to operational safety. Smart manufacturing is a broad idea that involves the direct implementation of tools and technologies during the production process. These encompass various technologies and solutions, which, if evaluated in a manufacturing industry, are termed as "smart manufacturing." We also call these technologies "problem solvers" as they assist in boosting the whole manufacturing process and, in turn, enhance profitability.

STs collect manufacturing site data, analyze it, and then provide summarized, better decision-oriented, and optimized results. The benefit of IoT is to enhance production. The core concept behind STs in manufacturing aims to establish a smart factory to derive accurate and efficient results. To acquire productive, energetic, and efficient manufacturing, some steps should be followed. Consider the following steps as shown in the given diagram (see Figure 5.1).

It has been nearly 260 years since the start of the initial age, which is thought to have started around 1760 in the United States. The most recent iteration of this process, the fourth technological revolution, has been called "smart manufacturing," while in Europe it is referred to as "Industry 4.0." IoT, 5G, AI, blockchain, edge computing, predictive analysis, and digital twins (DT) are grouped together as STs that yield long-term savings, safety, security, and increased productivity, as shown in Table 5.2.

*Topics in Artificial Intelligence Applied to Industry 4.0*, First Edition. Edited by Mahmoud Ragab AL-Refaey, Amit Kumar Tyagi, Abdullah Saad AL-Malaise AL-Ghamdi, and Swetta Kukreja.
© 2024 John Wiley & Sons Ltd. Published 2024 by John Wiley & Sons Ltd.

**Table 5.1** Abbreviation of technologies used in manufacturing sectors.

| Terms used in ST technology | Abbreviation |
| --- | --- |
| CPPS | Cyber-physical production system |
| ML | Machine learning |
| CAPEX | Capital expenditures |
| OPEX | Operating expense |
| CNC | Computer numeral control |
| CAD | Computer-aided design |
| CAM | Computer-aided manufacturing |
| IoT | Internet of Things |
| IIoT | Industrial Internet of Things |
| RFID | Radio frequency identification |
| MEMs | Micro-electrical-mechanical sensor |
| SNA | Social network analysis |
| CCT | Computer and communication technology |
| AI | Artificial intelligence |
| MBE | Model-based enterprise |
| MIOT | Manufacturing internet of things |

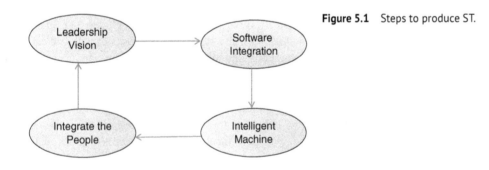

**Figure 5.1** Steps to produce ST.

**Table 5.2** Survey theme.

| S. no. | Theme of survey |
| --- | --- |
| 1 | Article from Kusiak [2] shows the industrial internet of things, recent advances, enabling technologies and open challenges |
| 2 | Massive internet of things for industrial applications: addressing wireless IoT connectivity challenges and ecosystem fragmentation |
| 3 | Articles shows the brief description of smart manufacturing |
| 4 | This article presents a systematic review determinants of information and digital technology implementation for smart manufacturing |
| 5 | This articles show technology using graphs to link data across the product life cycle for enabling smart manufacturing digital threads |
| 6 | This articles shows complete literature review of a smart manufacturing adoption framework for SMEs |
| 7 | This article shows a digital-driven smart technology prioritization challenges toward development of smart manufacturing using BWM method |
| 8 | This paper shows the fundamental of smart manufacturing, a multi-thread perspective |
| 9 | This article show the literature review to conceptual framework of enablers for smart manufacturing tools |
| 10 | This article shows smart manufacturing based on cyber-physical systems and beyond |

The structure of this chapter includes the following:

- Introduction
- A new definition of ST
- State-of-the-art research literature review
- Structure of any smart manufacturing technology
- Various technologies related to the manufacturing sector
- Challenges in ST
- Conclusion

"ST" is the term used to reduce human workload, ensure error-free and highly productive industries, and enable productive and predictive maintenance using AI and the Internet of Things (IoT). Industry 4.0 is the recent technology used to efficiently collect and maintain data. Technical and skillful staff are required to operate these computer-sensing machines and communication networks.

## 5.2 Literature Review

Khan et al. [3] show that IIoT is the latest technology in the manufacturing sector. This includes the latest framework communication protocols and highlights the different heterogeneous technologies involved in meeting the challenges of ST (refer Table 5.2). Andrew [1] also elaborated on the solution for making smart manufacturing more effective and more productive. He discussed the pillar of smart manufacturing. Kute et al. and Yuqian et al. [4, 5] also discussed the issues related to digital twin–deriving techniques in the context of Industry 4.0. They explored future challenges in smart manufacturing and defined industrial communication twining tools. The literature review of smart manufacturing [2] indicates that smart manufacturing is not a single domain but a multi-domain concept that shows different perspectives of smart manifesting using hardware and communication tools. It highlights the difference between resilient manufacturing and sustainable manufacturing [6–8]. Researchers also elaborate on the graph cycles and the linking of different diagrams to access various technologies for designing and quality domains for increasing the life cycle of the product. The term "model-based enterprise" (MBE) is also defined. West (2017) elaborates on the research about the cost analysis of smart manufacturing technology, addressing questions such as whether digital twin is affordable [9]. Attention is given to trending technologies in smart manufacturing, such as CPPS, and the definition of SCPS (space communication protocol standards), which includes eight tuples of CPS-based smart manufacturing for society [10]. Additionally, terms such as "cloud computing," "fog computing," and "edge computing" are also defined.

Over the past decade, we have been boosting our technologies and techniques with the assistance of communication networks and AI to reduce costs in capital expenditures (CAPEX) and operating expenses (OPEX). Advances in communication and computer intelligence in the industry contribute to the accuracy and perfection of automation in the manufacturing sector. In 1990, Kusaik published a journal on automatic manufacturing, and in 1995, the work on intelligent manufacturing systems started to support the industrial industry. Well-reputed companies from different countries, such as Japan, Korea, and the United States, started intelligent manufacturing systems to enhance their industrial sectors. Some of the STs are discussed in the next section.

## 5.3 Materials and Methods

### 5.3.1 Manufacturing-Led Design

In the world of product development, the concept of manufacturing-led design (MLD) emerges as a pivotal bridge between imagination and realization. MLD is a strategic approach that prioritizes the integration of manufacturing considerations during the design phase of a product's life cycle. This chapter delves into the essence of MLD, highlighting its significance, benefits, challenges, and the paradigm shift it brings to the design process as shown in Figure 5.2.

At its core, MLD embodies a proactive collaboration between design and manufacturing teams. Unlike traditional design practices, where aesthetics and functionality are often pursued without a deep understanding of production feasibility, MLD encourages a holistic approach. Here, design decisions are informed by the practical considerations of manufacturing processes, materials, and cost implications. This alignment between creative vision and real-world constraints ensures that the end product is not only innovative and appealing but also manufacturable without compromise.

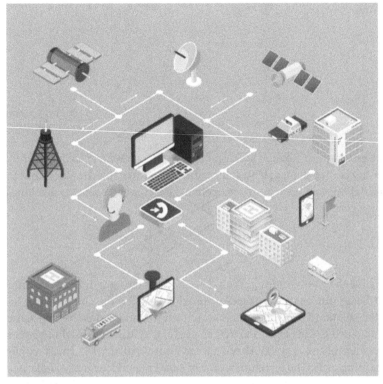

**Figure 5.2** Smart technology.

The advantages of MLD are manifold. First and foremost, it streamlines the development process, minimizing costly iterations and revisions. By addressing potential manufacturing bottlenecks early on, MLD averts production delays and unexpected expenses. Additionally, MLD facilitates better resource utilization as designers work in tandem with manufacturing experts, optimizing material usage and minimizing waste. Moreover, the collaboration inherent in MLD nurtures cross-functional teams, enhancing communication and knowledge sharing across departments.

However, the journey toward MLD is not without its challenges. One of the primary hurdles is fostering a culture of interdisciplinary collaboration. Designers and engineers accustomed to working within their respective silos must transition to a mindset of mutual understanding and partnership. Bridging this gap requires not only technical acumen but also effective communication and a shared commitment to the end goal.

MLD also marks a paradigm shift in the traditional design process. Instead of treating manufacturing as a mere executor of design concepts, it positions manufacturing as an active contributor to the creative process. This shift demands a deeper integration of design and manufacturing knowledge, enabling a harmonious fusion of aesthetic innovation and practical execution.

In conclusion, MLD signifies the evolution of product development from a linear and compartmentalized process to a collaborative and integrated approach. By uniting design creativity with manufacturing feasibility, MLD paves the way for more efficient, cost-effective, and sustainable product development. While it demands a departure from traditional practices and the cultivation of interdisciplinary collaboration, the potential benefits are vast. As industries continue to strive for innovation and efficiency, MLD emerges as a catalyst for achieving this delicate balance between inspiration and implementation.

### 5.3.2 3D Printing

The 3D printing technology, also known as "additive manufacturing," has emerged as a revolutionary innovation with the potential to revolutionize numerous industries. By layering materials based on digital designs, this cutting-edge technology permits the creation of three-dimensional objects. Over the past few decades, 3D printing has evolved from a niche concept to a mainstream technology, offering unprecedented design flexibility, cost-effectiveness, and personalization. This chapter examines the far-reaching impact of 3D printing technology across various industries and its potential to transform the future of manufacturing as shown in Figure 5.3.

Historically, manufacturing processes employed subtractive techniques, in which materials were carved or molded into the desired shapes. By facilitating additive manufacturing, however, 3D printing has ushered in a paradigm shift. It enables manufacturers to produce intricate geometries, complex structures, and customized components with less waste and greater efficiency. This technology has drastically sped up the prototyping phase, enabling rapid iteration and design optimization. Companies can now bring their products to market more quickly, lowering production costs and increasing their overall competitiveness.

The health-care industry has been revolutionized by 3D printing, particularly in the fields of medical devices and prosthetics. This technology has enhanced the quality of care and patient outcomes by enabling the production of patient-specific implants, prosthetic limbs, and surgical tools. Surgeons can now construct intricate 3D models of organs or bones in order to plan complex operations, resulting in more precise interventions. In addition, bio-printing techniques hold promise for tissue engineering, regenerative medicine, and drug development, which could lead to advances in personalized health care.

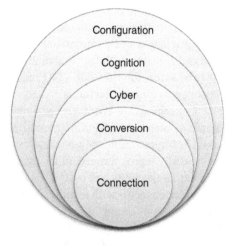

**Figure 5.3** Five architecture of cyber-physical production system.

The aerospace and automotive industries have also adopted 3D printing technology due to its unique benefits. Using lightweight materials and intricate designs, 3D printing permits the production of complex components with reduced weight and enhanced fuel efficiency. Aerospace companies are utilizing this technology to manufacture lighter aircraft components, thereby reducing maintenance costs and improving aircraft performance. Similarly, 3D printing in the automotive industry enables the creation of customized vehicle parts, expanding design options and streamlining production.

The availability of 3D printers has expanded educational and scientific horizons. Now, educational institutions can provide students with hands-on experiences, enabling them to materialize their concepts. From engineering and architecture to art and design, 3D printing encourages student creativity and innovation. In addition, scientists use this technology to fabricate prototypes, conduct experiments, and develop complex models for scientific analysis. The democratization of 3D printing has enabled individuals to investigate and materialize their ideas.

One of the most notable advantages of 3D printing is its sustainability potential. This technology minimizes waste compared to conventional manufacturing processes by utilizing only the necessary materials for production. In addition, 3D printing allows for the localization of production, which reduces transportation requirements and associated carbon emissions. In addition, the ability to repair or construct spare parts using 3D printing increases the longevity of products, thereby reducing the consumption of new resources. As the importance of sustainability rises, 3D printing offers a promising way to resolve environmental concerns.

The introduction of 3D printing technology has disrupted conventional manufacturing techniques and opened up a universe of opportunities across industries. Beyond manufacturing, its influence has reached health care, aerospace, education, and more. As the technology continues to develop, improvements in materials, scalability, and cost will expand its applications. To completely realize the potential of 3D printing, challenges such as intellectual property concerns and regulatory frameworks must be addressed. With continued innovation and widespread adoption, 3D printing will unquestionably revolutionize the future of manufacturing, allowing for greater customization and sustainability.

### 5.3.3 CNC Machining

Computer numerical control (CNC) machining is a revolutionary manufacturing process whose precision, efficiency, and adaptability have revolutionized numerous industries. By employing computerized controls, CNC machines are able to automate the production of intricate parts and components with exceptional precision and consistency. This chapter examines the primary characteristics, benefits, and applications of CNC machining, highlighting its significant impact on contemporary manufacturing.

CNC machining is distinguished by its ability to attain unparalleled precision and accuracy. Computerized controls eliminate human error and guarantee consistency. CNC machines are capable of performing intricate cutting, drilling, milling, and turning operations with micron-level precision, ensuring close tolerances and superior surface finishes. This precision renders CNC machining indispensable for industries that require precise specifications, such as the aerospace, medical, and automotive industries. Compared to traditional manual machining methods, CNC machining offers remarkable gains in efficiency and productivity. Once the CNC program is created, the machine can operate nonstop, 24 hours a day, seven days a week, minimizing interruption and maximizing output. In addition, CNC machines can swiftly change tools and execute multiple operations in a single configuration, eliminating the need for manual intervention and decreasing production time overall. CNC machining's automation and speed contribute to increased productivity and cost-effectiveness for manufacturers.

CNC machines are extremely versatile and adaptable to a variety of manufacturing needs. They are able to work with a variety of materials, including metals, plastics, composites, and wood. Complex geometries and intricate designs that would be impracticable or impossible to achieve manually can be handled by CNC machining. In addition, CNC machines can be easily reprogrammed, allowing manufacturers to easily produce a variety of components and prototypes, thereby augmenting flexibility and customization.

Complex elements with intricate features and geometries can be manufactured using CNC manufacturing. It is capable of precisely duplicating intricate designs on a large scale, ensuring consistency throughout a production run. This capability is especially advantageous for the aerospace and medical industries, where intricate components and custom-designed parts are essential. CNC machining enables the production of highly specialized and customized parts, satisfying the requirements of contemporary engineering and design.

With the integration of computer-aided design (CAD) and computer-aided manufacturing (CAM) software, CNC machines can convert digital designs into physical objects in a seamless fashion. This integration expedites the manufacturing process because CAD/CAM software generates precise tool trajectories and CNC machine instructions. In addition, CNC machines can be integrated into automated production lines, thereby increasing productivity and decreasing labor-intensive duties. Smart factories and Industry 4.0 are evolving due to the combination of CNC machining and automation technologies.

Modern manufacturing has been revolutionized by CNC machining, which enables the production of precise, efficient, and versatile parts and components. It has applications in industries spanning from aerospace and automotive to health care and electronics, thanks to its unparalleled precision, efficiency, and adaptability. As technology advances, CNC machining will continue to evolve, integrating features such as multi-axis machining, additive manufacturing, and increased automation. The future of CNC machining offers manufacturers the means to press boundaries, meet stringent demands, and fuel innovation in an ever-changing manufacturing environment.

### 5.3.4 Cloud Computing and Storage

Cloud computing storage devices are involved in the handling of data, which also entails the collision of cybersecurity intelligence and many other machine-related smart techniques to be effective and reduce costs by using many efficacious data-storage techniques. The main idea behind cloud computing is creating bonding between chain suppliers to facilitate product creation. Distributed devices are used for cloud computing, and three formats are predominantly used: PAAS, SAAS, and AAS.

Computing and storage in the cloud have emerged as potent technologies that are reshaping numerous industries, including manufacturing. In the manufacturing industry, where data management, collaboration, and scalability are essential, cloud-based solutions provide a number of advantages. This chapter examines the significance, benefits, and applications of cloud computing and storage in the manufacturing sector, emphasizing their potential to revolutionize efficiency, collaboration, and overall business operations.

Manufacturing processes generate enormous quantities of data from numerous sources, such as equipment sensors, supply chain management systems, and monitoring of production lines. Cloud computing provides a centralized platform for efficiently collecting, storing, and managing this data. Using cloud-based databases and data lakes, manufacturers can store and process enormous datasets, enabling real-time analytics, predictive maintenance, and quality control. This streamlined data administration improves decision-making, process optimization, and operational effectiveness as a whole.

The dynamic nature of the manufacturing industry necessitates cloud computing's unparalleled scalability and adaptability. Manufacturers can simply scale up or down their computing resources based on demand, eliminating the need for significant up-front hardware and infrastructure investments. Cloud-based solutions enable manufacturers to rapidly adapt to market shifts, accommodate production growth, and facilitate the incorporation of new technologies. This adaptability increases agility, reduces costs, and enables manufacturers to remain competitive in a landscape that is constantly changing.

Cloud computing facilitates collaboration and connectivity between manufacturing ecosystem stakeholders. Cloud-based platforms allow for real-time data collaboration, document sharing, and project management across departments, locations, and even partner organizations. This connectivity facilitates communication, accelerates decision-making, and improves collaboration between engineers, designers, suppliers, and consumers. Remote access to data and applications fosters a collaborative and adaptable work environment.

Because manufacturers deal with sensitive and proprietary data, data security and disaster recovery are crucial concerns. The security measures provided by cloud computing, such as data encryption, authentication protocols, and access controls, ensure the confidentiality and integrity of sensitive data. Cloud-based backups and disaster recovery solutions provide an additional layer of protection against data loss and ensure business continuity during disruptions and natural disasters. Manufacturers are relieved of the responsibility of sustaining on-premises security infrastructure by virtue of the fact that cloud providers frequently have dedicated teams and resources for data security.

Cloud computation and storage offer manufacturers significant cost savings. Manufacturers can reduce capital expenditures and transition to a more predictable, pay-as-you-go model by eradicating the need for on-premises infrastructure and maintenance costs. In addition, cloud-based solutions reduce the need for extensive IT support and maintenance, enabling manufacturers to focus their resources on core business activities. In addition, the scalability of cloud services guarantees that manufacturers only pay for the computing resources they require, maximizing cost-effectiveness.

In the manufacturing industry, cloud computing and storage have emerged as game-changing technologies. Their ability to expedite data administration, improve collaboration, provide scalability, ensure data security, and reduce costs renders them indispensable for manufacturers seeking to remain competitive in the digital age. As cloud technology continues to develop, with advances in peripheral computing, machine learning (MI), and IoT integration, its impact on the manufacturing sector is anticipated to increase. Adopting cloud-based solutions will enable manufacturers to improve efficiency, innovation, and collaboration, thereby fueling their success in a digitally driven, fast-paced manufacturing environment.

### 5.3.5 Internet of Things

This ST includes sensor technology and incorporates many internet technologies such as AI, network, and immutable communication. There is a central system to which the data is delivered and then provided to various researchers' central systems. Many researchers use different protocols, topologies, and hardware and software designs tailored to different industries. The main issue is to address the various problems within the industry, providing connectivity between manufacturing software tools and hardware. This aims to support the company's terms related to wireless computer devices, reducing labor costs. This ST starts with the collection of raw data through the implementation of smart systems, using CAD and manual system management. This smart technique also requires a platform of technical staff to run the STs to meet the challenging tasks [3].

The IoT has emerged as a game-changing technology with the potential to revolutionize multiple industries, including the manufacturing sector. Often referred to as Industrial IoT or IIoT, IoT in manufacturing is reshaping conventional production processes, optimizing operations, and generating unprecedented levels of efficiency and productivity. This chapter examines the significance, advantages, obstacles, and future prospects of IoT in the manufacturing industry.

In manufacturing, the IoT enables seamless connectivity between machines, devices, and systems [11]. In-machine sensors and actuators acquire real-time data, allowing for a vast network of interconnected devices. This data is then transmitted to sophisticated analytics platforms, which provide manufacturers with priceless insights into production processes, equipment performance, and supply chain operations. This data-driven methodology permits improved decision-making, predictive maintenance, and optimal resource allocation.

With IoT-enabled devices and data analytics, manufacturing facilities can become smart factories. These factories utilize real-time data to make intelligent decisions, automate processes, and facilitate proactive actions. Predictive maintenance is a crucial component of smart manufacturing because it enables manufacturers to anticipate equipment failures. This proactive approach minimizes disruption, reduces maintenance costs, and maximizes machinery and equipment's life span.

The IoT facilitates thorough quality control throughout the manufacturing process. Continuous monitoring of critical parameters by sensors ensures that products meet stringent quality standards. Any deviations or anomalies activate instantaneous notifications, allowing for prompt intervention to resolve problems. In addition, IoT-enabled traceability systems provide complete visibility into the production process, allowing manufacturers to monitor raw materials, components, and finished goods throughout the supply chain.

By monitoring and analyzing supply chain operations in real time, IoT in manufacturing optimizes these processes. By using IoT devices to monitor inventory levels, production progress, and transportation, manufacturers can improve supply chain efficiency and shorten lead times. Real-time data insights facilitate demand forecasting, inventory management, and improved supplier-manufacturer coordination. This optimization results in cost savings, fewer stock-outs, and enhanced customer satisfaction.

In manufacturing facilities, the IoT plays a vital function in enhancing workplace safety. Connected devices are able to monitor environmental conditions, identify potential dangers, and ensure compliance with safety regulations. Wearable devices containing IoT sensors provide employees with real-time feedback, thereby reducing the risk of accidents and enhancing overall safety standards. In addition, IoT in manufacturing facilitates the incorporation of ergonomic enhancements that boost worker comfort and productivity.

Despite its numerous advantages, implementing IoT in the manufacturing industry presents a number of obstacles. Interoperability, data security, and the high cost of IoT infrastructure continue to be significant issues for many manufacturers. To address these obstacles, collaboration among industry stakeholders and the creation of comprehensive IoT standards are required.

The future of the IoT in manufacturing bears tremendous promise. Edge computing, 5G technology, and AI advancements will further improve IoT capabilities. Integrating IoT with other emergent technologies, such as blockchain and augmented reality, will create new opportunities for enhanced supply chain transparency, product customization, and virtual worker training.

The IoT has opened a new era of opportunities for the manufacturing industry. By leveraging the connectivity and data-driven insights of the IoT, manufacturers can construct smart factories, optimize production processes, and enhance product quality. As IoT technology continues to evolve and become more accessible, its transformative effect on the manufacturing sector will only intensify. Adopting IoT in manufacturing is not a choice but rather a necessity for remaining competitive, resilient, and adaptable in a swiftly evolving industrial environment.

### 5.3.6 Cyber-Physical Production Systems

In the ever-evolving landscape of manufacturing, a remarkable transformation has been set in motion with the advent of CPPS. These systems represent a significant leap forward, intertwining the realms of digital technology and physical production processes. This fusion has ushered in a new era of manufacturing, characterized by unprecedented levels of efficiency, adaptability, and innovation as given in Figure 5.4.

At its core, CPPS embodies the seamless integration of the virtual and the physical. It orchestrates a symphony of computer simulations, real-time data analytics, and sensor-derived insights, all harmonizing with the tangible machinery and equipment of the production floor. This convergence yields multifaceted advantages that ripple through the manufacturing industry.

Connectivity and the IoT lie at the heart of CPPS. Devices, machines, and systems communicate through interconnected networks, creating a web of data streams that provide real-time insights. This data-driven foundation empowers manufacturers to make informed decisions and predictions, paving the way for proactive measures, such as predictive maintenance. Through constant monitoring, CPPS can forecast when equipment might falter, preempting costly downtime and elongating the life span of machinery.

The virtuous cycle of real-time monitoring and control marks a cornerstone of CPPS's impact. Sensors embedded in the manufacturing process generate a steady stream of data that unveils the intricate details of each production step. Manufacturers gain an unprecedented vantage point, enabling them to address deviations and challenges as they arise. Defects can be nipped in the bud, and quality is upheld at every stage, minimizing waste and elevating the overall product standard.

A hallmark feature of CPPS is its flexibility and customization. The amalgamation of digital and physical realms grants manufacturers the power to swiftly reconfigure their production processes. This newfound agility meets the demands of a dynamic market, facilitating on-demand production and responding adeptly to shifts in consumer preferences. The result is not only greater customer satisfaction but also an optimization of resources, minimizing excess inventory and waste.

CPPS is an embodiment of innovation and efficiency. The data deluge it produces forms the raw material for advanced data analytics and ML algorithms. This treasure trove of information can uncover hidden patterns, refine processes, and predict future trends. The manufacturing process becomes a canvas for continuous improvement, where data is the artist's brushstroke and innovation the masterpiece.

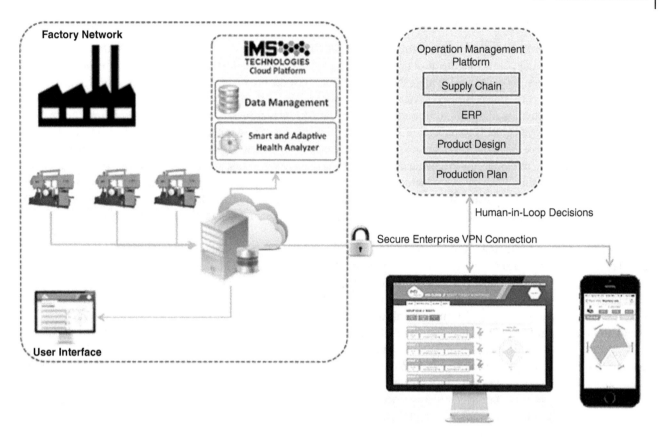

**Figure 5.4**   How CPPS system works. Adapted from Behrad Bagheri, 2015.

Beyond the technological realm, CPPS fosters collaboration and distributed manufacturing. Geographical boundaries blur as information flows seamlessly across different locations. This interconnectedness paves the way for collaborative production, optimizing resource allocation and nurturing a global approach to manufacturing.

In terms of cost savings, CPPS operates on multiple fronts. Predictive maintenance eliminates the specter of unplanned downtime and its associated expenses. Efficiency enhancements and waste reduction lead to fiscal gains, while quality improvements curb the expenses incurred from rework and recalls. The cost-effectiveness of CPPS extends beyond monetary realms, as it prioritizes worker safety by automating hazardous tasks and providing real-time insights to avert accidents.

In conclusion, CPPS has emerged as a transformative force in the manufacturing landscape. CPPS embodies the symbiosis of technology and industry, where data dances with machinery, and insights are married to action. As manufacturing processes embrace this digital embrace, industries stand poised to experience heightened efficiency, greater customization, and enhanced innovation. The CPPS revolution is not just about products rolling off assembly lines; it is about ideas coming to life, markets responding to the rhythm of demand, and industries marching confidently into a future of boundless possibilities.

CPPS is an ST proposed for AI. This system is a collection of collaborative technologies within the global context, and its ongoing processes are related to the further development of computer and communication technologies (CCT). This has led to the emergence of the fourth-generation industry, referred to as "4.0." The heap of raw data collected for industrial manufacturing from different sensors is very helpful for the detection of faults or the prognosis of equipment wear. There are five different levels of architectures of CPPS that produce an efficient productive system. The 5C levels of architecture are described in Figure 5.3.

The five 5C surface levels of CPS show a close bonding from the first level to the fifth level. All levels start from C, and CPPS starts from the connection of self-sensors and the collection of data that has to be obtained from this system, which extends to network implementation to produce cost-reducing products. The main architecture of CPPS is given in the subsequent section.

### 5.3.7 Sensors and Automatic Identification

Different types of sensors are used for the automatic identification of various factors such as cost, temperature, motion, and environmental conditions. However, the most popular sensor that is used in automation is called "radio frequency identification" (RFID). These sensors are used to meet the complete quality of the products in manufacturing sectors. Tags are used in the RFID technology to control wireless devices; every product is labeled with different tags. However, these tags are used to perform specific activities, including storing and retrieving data. RFID is always associated with cybersecurity and privacy issues. The tags employed in RFID, which is used in these sensors, have a certain range and can be easily hacked from other RFID sensor tags. There are many sensors that are used in the industrial and automation industry, such as temperature sensors, pressure sensors, micro-electro-mechanical sensors (MEMs), motion sensors, and torque sensors (Figure 5.5).

In the ever-evolving landscape of manufacturing, a remarkable transformation has been set in motion with the advent of CPPS. These systems represent a significant leap forward, intertwining the realms of digital technology and physical production processes. This fusion has ushered in a new era of manufacturing, characterized by unprecedented levels of efficiency, adaptability, and innovation.

At its core, CPPS embodies the seamless integration of the virtual and the physical. It orchestrates a symphony of computer simulations, real-time data analytics, and sensor-derived insights, all harmonizing with the tangible machinery and equipment on the production floor. This convergence yields multifaceted advantages that ripple through the manufacturing industry.

In the ever-evolving landscape of manufacturing industries, the integration of cutting-edge technologies has fundamentally reshaped production processes. Central to this transformation is the widespread adoption of sensors – intricate devices that detect and quantify physical, chemical, and biological phenomena, translating them into actionable data. In manufacturing, sensors serve as the bedrock of data-driven decision-making, enabling precise monitoring, adaptive control, and optimized resource allocation. This chapter explores the various types of sensors employed in manufacturing industries, their pivotal role in enhancing efficiency and quality, and the challenges and future prospects of sensor integration.

The diverse array of manufacturing processes necessitates a variety of sensors to monitor, analyze, and control different parameters. Some prominent sensor types and their applications are given here.

Temperature sensors are used for processes involving heat, such as metal casting or chemical reactions. They ensure optimal conditions and prevent overheating, which could compromise product quality. Pressure sensors are crucial in maintaining stability within pneumatic and hydraulic systems, ensuring consistent performance and avoiding potential leaks or failures. Proximity sensors are employed in automated assembly lines and robotics; these sensors detect the presence or absence of objects, facilitating efficient material handling and quality checks. Level sensors are used in monitoring material levels in storage tanks; these sensors prevent overflow or depletion, optimizing inventory management.

Despite the benefits, sensor integration poses challenges like initial costs, complexity in integration, and data management. Cybersecurity concerns must also be addressed to protect sensitive data from breaches. Looking forward, the synergy between sensors and the IIoT holds immense potential. The advent of Industry 4.0 promises to usher in an era of interconnected smart factories, where sensors collaborate seamlessly with AI-driven analytics, enabling autonomous decision-making and optimizing the entire production ecosystem.

Sensors have become the backbone of modern manufacturing, transforming traditional processes into adaptive, efficient, and quality-driven systems. Their ability to capture, analyze, and respond to data in real time empowers manufacturers to stay competitive in an ever-changing market. As technology continues to advance, the role of sensors in reshaping the

**Figure 5.5** Some types of sensors. *Sources:* Ochre Digi Media Pvt Ltd; Ochre Digi Media Pvt Ltd; Transcat, Inc; Agilent.

manufacturing landscape is poised to grow, heralding an era of intelligent factories that stand as a testament to human ingenuity and innovation.

### 5.3.8 Big Data Analytics

There is a colossal amount of data that also requires a highly productive computing system. The demand for a high amount of data that is collected but not analyzed poses challenges in how to handle and work with this data. Some specific data-collection techniques are used in smart manufacturing techniques to reduce costs, time constraints, and budget issues. There is also a need for the collection of correct raw data. Big data analytic systems play a key role in the manufacturing sector. Hong-Ning Dai (2018) explored the big data analytic industries and their challenges in his research. He also defined the term "Manufacturing Internet of Things" (MIOT). Shan Ren (2019) performed research on big analytic data technologies and gave comprehensive reviews of different big analytic data in smart manufacturing and offered an enhanced briefing on how to get productive results.

In the realm of smart manufacturing, the integration of big data analytics has set in motion a revolutionary transformation, reshaping the very core of how factories and production processes operate in the digital era. Within this landscape, where interconnected devices and sensors continuously generate vast streams of data, big data analytics emerges as the orchestrator of efficiency and innovation. This technology, adept at swiftly deciphering and making sense of these intricate data flows, becomes the catalyst for agile and informed decision-making. This newfound capability to not only predict but also adapt in real-time lends an unprecedented level of precision to operations, enhancing both overall efficiency and the quality of end products.

One of the remarkable implications of this integration is the realization of predictive maintenance. By meticulously analyzing the data churned out by the smart manufacturing environment, manufacturers can pinpoint potential issues and address them proactively. This capability minimizes unplanned downtimes, reduces maintenance costs, and keeps production processes uninterrupted, ensuring a smoother operational flow. In this way, big data analytics serves as a proactive guardian of operational continuity.

Yet the significance of big data analytics in smart manufacturing extends beyond immediate gains. It propels industries toward broader objectives such as sustainability, customization, and competitive advantage. The wealth of data collected enables the identification of energy consumption patterns, fostering efficient resource allocation and bolstering sustainability efforts. Moreover, the customization of products based on real-time data insights amplifies customer satisfaction and market responsiveness.

In this evolving landscape, manufacturing is not merely a static process but a dynamic, data-rich ecosystem that thrives on innovation and continuous improvement. As the amalgamation of big data analytics and smart manufacturing reaches new heights, industries stand on the precipice of a paradigm shift. With data-driven precision as its hallmark, this fusion empowers manufacturers to navigate complexity with agility, transforming challenges into opportunities and redefining the boundaries of what is achievable in the realm of production.

### 5.3.9 Blockchain Technology

In the latest STs, blockchain plays a key role in IIoT, leading toward the next generation in the manufacturing industrial sector. Blockchain technology is involved not only in the manufacturing sector but also in health care, finance, supply chain, and car insurance. The unique characteristics of blockchain technology, such as its decentralized nature, discoverability, durability, trust, security, and cost-effectiveness, make it the trending ST in IoT. The term "Ethereum" was defined by the researcher Arshdeep Bahga in 2016. On the blockchain platform, users can sign the Ethereum, which is decentralized and run not only by one person but also by peers. The Ethereum virtual machine is also established, mainly with nodes present in the network. The latest research shows how blockchain can be developed for DT to produce authentic, efficient, and secure manufacturing.

### 5.3.10 Artificial Intelligence

AI is a key element of ST. Due to AI, Industry 4.0 has come into existence. Without AI, there may be no concept of ST; the success of any ST depends on the level of AI involved in that technique. The terms "AI" and "ML" are related to each other (refer Table 5.3). The motivation behind AI and ML in creating ST is to increase the speed of the analysis and

**Table 5.3** Survey report of MHI in 2018.

| Name of ST | Rate of adoption in manufacturing sector (%) |
|---|---|
| Cloud computing | 57 |
| Inventory and network optimization | 44 |
| Sensors and automatic detection | 45 |
| Predictive analysis | 20 |
| Internet of Things | 22 |
| Robotics and automation | 34 |
| Blockchain | 6 |
| Driverless vehicle and drones | 11 |
| 3D printing | 16 |
| Artificial intelligence | 6 |

decision-making process. Social network analysis (SNA) is derived from social network theories. In 2018, material handling industry (MHI) and his team conducted a survey of different STs used worldwide. This survey took a long time, spanning five years.

## 5.4 Discussion

### 5.4.1 Present and Future Challenges

Due to the diverse and complex nature of ST, stemming from the diversity of communication network intelligence, there are many challenges present in this sector that should be resolved. Some of these challenges are given in the following section.

#### 5.4.1.1 Technical Staff
The technical staff is required to run this complex and compatible architecture. This includes processing of efficient ML, understanding communication networks, and ensuring smooth operation of these topologies over the network. A better understanding of fast data-driven techniques and operating systems is crucial. With a technical team, it becomes easier to handle data management schemes and machine integration with software strategies to gain efficient results in ST *within* the manufacturing sector – to meet the security and sustainability of computer-related results.

#### 5.4.1.2 Difficult to Handle Huge Data Analytic and Management Techniques
There are many data analytic techniques to address the collection, sensing, and processing of data for making future decisions. These data management techniques are essential for the operation of various IIoT systems to handle and retrieve large amounts of data effectively.

#### 5.4.1.3 System Integration
The compatibility of various ST platforms with ML is crucial, and the new manufacturing system requires IPV6 connectivity for the smooth operation of different interfaces. Over the previous decades, many technologies and platforms have been used to attain productive results in the manufacturing sector. However, integrating all these platforms poses a challenge. The evolution of technology has moved from mechanical manufacturing to manual production and further progressed to information technology (IT), and nowadays, CPPS is used.

#### 5.4.1.4 Big Data Analytic Tools
It is challenging to handle a large amount of data and then analyze it. Various data analytic tools are used, and the significant issue behind these tools is the sharing and maintenance of records across different networks.

### 5.4.1.5 Robustness and Security Issue

To address security issues, various cyber and AI tools are used, but a lack of security persists in different technologies across networks. The issue of trust in using ST arises due to the transmission of data between different resources through various communication networks.

### 5.4.1.6 Use of Wireless Technologies and Different Protocols

It is challenging to use various technologies through wireless means, as all the communication between machines and humans occurs through the use of networks by using different topologies, whether centralized or distributed. Numerous communication networks and wireless technologies are employed, making it challenging to decide which communication network is better for a productive manufacturing system. Various technical issues that are related to communication networks include latency, bandwidth, and many more factors.

### 5.4.1.7 Invention of Specific Operating System

There is a need for specific operating systems in the manufacturing sectors, and TinyOS and ConTiki are the most commonly used operating systems that meet the requirements of smart manufacturing techniques. There is a need to design an operating system with characteristics such as smooth traffic flow, a smart grid, an intelligent communication framework, efficient bandwidth consumption, and interoperability.

### 5.4.1.8 Supply Chain Is Complex

Smart manufacturing industries employ numerous heterogeneous systems that are interconnected. There is a long chain of stakeholders, systems, and suppliers. The challenge behind the smart manufacturing systems lies in handling large supply chains among different stakeholders and technologies that are involved in ST all over the world. It is essential to address and resolve conflicts in communication between various platforms that are used in smart industrial manufacturer.

### 5.4.1.9 Customer Trust Involvement

The products emerging from smart manufacturing should be system-integrated, reliable, and durable. Various techniques are used in this context; there are many wireless technologies, presenting many challenges in IoT and Industrial 4.0. Customers should be informed about these technologies through the development of effective software models that are oriented toward human-to-machine interaction.

## 5.5 Conclusion

This chapter gives a perspective review of STs that produce energetic and profitable products, involving wireless communication by using various operating systems with reliable and effective systems and communication protocols. Additionally, STs are also emerging toward robotic technology to reduce human involvement and workload. DT utilize sensor technologies and offer numerous benefits such as cost reduction in products and enhanced product quality.

## References

1 Deshmukh, A., Patil, D.S., Soni, G., and Tyagi, A.K. (2023). Cyber security: new realities for Industry 4.0 and society 5.0. In: *Handbook of Research on Quantum Computing for Smart Environments* (ed. A. Tyagi), 299–325 IGI Global. 10.4018/978-1-6684-6697-1.ch017.

2 Hedberg, T.D., Bajaj, M., and Camelio, J.A. (2019). Using graphs to link data across the product lifecycle for enabling smart manufacturing digital threads. *ASME Journal of Computing and Information Science in Engineering* 1: 213–224.

3 Kusiak, A. (2018). Smart manufacturing. *International Journal of Production Research* 56 (2): 508–517.

4 Kusiak, A. (2019). Fundamentals of smart manufacturing: a multi-thread perspective. *Annual Reviews in Control* 47 (2019): 214–220.

5 Khan, W.H., Rehman, M.H., Zangoti, H.M. et al. (2020a). Industrial internet of things: recent advances, enabling technologies and open challenges. *Computers and Electrical Engineering* 81: 1–13.

**6** Kute, S.S., Tyagi, A.K., and Aswathy, S.U. (2022). Industry 4.0 challenges in e-Healthcare applications and emerging technologies. In: *Intelligent Interactive Multimedia Systems for e-Healthcare Applications* (ed. A.K. Tyagi, A. Abraham, and A. Kaklauskas). Singapore: Springer http://dx.doi.org/10.1007/978-981-16-6542-4_1.

**7** Kathirvel, A., Sudha, D., Naveneethan, S., Subramaniam, M., Das, D. and Kirubakaran, S. (2022). AI Based Mobile Bill Payment System using Biometric Fingerprint. *American Journal of Engineering and Applied Sciences* 15 (1): 23–31. https://doi.org/10.3844/ajeassp.2022.23.31

**8** Nair, M.M., Tyagi, A.K., and Sreenath, N. (2021). The future with industry 4.0 at the core of society 5.0: open issues, future opportunities and challenges, *2021 International Conference on Computer Communication and Informatics (ICCCI).* 1–7. http://dx.doi.org/10.1109/ICCCI50826.2021.9402498.

**9** Qi, Q. and Tao, F. (2019). A smart manufacturing service system based on edge computing, fog computing, and cloud computing. *IEEE Access* 7: 86769–86777.

**10** Yao, X., Zhou, J., Lin, Y. et al. (2019). Smart manufacturing based on cyber-physical systems and beyond. *Journal of Intelligent Manufacturing* 30: 2805–2817.

**11** Yuqian, L., Liu, C., Kevin, I. et al. (2020). Digital twin-driven smart manufacturing: connotation, reference model, applications and research issues. *Robotics and Computer-Integrated Manufacturing* 61: 1–14.

# 6

# Blockchain Technology for Industry 4.0

*Rajiv Kumar Berwer[1], Sanjeev Indora[1], Dinesh Kumar Atal[2], and Vivek Yadav[3]*

[1] *Department of Computer Science and Engineering, Deenbandhu Chhotu Ram University of Science and Technology, Sonipat, Haryana, India*
[2] *Department of Biomedical Engineering, Deenbandhu Chhotu Ram University of Science and Technology, Sonipat, Haryana, India*
[3] *Expresslending Pty Ltd, Melbourne, Victoria, Australia*

## 6.1 Introduction

### 6.1.1 Definition and Overview of Industry 4.0

The Fourth Industrial Revolution, or Industry 4.0, is identified by the transformative paradigm shift in the manufacturing and industrial sectors due to the combination of information technology and automation in various aspects of production, logistics, and service delivery [1–3]. The term "Industry 4.0" originated in Germany as an element of a strategic initiative to establish the country as a global leader in advanced manufacturing technologies. However, it has gained widespread recognition and adoption worldwide as a concept that goes beyond national boundaries. Figure 6.1 illustrates how Industry 4.0 builds on the achievements of its predecessors, including Industry 1.0 with mechanization, Industrial 2.0 with industrial production, and Industrial 3.0 with digitalization [4]. At its origin, Industry 4.0 represents a holistic and interconnected approach to industrial processes where intelligent machines, systems, and products communicate, collaborate, and make autonomous decisions. The integration of physical and digital technologies creates a digital ecosystem that spans the entire value chain, from the creation and manufacture of products to distribution and customer service [5].

The main concepts of Industry 4.0 are illustrated in Figure 6.2 [6]:

1) IoT: The IoT enables the interconnectivity of physical objects and devices, which enables real-time data collection and sharing. It forms the foundation for creating smart and connected systems in Industry 4.0.
2) Machine Learning: AI technologies, including machine learning (ML) algorithms, enable machines and systems to learn from data, recognize patterns, and make intelligent decisions. ML facilitates automation, predictive analytics, and adaptive processes.
3) Big Data Analytics: In Industry 4.0, the capacity to gather, store, and evaluate an enormous amount of information produced through interconnected devices and systems is crucial. Big data analytics provides valuable insights for decision-making, process optimization, and predictive maintenance.
4) Additive Manufacturing (3D Printing): 3D printing allows for the production of complex and customized parts on demand. It offers greater design flexibility, reduced costs, and shorter lead times.
5) Cyber-Physical Systems: Cyber-physical systems (CPS) facilitate the easy integration of the online and offline worlds by fusing physical and digital capabilities. The foundation for continuous monitoring, management, and process enhancement in industrial settings is CPS.

Benefits of Industry 4.0 [7]

1) Increased productivity and efficiency through automation, real-time data analysis, and optimization of processes
2) Cost reduction and waste minimization through improved resource allocation and predictive maintenance
3) Enhanced product quality and customization through continuous monitoring and management of production processes

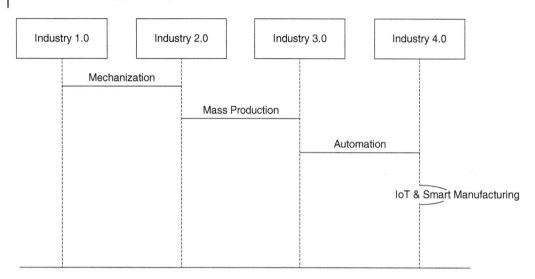

**Figure 6.1** Evolution of Industry 4.0.

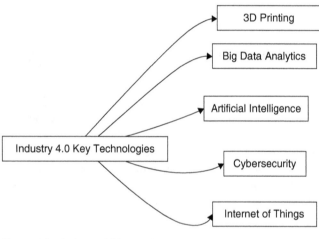

**Figure 6.2** Industry 4.0 key technologies.

4) Improved supply chain management with real-time visibility, traceability, and demand forecasting
5) Accelerated innovation and product development through digitalization and collaboration.

Challenges and Considerations

Industry 4.0 has a lot of potential advantages, but there are also obstacles and issues to consider.

1) Cybersecurity and data privacy concerns as digital systems become more interconnected.
2) The need for upskilling and reskilling of the workforce to adapt to changing job requirements.
3) Interoperability and standardization issues between different technologies and systems.
4) Ethical and social implications, including the impact on employment and societal well-being.
5) The investment required for infrastructure upgrades and technology adoption.

Industry 4.0 represents an innovative era in production and trade, driven by the combination of digital technologies, automation, and data-driven decision-making. It offers unprecedented opportunities for organizations to improve productivity, efficiency, and innovation. However, it also poses challenges that must be resolved to realize its true capacity. Embracing Industry 4.0 is crucial for organizations to remain competitive and thrive in the digital age [8].

### 6.1.2 Introduction to Blockchain Technology and Its Key Characteristics

In the past decade, the blockchain has drawn significant curiosity as a revolutionary method of handling data and performing transactions. Blockchain was first created as the foundational technology for digital currencies like Bitcoin, but it has since matured into a flexible platform with uses in a number of other sectors, including banking, supply chain management, health care, and more (Figure 6.3). Its unique characteristics offer several advantages that have sparked widespread interest and exploration [9]. A blockchain is fundamentally a decentralized, distributed file system that securely and openly stores and records transactions. Blockchain operates on a hope-to-hope network, which allows participants to jointly update and validate the ledger. It is in contrast to typical systems that are centralized, in

**Figure 6.3** Various applications of blockchain.

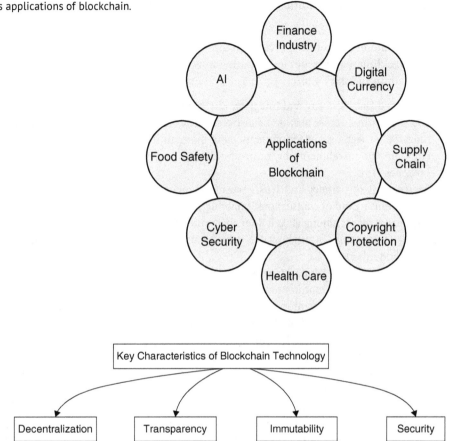

**Figure 6.4** Key characteristics of blockchain technology.

which a governing body manages the records [10]. The decentralized structure decreases reliance on trust, removes the need for middlemen, and increases efficiency and security.

Key characteristics of blockchain technology (Figure 6.4):

- Decentralization: Blockchain uses nodes, distributed systems that keep the full ledger. None of the organizations will have complete control over or ownership of the data thanks to its decentralized design. Decentralization enhances transparency, eliminates single points of failure, and makes the system resistant to tampering or manipulation.
- Transparency and immutability: All network members may see and access transactions that have been logged on the blockchain. A transaction can almost never be changed or removed after it is uploaded to the blockchain. The immutability of blockchain records, which offers a high level of data integrity, produces an auditable trail of transactions.
- Security and cryptography: Blockchain uses advanced cryptographic methods to guarantee the confidentiality and security of transactions. Each transaction is digitally signed, rendering it verifiable and impervious to tampering.
- Smart contracts: Self-executing contracts are those that have been encoded into the blockchain. When particular requirements are satisfied, these contracts automatically carry out the prescribed activities. Smart contracts reduce the need for middlemen, improve automation, and allow for trustless transactions. They have the potential to revolutionize various industries by automating complex processes and reducing costs.

The distinctive characteristics of blockchain technology offer multiple advantages, including increased security, accountability, and effectiveness across a variety of businesses. Blockchain technology has the potential to change established processes and open up new possibilities for innovation and cooperation in a variety of industries, including finance, logistics management, and distributed applications [11]. Explore the different uses of blockchain in many sectors and the effects it may have on their practices in the sections that follow.

### 6.1.3  Significance and Rationale for Integrating Blockchain in Industry 4.0

Industry 4.0's use of blockchain technology is important and has numerous strong justifications [12]. As the industrial and manufacturing sectors go through a digital transition, blockchain has the potential to be a fundamental tool for tackling major issues and creating new possibilities. Here are some of the significant reasons for integrating blockchain in Industry 4.0:

1) Enhanced Data Security and Integrity: Ensuring data security and integrity is crucial in Industry 4.0, as data is created and transmitted across networked devices and systems. The distributed and unchangeable characteristics of blockchain offer the perfect answer for protecting and maintaining the integrity of data. Blockchain promotes trust and transparency among stakeholders by recording data in a public ledger that is distributed and reducing the risk of data breaches [13].

2) Increased Transparency and Trust: Trust is a crucial aspect of any industrial ecosystem. Blockchain offers an open and transparent record of transactions, enabling all participants to have visibility into the data and processes. The transparency fosters trust among stakeholders by removing the requirement for a mediator and reducing disputes and fraud. It enables traceability throughout the value chain, enhancing accountability and ensuring compliance with regulations and standards [14].

3) Streamlined Supply Chain Management: Supply chains in Industry 4.0 are complex and involve multiple parties, making it challenging to monitor or verify the flow of products, components, and metadata. Blockchain helps prevent counterfeiting, improves inventory management, and enhances the efficiency of logistics and product recalls [15].

4) Secure and Efficient Digital Identity Management: With the increasing digitalization of Industry 4.0, managing digital identities becomes crucial. Decentralized and self-sovereign identity systems can be built using blockchain technology. Identity management and selective information sharing enable people and organizations to lower the risk of identity theft and illegal access. Blockchain-based identity management can streamline authentication processes, enhance privacy, and facilitate secure access to digital systems and services [16].

5) Collaborative Networks and Interoperability: Industry 4.0 emphasizes collaboration and interoperability among different systems and stakeholders. Blockchain provides a platform for creating trusted networks and collaborations. It enables secure and efficient sharing of data, assets, and value, eliminating the requirement for mediators and reducing handling overheads. Blockchain-based smart contracts automate and enforce agreements, facilitating seamless transactions and interactions among participants [17].

6) Tokenization and New Business Models: Blockchain enables the tokenization of assets, which can represent physical assets, intellectual property, or even fractional ownership rights. Tokenization opens up new possibilities for creating and exchanging value in Industry 4.0. Blockchain enables the development of decentralized marketplaces, hope-to-hope transactions, and innovative business models. Tokens can be used to incentivize collaboration, reward stakeholders, and streamline payment processes [18].

7) Auditing and Compliance: Compliance with regulations and standards is crucial in highly regulated industries. Due to its transparency and immutability, blockchain facilitates auditing procedures and ensures compliance by offering a trustworthy and secure log of activities. Blockchain can simplify regulatory reporting, lower the cost of compliance, and make real-time audits and monitoring easier [19].

Integrating blockchain technology in Industry 4.0 has the ability to transform manufacturing and industrial processes. It offers increased security, transparency, efficiency, and trust among stakeholders, while unlocking new business models and opportunities for collaboration. As the technology continues to mature and industry-specific use cases emerge, blockchain's significance in Industry 4.0 is expected to grow, driving innovation and transformation across various sectors [20].

## 6.2  Key Concepts of Blockchain

### 6.2.1  Distributed Ledger Technology and Decentralized Consensus Mechanisms

Distributed ledger technology (DLT) and consensus-based decentralized methods are key elements of blockchain technology. DLT refers to a system that enables the recording, sharing, and synchronization of data across multiple nodes or computers within a network. A centralized authority is not required to authenticate and maintain the ledger since it offers a visible and unchangeable record of all transactions. The decentralized nature of DLT is one of its core features.

DLT distributes the ledger across several participants, or nodes, in a network, as opposed to conventional centrally controlled systems, in which a single organization holds control over the ledger [21]. The decentralized architecture offers several benefits, including the following items:

1) Increased Security: Distributed ledgers are more secure than centralized systems. Because the data is spread among several nodes, it is challenging for hostile parties to attack the entire system.
2) Improved Transparency: DLT provides transparency by allowing all nodes in the network to provide similar data. Transactions logged on the ledger are openly accessible to all members, promoting trust and accountability. The transparency reduces the reliance on intermediaries and increases the efficiency of processes that require multiple parties to access and verify information.
3) Enhanced Efficiency: DLT enables node-to-node interactions, eliminating the requirement for mediators and reducing administrative overheads. Transactions can be executed directly between participants, streamlining processes and reducing costs. Moreover, the decentralized nature of DLT allows for faster transaction settlement and real-time updates across the network.
4) Resilience and Scalability: The distributed nature of DLT provides resilience for the network. Operations can continue despite the fact that certain nodes malfunction or are hacked since the data is still available from other nodes. Additionally, DLT can be designed to scale horizontally by adding more nodes to the network, accommodating a larger number of transactions and participants without sacrificing performance.

### 6.2.2 Smart Contracts and Their Role in Automating Processes

Within blockchain technology, smart agreements are essential for automating procedures. The provisions of the contract are explicitly written into each segment of the program, making them self-executing contracts (Figure 6.5). When established circumstances are satisfied, these contracts automatically carry out activities and enforce commitments. Smart contracts are designed to eliminate the requirement for mediators as they enable direct peer-to-peer transactions and automate various business processes [22].

Here are some key roles of smart contracts in automating processes:

1) Automation of Agreement Execution: Once the required criteria are satisfied, smart contracts carry out programmed activities automatically. This eliminates the need for manual intervention and ensures that agreements are enforced without reliance on intermediaries. For instance, once the parties involved have confirmed the delivery of the items, a smart contract in the supply network can automatically release payment to a supplier.
2) Streamlining Complex Workflows: Complex workflows involving multiple parties and dependencies can be automated using smart contracts. These contracts define the sequence of actions and trigger the next step once the previous one is completed. This streamlines the workflow and reduces administrative overhead, as each participant can trust that the smart contract will automatically progress the process.
3) Enhancing Efficiency and Speed: By automating processes, smart contracts eliminate manual tasks and reduce human errors. This leads to improved efficiency and faster transaction processing. Smart contracts, for instance, can automate the reimbursement process in the insurance sector, enabling quicker evaluation, verification, and settlement of claims.
4) Immutable and Transparent Record-Keeping: A blockchain, which offers a permanent and public log of all transactions and contract states, is where smart contracts are executed. This ensures that the history and current state of a contract are readily accessible and cannot be tampered with. The transparency and immutability of smart contracts enhance trust among participants and facilitate auditing and compliance processes.
5) Cost Reduction: By automating processes and eliminating intermediaries, smart contracts can significantly reduce costs associated with administrative tasks, intermediation fees, and potential disputes. For example, in real estate

**Figure 6.5** Smart contracts.

Predefined Contract     Events     Execution     Settlements

transactions, smart contracts may automate the handover of land rights and eliminate the necessity for mediators like brokers and lawyers.

6) Trust and Security: Smart contracts operate on a blockchain, which is based on decentralized consensus mechanisms. This ensures that the execution of smart contracts is transparent, secure, and resistant to tampering. The use of cryptographic techniques in smart contracts provides a high level of security and trust among participants.

Overall, smart contracts enable the automation of processes, improve efficiency, enhance transparency, reduce costs, and foster trust in various industries. As blockchain technology further develops, the capabilities and applications of smart contracts are expected to expand, leading to further automation and innovation in business processes [23].

### 6.2.3 Public Versus Private Blockchains: Trade-Offs and Considerations

There are two basic kinds of blockchain networks: public and private (Figure 6.6). Each has its own trade-offs and considerations to take into account. Understanding these differences is crucial when deciding which type of blockchain is appropriate for a specific use case [24]. Let's explore the trade-offs and considerations between public and private blockchains [25].

1) Access and Permission: Public blockchains are open to anyone, allowing anyone to participate, validate transactions, and maintain the distributed ledger. Contrarily, private blockchains limit usage to a select number of users who are given authorization to join the network [26]. Public blockchains offer more decentralized and permissionless access, while private blockchains provide more control and privacy.
2) Security and Trust: Public blockchains achieve security with consensus mechanisms like proof of work (PoW) or proof of stake (PoS) [27], which require users to contribute processing power or stake currency to validate transactions. This decentralized security model makes public blockchains highly resilient against attacks. Contrarily, private blockchains rely on a smaller group of trusted participants for validation, which can be more efficient but may raise concerns about centralization and trust.
3) Scalability: Public blockchains, due to their decentralized nature, often face scalability challenges. The consensus mechanisms and the need to replicate the entire blockchain across multiple nodes can limit the transaction throughput [28]. Private blockchains, being more centralized, can adopt different consensus algorithms and scalability solutions that suit the specific use case, allowing for higher transaction speeds and scalability.
4) Data Privacy: Public blockchains provide transparency, as all transactions are accessible to anyone on the chain. While this transparency enhances trust and accountability, it may not be suitable for scenarios where data privacy is a concern. Contrarily, private blockchains can provide greater security by restricting utilization to authorized participants and implementing encryption techniques to protect sensitive data.

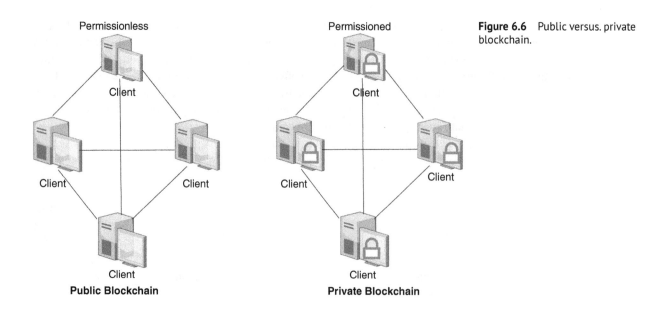

**Figure 6.6** Public versus. private blockchain.

**Public Blockchain**

**Private Blockchain**

5) Governance: Public blockchains typically operate under decentralized governance models, where decisions regarding protocol upgrades and network rules are made through community consensus. Private blockchains, being controlled by a select group of participants, can adopt more centralized governance models, enabling faster decision-making but potentially raising concerns about concentration of power [29].

6) Regulatory Compliance: Public blockchains operate in a global and open environment, which can pose challenges in terms of regulatory compliance, in particular when it concerns data privacy, data residency, and compliance with specific industry regulations. Private blockchains, with restricted access and a defined set of participants, may offer greater flexibility in complying with regulatory requirements [30].

7) Cost and Resources: Public blockchains often require significant computational resources and energy consumption for consensus mechanisms like PoW. Private blockchains, being more centralized, can be more resource-efficient and cost-effective, as they don't require extensive computational power for consensus [31]. However, private blockchains may incur additional costs for infrastructure setup and maintenance.

The selection between public and private blockchains is based on the particular requirements of the use case, including factors such as desired levels of transparency, privacy, scalability, governance, regulatory compliance, and resource considerations [28].

### 6.2.4 Interoperability and Standardization Challenges

Interoperability and standardization are critical challenges in the blockchain organization. While blockchain technology offers the ability to revolutionize various sectors, the lack of interoperability and standardized protocols can hinder its widespread adoption and limit its full potential [32]. Here are some key challenges related to interoperability and standardization in blockchain [33]:

1) Fragmented Blockchain Ecosystem: The blockchain ecosystem is highly fragmented, with numerous blockchain platforms, each with its own unique protocols, consensus mechanisms, and smart contract languages. This fragmentation creates interoperability challenges, as different blockchains struggle to communicate and share data seamlessly [34]. The lack of interoperability hinders collaboration and limits the ability to build comprehensive and interconnected blockchain solutions.

2) Siloed Data and Incompatible Formats: Each blockchain network maintains its own data structure and format, making it difficult to share and integrate data across different blockchains. Data interoperability is crucial for industries that require cross-platform collaboration and seamless data exchange. The absence of standardized data formats and protocols makes it challenging to achieve interoperability and efficient data sharing [35].

3) Divergent Consensus Mechanisms: Different blockchains employ varying consensus algorithms, PoW, PoS, and practical Byzantine fault tolerance (PBFT) [36]. The lack of standardized consensus mechanisms makes it difficult for different blockchains to interact and validate transactions across networks. Achieving consensus among diverse blockchains is a complex task that requires bridging the gaps in consensus protocols.

4) Smart Contract Interoperability: The automatic agreements found in blockchains are often specific to the platform and programming language. This limits their interoperability across different blockchain networks [37]. The inability to execute smart contracts across multiple blockchains restricts the seamless automation of processes and the exchange of value between different platforms.

5) Governance and Standardization Efforts: The distributed structure of blockchain networks makes establishing uniform governance standards difficult. Blockchain communities and industry consortia are working toward developing common standards and protocols, but achieving consensus among diverse stakeholders is a complex process. The lack of standardized governance models and protocols further impedes interoperability efforts [38].

6) Security and Privacy Concerns: Interoperability between blockchains raises security and privacy concerns. Sharing data and assets across different blockchain networks may introduce vulnerabilities and risks. Ensuring secure communication and data privacy while maintaining interoperability is a significant challenge that requires robust cryptographic techniques and privacy-preserving protocols [32].

Addressing these challenges requires collaborative efforts from industry participants, standardization bodies, and regulatory entities. Some initiatives are already underway to tackle interoperability, such as the development of interoperability protocols, cross-chain bridges, and blockchain interoperability frameworks. These efforts aim to establish common standards, protocols, and governance models to enable seamless communication, data exchange, and value transfer between different blockchain networks [39]. Overcoming these challenges will enable blockchain networks to work together effectively, unlock new use cases, and foster innovation in a decentralized and interconnected ecosystem [40].

## 6.3 Blockchain in Data Privacy and Security

### 6.3.1 Ensuring Data Privacy in a Decentralized Ecosystem

Ensuring data privacy in a decentralized ecosystem is a critical aspect of maintaining trust and security. While blockchain technology offers transparency and immutability, it also presents challenges when it comes to protecting sensitive data. Here are some key considerations for ensuring data privacy in a decentralized ecosystem:

1) Encryption: Implementing encryption techniques is essential to protect sensitive data in a decentralized ecosystem. Encryption assures that information is securely stored and transmitted, and only authorized parties with the decryption keys can access and interpret the data [41]. Data should be encrypted both in transmission and at rest to give an extra degree of security against unwanted access.
2) Data Minimization: Applying a data minimization strategy includes gathering and storing the important data required for specific purposes. By minimizing the amount of personal or sensitive information stored on the blockchain, the chance of exposure is decreased. Implementing privacy by design principles can help ensure that only critical information is recorded on the blockchain, while sensitive information could be kept off-chain or encrypted [42].
3) Private Transactions: Public blockchains inherently provide transparency, as every transaction is available to all users. To ensure data privacy, private transactions can be implemented using techniques like zero-knowledge proofs or ring signatures. These methods allow for the verification of transactions without revealing the specific details, ensuring confidentiality while still maintaining the integrity of the blockchain [24].
4) Private Blockchains: In a permissioned blockchain, connectivity to the network and involvement in the consensus process are restricted to authorized participants. This allows for greater control over data privacy and ensures that only trusted entities are involved in the validation and storage of sensitive data. Permissioned blockchains are suitable for scenarios where privacy and confidentiality are paramount [25].
5) Off-Chain Storage and Hashing: Storing sensitive data off-chain while maintaining a reference or hash on the blockchain is another method to enhance data privacy. The aforementioned method reduces the exposure of sensitive information on the blockchain while still maintaining the integrity and transparency of the transaction. Off-chain storage solutions, such as decentralized storage networks or encrypted cloud storage, can be utilized to securely store sensitive information [43].
6) Consent and Identity Management: Implementing robust consent management mechanisms allows individuals to control how their data is used and shared within a decentralized ecosystem. Blockchain-based identity management systems can provide individuals with control over their personal information, allowing them to selectively share data while maintaining privacy [44]. Consent-based data sharing guarantees that data is only used for the planned purpose and with the explicit consent of the data owner.
7) Regulatory Compliance: Adhering to relevant data protection regulations and privacy frameworks is crucial in a decentralized ecosystem. Compliance with regulations such as the General Data Protection Regulation (GDPR) requires enterprises to deploy proper administrative and technological precautions to secure private information. By aligning with regulatory requirements, organizations can ensure that data privacy is upheld within the decentralized ecosystem [45].

Remember that maintaining the privacy of information in a distributed system is challenging. The implementation of privacy-enhancing technologies, adherence to privacy best practices, and continuous evaluation of emerging standards and regulations are essential to address the evolving challenges associated with data privacy in a decentralized ecosystem.

### 6.3.2 Immutable and Tamper-Proof Records for Enhanced Data Integrity

Immutable and tamper-proof records are fundamental characteristics of blockchain technology that contribute to enhanced data integrity. These characteristics guarantee that the information stored on a distributed ledger is unable to be changed or tampered with, providing faith in the reliability of the information [46]. The consistency and tamper-proof qualities of blockchain records improve data integrity in the following ways:

1) Data Immutability: In a blockchain, once a transaction or data record is added to the chain, it becomes a permanent part of the ledger. The data is stored in blocks that are connected by means of encrypted hashes, creating an immutable and chronological chain of transactions. Immutability means that past records cannot be modified, deleted, or tampered with without consent from the blockchain system participants [47]. This feature maintains the consistency and accuracy of historical data as it remains unchanged over time.

2) Cryptographic Hash Functions: Blockchain uses cryptographic hash functions to create a unique digital fingerprint, or hash, for every block of information. The hash, which is a fixed-length alphanumeric string, is created from the contents of the block. Any change to the data, no matter how slight, would change the hash value. This property makes it almost impossible to change the data within a block without detection, as the hashes of subsequent blocks would also change, breaking the chain's integrity [48].

3) Consensus Mechanisms: Blockchain networks rely on consensus mechanisms to verify and approve the order and accuracy of transactions. Consensus methods [27] ensure that the majority of network participants agree on the verification of records before putting them on the blockchain. This distributed consensus process adds an extra layer of security and prohibits intruders from altering the record by requiring a majority agreement [21, 31].

4) Transparency and Auditing: Blockchain's transparent nature enables all participants to have visibility into the data and transactions recorded on the blockchain. This transparency facilitates auditing and verification of data integrity by allowing anyone to independently verify the transactions [49]. The ability to monitor the source and history of information on the blockchain enhances accountability and fosters trust among participants, as any discrepancies or tampering attempts can be easily identified.

By leveraging these features, blockchain technology provides a robust and tamper-proof framework for maintaining data integrity [50]. Immutable and tamper-proof records on the blockchain offer assurances of the accuracy, consistency, and security of data, making it a compelling solution for applications where data integrity is critical, like financial transactions, logistics management, and health-care records [51].

### 6.3.3 Secure Peer-to-Peer Communication and Encryption Protocols

Secure hope-to-hope communication and encryption protocols are essential for ensuring the confidentiality, integrity, and authenticity of data exchanged between participants in a decentralized network. These protocols provide a secure framework for communication, protecting sensitive information from unauthorized access and interception [52]. Here are some common protocols used for secure peer-to-peer communication:

1) Transport Layer Security/Secure Sockets Layer: Transport layer security (TLS)/secure sockets layer (SSL) protocols are widely used to secure communication over the Internet. They establish an encrypted connection between two endpoints, ensuring that data transmitted between them remains confidential and protected from eavesdropping or tampering. TLS/SSL protocols provide authentication, encryption, and integrity verification, making them a standard choice for secure communication in various applications, including peer-to-peer networks [53].

2) Internet Protocol Security: Internet protocol security (IPsec) is a protocol suite that provides security services at the Internet protocol (IP) layer. It can be used to establish secure communication between networked devices, including peer-to-peer connections. IPsec encrypts and authenticates IP packets, ensuring confidentiality and integrity of data. It can operate in transport mode (protecting the payload of IP packets) or tunnel mode (protecting entire IP packets) [54].

3) Off-the-Record Messaging: Off-the-record messaging (OTR) is a cryptographic protocol specifically designed for secure instant messaging. It provides end-to-end encryption and deniable authentication, ensuring that messages exchanged between participants are confidential and cannot be attributed to specific senders or recipients. OTR supports perfect forward secrecy, where even if an encryption key is compromised, past messages remain secure [55].

4) Pretty Good Privacy: Pretty good privacy (PGP) is a widely used encryption protocol for secure email communication. It uses public-key cryptography to encrypt and digitally sign emails, ensuring confidentiality and authenticity [56]. PGP enables users to exchange encrypted messages without relying on a centralized authority, making it suitable for secure peer-to-peer communication.

5) Signal Protocol: The signal protocol is an open-source protocol used for secure messaging and voice/video calls. It provides end-to-end encryption, secure key exchange, and forward secrecy [57]. Signal protocol is known for its strong security and privacy features and is implemented in various messaging applications, including the Signal app itself and WhatsApp.

6) Datagram Transport Layer Security: Datagram transport layer security (DTLS) is a variant of TLS designed for securing datagram protocols, such as user datagram protocol (UDP). It provides similar security features as TLS, including encryption, authentication, and integrity verification, but is tailored for unreliable and connectionless communication channels. DTLS is commonly used for securing real-time communications and IoT devices [58].

Implementing these secure peer-to-peer communication and encryption protocols ensures that data exchanged between participants in a decentralized network remains private and protected from unauthorized access. These protocols establish secure channels, verify the identities of participants, encrypt data, and provide mechanisms for detecting and preventing tampering. By leveraging these protocols, decentralized systems can maintain a high level of security and enable trusted interactions among participants [59].

### 6.3.4 Self-Sovereign Identity and User-Controlled Data Sharing

Self-sovereign identity (SSI) allows individuals to manage identity attributes and selectively share them with different parties as desired. This concept empowers individuals with ownership and agency over their personal information, reducing reliance on centralized authorities and enhancing privacy and security. SSI is closely related to the idea of user-controlled data sharing, where individuals can determine who has access to their data and under what conditions [60]. The traditional identity systems often rely on centralized authorities or intermediaries to validate and authenticate identities. This approach can result in privacy concerns, data breaches, and a lack of control for individuals over their personal information [61]. SSI addresses these challenges by leveraging decentralized technologies, such as blockchain, to create a secure and privacy-enhancing identity framework. In an SSI system, individuals have a digital wallet that holds their identity credentials, such as government-issued IDs, educational certificates, or financial records [62]. These credentials are cryptographically secured and can be selectively shared with others.

The concept of user-controlled data sharing complements SSI by allowing individuals to determine when and with whom their personal data is shared. Rather than having data stored in centralized databases, SSI enables data to be stored in individual wallets or decentralized storage systems. Individuals can then grant permission to specific entities or applications to access their data for specific purposes, all while maintaining control over their information. This approach ensures that individuals have full visibility and control over their data and can revoke access at any time. The integration of SSI and user-controlled data sharing brings several benefits [63]. It enhances privacy by reducing the reliance on centralized databases and limiting the exposure of personal information. Additionally, it enables more efficient and trusted digital interactions, as individuals can easily share verified credentials without relying on lengthy verification processes.

Interoperability and standardization are critical to ensure that different SSI systems can interact seamlessly. Additionally, addressing legal and regulatory frameworks around data protection, consent, and liability is necessary to establish a supportive environment for SSI adoption. Overall, SSI and user-controlled data sharing provide individuals with greater control, privacy, and security over their personal information. By shifting the control of identity and data to the individuals themselves, SSI promotes a more decentralized and user-centric approach to digital identity management [64].

## 6.4 Cybersecurity in the Era of Industry 4.0

Cybersecurity in the era of Industry 4.0 is of paramount importance as the proliferation of interconnected devices and networks introduces new vulnerabilities and risks [65]. As organizations embrace the benefits of digital transformation, it becomes crucial to implement robust security measures to protect critical infrastructure, sensitive data, and ensure uninterrupted operations. In this context, blockchain technology is an important asset to improve cybersecurity. Here are some key aspects of blockchain's contribution to cybersecurity in Industry 4.0:

### 6.4.1 Blockchain as a Security Layer for IoT Devices and Networks

Blockchain technology has numerous crucial qualities, including decentralization, immutability, transparency, and consensus procedures, which have the ability to operate as a security layer over IoT devices and networks. Here are some points outlining how blockchain can enhance security in IoT [66, 67]:

1) Secure Communication: Blockchain can facilitate secure peer-to-peer communication between IoT devices. By using cryptographic techniques and decentralized consensus mechanisms, blockchain ensures that data exchanged between devices is encrypted and remains confidential [68].
2) Firmware Updates: Blockchain can enhance the security of firmware updates for IoT devices. By leveraging smart contracts and consensus mechanisms, blockchain ensures that only authorized updates are applied, preventing malicious or unauthorized modifications [69].
3) Supply Chain Security: Blockchain can improve the security of IoT devices throughout the supply chain [70]. By recording the origin, ownership transfers, and maintenance history of devices on the blockchain, it becomes easier to detect counterfeit or tampered devices.
4) Data Privacy: Users can grant selective access to their data through decentralized identity management systems, ensuring that sensitive information is shared only with authorized entities [71].

Blockchain technology offers several security benefits for IoT; it also presents challenges such as scalability, energy consumption, and integration complexity. Therefore, the adoption of blockchain as a security layer for IoT devices and networks requires careful consideration of these factors. Although specific references were not mentioned in this response, you can refer to the previous answers in this conversation for a list of research papers and resources that discuss the topic in detail.

### 6.4.2 Prevention of Unauthorized Access and Tampering of Data

Blockchain technology serves as a powerful tool for preventing unauthorized access and tampering of data in IoT devices and networks. Through decentralized consensus mechanisms, blockchain requires network-wide agreement for any modifications, effectively preventing unauthorized tampering. The immutability of blockchain guarantees that as data is recorded, it cannot be altered without consensus, establishing a reliable and trustworthy source of truth. By employing cryptographic security measures, including encryption and digital signatures, blockchain safeguards data in transit and verifies the authenticity of transactions, thwarting unauthorized access attempts [72]. Additionally, access control and permission mechanisms, implemented through blockchain-based smart contracts, enable granular control over device interaction and data access, ensuring that only authorized entities can engage with the IoT devices and access sensitive information. The distributed ledger architecture of blockchain further strengthens security by eliminating central points of failure and providing transparent visibility into all data transactions. With these capabilities, blockchain effectively prevents unauthorized access and tampering of data, bolstering the overall security posture of IoT devices and networks [73–75].

Additionally, blockchain can be utilized for secure identity management, financial transactions, and intellectual property protection, among other cybersecurity applications [76, 77]. By leveraging blockchain technology, Industry 4.0 can strengthen its cybersecurity posture and address the evolving challenges of a digital and interconnected world. As organizations embrace blockchain-based solutions and develop robust cybersecurity strategies, they can enhance the trust, reliability, and resilience of their systems in the era of Industry 4.0 [78].

## 6.5 Supply Chain Management and Traceability

Supply chain management and traceability are critical aspects of Industry 4.0, and blockchain technology offers several advantages in addressing these challenges. Here are some key aspects of blockchain's role in supply chain management and traceability:

1) Enhanced Supply Chain Visibility and Transparency: Blockchain provides a decentralized and transparent ledger that enables real-time monitoring in logistics. By documenting and verifying records on a blockchain, all stakeholders can access a single source of truth, eliminating information asymmetry and enhancing visibility across the entire supply chain network. This visibility enables organizations to monitor the location of products, monitor inventory levels, and identify bottlenecks or inefficiencies in the supply chain [79].

2) Immutable Records for Product Provenance and Traceability: Blockchain's immutable nature ensures that once data is kept in the block, it won't change. The immutable and traceability feature is particularly valuable in supply chain management, where maintaining an auditable record of product provenance and traceability is crucial [80]. By storing product-related information, such as origin, manufacturing processes, and quality-control measures, on a blockchain, organizations can establish an unalterable record of a product's journey from its source to the end customer. This helps in verifying the authenticity of products and ensuring compliance with regulatory standards [81].

3) Preventing Counterfeiting and Ensuring Product Authenticity: Counterfeiting is a significant challenge in supply chains, leading to financial losses and reputational damage for organizations. Blockchain technology can mitigate this risk by providing a transparent and tamper-proof record of product information [82]. By storing unique product identifiers, such as serial numbers or QR codes, on a blockchain, it becomes easier to verify the authenticity of products at various steps of the supply chain. Consumers can scan these identifiers to access information about the product's origin, manufacturing processes, and distribution, ensuring they are purchasing genuine goods [83].

4) Smart Contracts for Automated Supply Chain Processes: Smart contracts automatically execute predefined actions when specific conditions are met. For example, a smart contract can automatically execute the release of payment to a supplier when certain predefined conditions, such as delivery confirmation, are met [84, 85]. This reduces the need for intermediaries, minimizes errors, and increases operational efficiency in the supply chain.

These capabilities of blockchain technology contribute to more efficient and secure supply chain management in Industry 4.0. By leveraging enhanced visibility, traceability, and automation through blockchain and smart contracts, organizations can optimize their supply chain processes and mitigate risks [86].

## 6.6 Blockchain-Enabled Smart Manufacturing

Blockchain-enabled smart manufacturing refers to the combination of blockchain technology into the manufacturing processes to enhance efficiency, quality control, supply chain optimization, and data sharing. Here's an explanation of the terms mentioned:

1) Improving Process Efficiency Through Real-Time Data Sharing: Blockchain facilitates real-time data sharing among different stakeholders involved in the manufacturing process [87]. By securely and transparently recording and sharing data on a distributed ledger, blockchain eliminates the need for manual data reconciliation and enables faster and more accurate decision-making. This real-time data sharing improves process efficiency by reducing delays, errors, and information asymmetry.

2) Enhancing Quality Control and Supply Chain Optimization: Blockchain provides an unchangeable and tamper-proof record of manufacturing data, including information about materials, components, and production processes. This ensures transparency and traceability throughout the supply chain, enabling enhanced quality-control measures. With blockchain, manufacturers can track and verify the origin, authenticity, and quality of inputs; identify potential bottlenecks or quality issues; and optimize the supply chain to ensure smooth and efficient operations [88].

3) Integration with IoT devices for data collection: In smart manufacturing, IoT devices play a crucial role in gathering real-time data from various stages of the production process. Blockchain can be integrated with IoT devices, enabling secure and decentralized data collection [89]. The data collected from IoT devices can be recorded on the blockchain, ensuring its integrity and accessibility to relevant stakeholders. This integration enhances data accuracy, reliability, and availability, enabling manufacturers to make data-driven decisions and optimize their operations.

4) Overall, blockchain-enabled smart manufacturing offers several advantages, including improved process efficiency, enhanced quality control, optimized supply chain operations, and seamless integration with IoT devices. By leveraging blockchain technology, manufacturers can transform their operations, streamline processes, and drive innovation in the digital age of manufacturing [90].

## 6.7 Overcoming Challenges in Blockchain Implementation

Implementing blockchain technology comes with its own set of challenges that organizations need to overcome to ensure successful deployment. Here are some key challenges and ways to overcome them:

1) Scalability and Performance Limitations of Blockchain Technology: Blockchain technology, particularly public blockchains, often faces scalability and performance limitations due to factors like block size, transaction throughput, and consensus mechanisms. To overcome these challenges, various solutions are being explored, such as layer-two scaling solutions like payment channels and sidechains, sharding techniques, and the development of more efficient consensus algorithms [91]. Additionally, the use of private or consortium blockchains with controlled participation can provide better scalability and performance for specific use cases [92].

2) Interoperability Between Different Blockchain Platforms: As multiple blockchain platforms and protocols emerge, interoperability between them becomes crucial for seamless data and value transfer [93]. Efforts are underway to develop standards and protocols that enable interoperability, such as cross-chain bridges, interoperability layers, and standardized smart contract languages [94]. These initiatives aim to create an interconnected blockchain ecosystem where different platforms can communicate and share data securely.

3) Regulatory and Legal Considerations for Blockchain Adoption: Blockchain technology often operates in a regulatory gray area, and its adoption requires careful consideration of legal and regulatory frameworks. Governments and regulatory bodies are increasingly recognizing the potential of blockchain and are working to establish clear guidelines and regulations. It is important for organizations to stay updated on evolving regulations, ensure compliance, and engage with relevant stakeholders to address any legal concerns or uncertainties [95].

4) Addressing Skepticism and Resistance to Change: Blockchain implementation may face skepticism and resistance to change from various stakeholders. It is crucial to address these concerns by educating stakeholders about the benefits and potential of blockchain technology. Demonstrating successful use cases, conducting pilot projects, and providing clear business justifications can help alleviate skepticism [96]. Collaborating with industry partners, academia, and regulatory bodies can also build trust and foster an environment conducive to blockchain adoption [97].

Overcoming these challenges requires a collaborative approach involving technology developers, industry players, regulators, and other stakeholders. Continuous research, innovation, and open dialog are necessary to address scalability, interoperability, regulatory, and resistance-related challenges. As the blockchain ecosystem evolves, solutions and best practices will continue to emerge, paving the way for widespread blockchain implementation across various industries.

## 6.8 Real-World Applications of Blockchain in Industry 4.0

Blockchain technology has found numerous real-world applications in the context of Industry 4.0, revolutionizing various sectors and enabling new possibilities. Here are some key examples of how blockchain is being implemented in Industry 4.0:

1) Blockchain in Automotive Manufacturing and Autonomous Vehicles: Blockchain can enhance the automotive manufacturing process by ensuring transparency and traceability of components, streamlining supply chain management, and improving vehicle maintenance and safety [98]. Smart contracts can automate and enforce agreements between manufacturers, suppliers, and customers [99].

2) Supply Chain Traceability in the Food and Pharmaceutical Industries: Blockchain enables source to destination traceability and transparency in supply chains, which is crucial for industries such as food and pharmaceuticals. This helps prevent counterfeiting, improve quality control, and provide consumers with verifiable information about the products they consume [100].

3) Blockchain for Intellectual Property Rights and Provenance Tracking: Blockchain can revolutionize intellectual property rights management by offering decentralized and tamper-proof logs and authenticating ownership and usage rights [101]. Smart contracts can automate licensing agreements and royalty payments, ensuring fair compensation for creators. Additionally, blockchain can enable provenance tracking for artworks, luxury goods, and other high-value items, allowing consumers to verify their authenticity and ownership history.

4) Case Studies Showcasing Successful Blockchain Implementations: Numerous case studies highlight successful implementations of blockchain in various industries. For example, IBM's Food Trust platform uses blockchain to improve transparency and traceability in the food supply chain, while the MediLedger Project leverages blockchain to track and verify pharmaceutical supply chains [102]. The Everledger platform utilizes blockchain for tracking and verifying the provenance of diamonds [103]. These case studies demonstrate the practical applications of blockchain technology; its impact on efficiency, transparency, and trust; and the benefits it brings to different industries.

These real-world applications showcase the capabilities of blockchain technology in Industry 4.0. By leveraging blockchain's key characteristics like transparency, immutability, and automation through smart contracts, businesses can address critical challenges, enhance efficiency, and unlock new opportunities for collaboration and innovation [104].

## 6.9 Future Trends

As blockchain-based technology continues to evolve, several future trends and considerations are shaping its development and adoption in Industry 4.0:

1) Integration of Blockchain with Emerging Technologies (AI, IoT, and so on): The integration of blockchain with emerging technologies can create synergies and enable new capabilities. For example, combining blockchain with AI can enhance data analysis and decision-making processes, while integrating blockchain with IoT devices can enable secure and trusted data sharing and automation [105–106].
2) Evolution of Blockchain Standards and Governance Frameworks: As blockchain adoption expands, there is a growing need for standardization and governance frameworks to ensure interoperability, compatibility, and security [106]. Efforts are underway to develop industry standards and best practices for blockchain implementation. These standards will help streamline integration, facilitate collaboration among stakeholders, and ensure the scalability and security of blockchain networks [107].
3) Collaborative Blockchain Ecosystems and Industry Consortia: Blockchain technology often thrives in collaborative ecosystems and industry consortia. These partnerships bring together multiple stakeholders to collectively develop and implement blockchain solutions. Collaborative efforts foster innovation, enable knowledge sharing, and address common challenges. Industry-specific consortia are emerging in different fields like finance, logistics, and the medical industry to drive blockchain adoption and create shared value [108].
4) Ethical and Societal Implications of Blockchain Technology: As blockchain becomes more prevalent in Industry 4.0, there are ethical and societal considerations that need to be addressed. These include privacy concerns, data governance, and the potential impact on employment and economic systems. It is mandatory to guarantee that blockchain solutions are designed and developed in a manner that upholds ethical principles, protects individuals' privacy rights, and contributes to the overall well-being of society [109].

Considering these future trends and considerations will help shape the responsible and sustainable adoption of blockchain technology in Industry 4.0. By integrating blockchain with emerging technologies, establishing standards and governance frameworks, fostering collaboration, and addressing ethical implications, businesses, and industries can fully harness the potential of blockchain while ensuring its positive impact on society.

## 6.10 Conclusion

In conclusion, the integration of blockchain technology with Industry 4.0 has the ability to revolutionize manufacturing and industrial processes. The chapter explored the transformative impact of blockchain across different fields of Industry 4.0, including data security, transparency, supply chain management, smart contracts, cybersecurity, and more. Blockchain's decentralized and tamper-proof nature enhances data security and integrity, fosters trust and transparency among stakeholders, streamlines supply chain management, and enables secure peer-to-peer communication. Smart contracts automate processes, while immutable records ensure data integrity. Blockchain also addresses cybersecurity challenges and enhances supply chain visibility and traceability. Businesses and policymakers can leverage the potential of blockchain in Industry 4.0 by taking several recommendations into consideration. First, they should invest in research and development to understand the capabilities and limitations of blockchain technology. Second, fostering collaboration and partnerships with industry players and technology providers will enable the creation of innovative blockchain solutions. Thirdly, developing industry-specific standards and governance frameworks will ensure interoperability, security, and scalability. Looking ahead, the usage of blockchain in Industry 4.0 is expected to continue growing. However, there are potential challenges that need to be addressed. Scalability and performance limitations, interoperability issues, regulatory and legal considerations, and resistance to change are some of the key challenges that must be navigated. In the future, we can expect to see further integration of blockchain with developed technologies like AI and IoT, the establishment of blockchain standards and governance frameworks, the formation of collaborative ecosystems and industry consortia, and a focus on addressing the ethical and societal implications of blockchain technology.

Overall, blockchain has the ability to drive significant transformations in Industry 4.0, enabling increased security, transparency, efficiency, and trust among stakeholders. By embracing blockchain and its capabilities, businesses and policymakers can unlock new opportunities for innovation and collaboration, leading to a more resilient and advanced industrial ecosystem.

## Declarations

Conflict of interest: The authors declare that they have no conflict of interest.

## References

**1** Lu, Y. (2017). Industry 4.0: a survey on technologies, applications and open research issues. *Journal of Industrial Information Integration* 6: 1–10.

**2** Machado, C.G., Winroth, M.P., and Ribeiro da Silva, E.H.D. (2020). Sustainable manufacturing in Industry 4.0: an emerging research agenda. *International Journal of Production Research* 58 (5): 1462–1484.

**3** Lu, Y. (2021). The current status and developing trends of Industry 4.0: a review. *Information Systems Frontiers* 1–20.

**4** Nayernia, H., Bahemia, H., and Papagiannidis, S. (2022). A systematic review of the implementation of Industry 4.0 from the organisational perspective. *International Journal of Production Research* 60 (14): 4365–4396.

**5** Alcácer, V. and Cruz-Machado, V. (2019). Scanning the Industry 4.0: a literature review on technologies for manufacturing systems. *Engineering Science and Technology, an International Journal* 22 (3): 899–919.

**6** Singh, H. (2021). Big data, Industry 4.0 and cyber-physical systems integration: a smart industry context. *Materials Today Proceedings* 46: 157–162.

**7** Mohamed, M. (2018). Challenges and benefits of Industry 4.0: an overview. *International Journal of Supply and Operations Management* 5 (3): 256–265.

**8** Tambare, P., Meshram, C., Lee, C.C. et al. (2021). Performance measurement system and quality management in data-driven Industry 4.0: a review. *Sensors* 22 (1): 224.

**9** Duy, P.T., Hien, D.T.T., Hien, D.H. and Pham, V.H. (2018). A survey on opportunities and challenges of Blockchain technology adoption for revolutionary innovation. *Proceedings of the 9th International Symposium on Information and Communication Technology*, 200–207 (6–7 Dec 2018). Da Nang City, Vietnam: ACM dl publication.

**10** Li, X., Wang, Z., Leung, V.C. et al. (2021). Blockchain-empowered data-driven networks: a survey and outlook. *ACM Computing Surveys (CSUR)* 54 (3): 1–38.

**11** Law, A. (2017). Smart contracts and their application in supply chain management. Doctoral dissertation, Massachusetts Institute of Technology.

**12** Javaid, M., Haleem, A., Singh, R.P. et al. (2021). Blockchain technology applications for Industry 4.0: a literature-based review. *Blockchain: Research and Applications* 2 (4): 100027.

**13** Manogaran, G., Thota, C., Lopez, D., and Sundarasekar, R. (2017). Big data security intelligence for healthcare industry 4.0. *Cybersecurity for Industry 4.0: Analysis for Design and Manufacturing*, 103–126.

**14** Vafiadis, N.V. and Taefi, T.T. (2019). Differentiating blockchain technology to optimize the processes quality in industry 4.0. In: *2019 IEEE 5th World Forum on Internet of Things (WF-IoT)*, 864–869. IEEE.

**15** Reda, M., Kanga, D.B., Fatima, T., and Azouazi, M. (2020). Blockchain in health supply chain management: state of art challenges and opportunities. *Procedia Computer Science* 175: 706–709.

**16** Leng, J., Ruan, G., Jiang, P. et al. (2020). Blockchain-empowered sustainable manufacturing and product lifecycle management in Industry 4.0: a survey. *Renewable and Sustainable Energy Reviews* 132: 110112.

**17** Dos Santos, L.M.A.L., da Costa, M.B., Kothe, J.V. et al. (2021). Industry 4.0 collaborative networks for industrial performance. *Journal of Manufacturing Technology Management* 32 (2): 245–265.

**18** Esmaeilian, B., Sarkis, J., Lewis, K., and Behdad, S. (2020). Blockchain for the future of sustainable supply chain management in Industry 4.0. *Resources, Conservation and Recycling* 163: 105064.

**19** Yaqoob, I., Salah, K., Jayaraman, R., and Al-Hammadi, Y. (2021). Blockchain for healthcare data management: opportunities, challenges, and future recommendations. *Neural Computing and Applications* 34: 1–16.

**20** Aoun, A., Ilinca, A., Ghandour, M., and Ibrahim, H. (2021). A review of Industry 4.0 characteristics and challenges, with potential improvements using blockchain technology. *Computers & Industrial Engineering* 162: 107746.

**21** El Ioini, N. and Pahl, C. (2018). A review of distributed ledger technologies. *On the Move to Meaningful Internet Systems. OTM 2018 Conferences: Confederated International Conferences: CoopIS, C&TC, and ODBASE 2018* (October 22–26, 2018), Valletta, Malta. *Proceedings*, Part, 277–288. Springer International Publishing.

**22** Abdelhamid, M. and Hassan, G. (2019). Blockchain and smart contracts. *Proceedings of the 8th International Conference on Software and Information Engineering*, 91–95 (9–12 April 2019). Cairo, Egypt: ACM dl publication.

**23** Taherdoost, H. (2023). Smart contracts in Blockchain technology: a critical review. *Information* 14 (2): 117.

**24** Dinh, T.T.A., Liu, R., Zhang, M. et al. (2018). Untangling blockchain: a data processing view of blockchain systems. *IEEE Transactions on Knowledge and Data Engineering* 30 (7): 1366–1385.

**25** Zheng, Z., Xie, S., Dai, H. et al. (2017). An overview of blockchain technology: Architecture, consensus, and future trends. In: *2017 IEEE International Congress on Big Data (BigData Congress)*, 557–564. IEEE.

**26** Kaur, M. and Gupta, S. (2021). Blockchain technology for convergence: an overview, applications, and challenges. *Blockchain and AI Technology in the Industrial Internet of Things*, 1-17.

**27** Akbar, N.A., Muneer, A., ElHakim, N., and Fati, S.M. (2021). Distributed hybrid double-spending attack prevention mechanism for proof-of-work and proof-of-stake blockchain consensuses. *Future Internet* 13 (11): 285.

**28** Tasca, P., Thanabalasingham, T., and Tessone, C.J. (2017). *Ontology of Blockchain Technologies. Principles of Identification and Classification*, 10. SSRN Electronic Journal.

**29** Rikken, O., Janssen, M., and Kwee, Z. (2019). Governance challenges of blockchain and decentralized autonomous organizations. *Information Polity* 24 (4): 397–417.

**30** Salmon, J. and Myers, G. (2019). Blockchain and associated legal issues for emerging markets.

**31** Idrees, S.M., Aijaz, I., Jameel, R., and Nowostawski, M. (2021). Exploring the blockchain technology: issues, applications and research potential. *International Journal of Online & Biomedical Engineering* 17 (7): 48–69.

**32** Madine, M., Salah, K., Jayaraman, R. et al. (2021). Appxchain: application-level interoperability for blockchain networks. *IEEE Access* 9: 87777–87791.

**33** Belchior, R., Vasconcelos, A., Guerreiro, S., and Correia, M. (2021). A survey on blockchain interoperability: past, present, and future trends. *ACM Computing Surveys (CSUR)* 54 (8): 1–41.

**34** Worley, C. and Skjellum, A. (2018). Blockchain tradeoffs and challenges for current and emerging applications: generalization, fragmentation, sidechains, and scalability. *2018 IEEE International Conference on Internet of Things (iThings) and IEEE Green Computing and Communications (GreenCom) and IEEE Cyber, Physical and Social Computing (CPSCom) and IEEE Smart Data (SmartData)*, 1582–1587. IEEE.

**35** Zhang, P., Schmidt, D.C., White, J., and Lenz, G. (2018). Blockchain technology use cases in healthcare. In: *Advances in Computers*, vol. 111 (ed. P. Raj and G.C. Deka), 1–41. Elsevier.

**36** Dugan, M. and Wilkins, W. (2018). Blockchain and cryptography for secure information sharing. In: *Artificial Intelligence for Autonomous Networks* (ed. M. Gilbert), 69–82. Chapman and Hall/CRC.

**37** Schulte, S., Sigwart, M., Frauenthaler, P. and Borkowski, M. (2019). Towards blockchain interoperability. *Business Process Management: Blockchain and Central and Eastern Europe Forum: BPM 2019 Blockchain and CEE Forum* (September 1–6, 2019), Vienna, Austria. Proceedings 17, 3–10. Springer International Publishing.

**38** Liu, Y., Lu, Q., Yu, G. et al. (2022). Defining blockchain governance principles: a comprehensive framework. *Information Systems* 109: 102090.

**39** Chituc, C.M., Toscano, C., and Azevedo, A. (2008). Interoperability in collaborative networks: independent and industry-specific initiatives–the case of the footwear industry. *Computers in Industry* 59 (7): 741–757.

**40** Meunier, S. (2018). Blockchain 101: what is blockchain and how does this revolutionary technology work? In: *Transforming Climate Finance and Green Investment with Blockchains* (ed. A. Marke), 23–34. Academic Press.

**41** Shafagh, H., Burkhalter, L., Hithnawi, A., and Duquennoy, S. (2017). November. Towards blockchain-based auditable storage and sharing of IoT data. *Proceedings of the 2017 on Cloud Computing Security Workshop*, 45–50 (3 November 2017). Dallas, Texas, USA: ACM dl publication.

**42** Kodym, O., Kubáč, L., and Kavka, L. (2020). Risks associated with logistics 4.0 and their minimization using blockchain. *Open Engineering* 10 (1): 74–85.

**43** George, G.M. and Jayashree, L.S. (2021). A survey on user privacy preserving blockchain for health insurance using ethereum smart contract. *International Journal of Information Privacy, Security and Integrity* 5 (2): 111–137.

**44** Liu, Y., He, D., Obaidat, M.S. et al. (2020). Blockchain-based identity management systems: a review. *Journal of Network and Computer Applications* 166: 102731.

**45** Giannopoulou, A. and Ferrari, V., (2019). Distributed data protection and liability on blockchains. *Internet Science: INSCI 2018 International Workshops* (October 24–26, 2018), St. Petersburg, Russia. Revised Selected Papers 5, 203–211. Springer International Publishing.

**46** Idrees, S.M., Nowostawski, M., Jameel, R., and Mourya, A.K. (2021). Security aspects of blockchain technology intended for industrial applications. *Electronics* 10 (8): 951.

**47** Ali, S., Wang, G., White, B., and Cottrell, R.L. (2018). A blockchain-based decentralized data storage and access framework for pinger. *2018 17th IEEE international conference on trust, security and privacy in computing and communications/12th IEEE international conference on big data science and engineering (TrustCom/BigDataSE)*, 1303–1308. IEEE.

**48** Saini, K. (2021). Blockchain foundation. In: *Essential Enterprise Blockchain Concepts and Applications* (ed. P.-A. Champin, F.L. Gandon, and L. Médini), 1–14. Auerbach Publications.

**49** Nguyen, H.L., Ignat, C.L., and Perrin, O. (2018). Trusternity: Auditing transparent log server with blockchain. *Companion Proceedings of the The Web Conference 2018*, 79–80.

**50** Rejeb, A., Keogh, J.G., and Treiblmaier, H. (2019). Leveraging the internet of things and blockchain technology in supply chain management. *Future Internet* 11 (7): 161.

**51** Ahmad, R.W., Salah, K., Jayaraman, R. et al. (2021). The role of blockchain technology in telehealth and telemedicine. *International Journal of Medical Informatics* 148: 104399.

**52** Taylor, P.J., Dargahi, T., Dehghantanha, A. et al. (2020). A systematic literature review of blockchain cyber security. *Digital Communications and Networks* 6 (2): 147–156.

**53** Pohlmann, N. (2022). Transport layer security (TLS)/secure socket layer (SSL). In: *Cyber Security: The Textbook on Concepts, Principles, Mechanisms, Architectures and Properties of Cyber Security Systems in Digitization* (ed. N. Pohlmann), 439–473. Wiesbaden: Springer Fachmedien Wiesbaden.

**54** Alshamrani, H. (2014). Internet protocol security (IPSec) mechanisms. *International Journal of Scientific and Engineering Research* 5 (5): 2229–5518.

**55** Mrsic, L., Adamek, J., and Cicek, I. (2019). Off-the-Record (OTR) Security Protocol Application in Cloud Environment. *International Conference on Communication and Intelligent Systems*, 195–206. Singapore: Springer Singapore.

**56** Roh, C.H. and Lee, I.Y. (2020). A study on PGP (pretty good privacy) using blockchain. In: *Advances in Computer Science and Ubiquitous Computing: CSA-CUTE 2018* (ed. V.R.Q. Leithardt), 316–320. Springer Singapore.

**57** Younes, O. and Albalawi, U. (2022). Securing session initiation protocol. *Sensors* 22 (23): 9103.

**58** Rescorla, E., Tschofenig, H., and Modadugu, N. (2022). RFC 9147: The Datagram Transport Layer Security (DTLS) Protocol Version 1.3.

**59** Yadav, U. and Sharma, A. (2023). Network security in evolving networking technologies: developments and future directions. In: *Evolving Networking Technologies: Developments and Future Directions* (ed. V.R.Q. Leithardt), 75–95. MDPI.

**60** Naik, N. and Jenkins, P. (2020). Self-Sovereign Identity Specifications: Govern your identity through your digital wallet using blockchain technology. *2020 8th IEEE International Conference on Mobile Cloud Computing, Services, and Engineering (MobileCloud)*, 90–95. IEEE.

**61** Diro, A., Lu Zhou, L., Saini, A. et al. Leveraging Blockchain for Zero Knowledge Identify Sharing: A Survey of Advancements, Challenges and Opportunities. Akanksha and Kaisar, Shahriar and Pham, Hiep, Leveraging Blockchain for Zero Knowledge Identify Sharing: A Survey of Advancements, Challenges and Opportunities.

**62** Schlatt, V., Sedlmeir, J., Feulner, S., and Urbach, N. (2022). Designing a framework for digital KYC processes built on blockchain-based self-sovereign identity. *Information & Management* 59 (7): 103553.

**63** Stockburger, L., Kokosioulis, G., Mukkamala, A. et al. (2021). Blockchain-enabled decentralized identity management: the case of self-sovereign identity in public transportation. *Blockchain: Research and Applications* 2 (2): 100014.

**64** Schardong, F. and Custódio, R. (2022). Self-sovereign identity: a systematic review, mapping and taxonomy. *Sensors* 22 (15): 5641.

**65** Culot, G., Fattori, F., Podrecca, M., and Sartor, M. (2019). Addressing Industry 4.0 cybersecurity challenges. *IEEE Engineering Management Review* 47 (3): 79–86.

**66** Latif, S., Idrees, Z., e Huma, Z., and Ahmad, J. (2021). Blockchain technology for the industrial internet of things: a comprehensive survey on security challenges, architectures, applications, and future research directions. *Transactions on Emerging Telecommunications Technologies* 32 (11): e4337.

**67** Da Xu, L., Lu, Y., and Li, L. (2021). Embedding blockchain technology into IoT for security: a survey. *IEEE Internet of Things Journal* 8 (13): 10452–10473.

**68** Rana, A., Sharma, S., Nisar, K. et al. (2022). The rise of blockchain internet of things (BIoT): secured, device-to-device architecture and simulation scenarios. *Applied Sciences* 12 (15): 7694.

**69** Choi, S. and Lee, J.H. (2020). Blockchain-based distributed firmware update architecture for IoT devices. *IEEE Access* 8: 37518–37525.

**70** Dutta, P., Choi, T.M., Somani, S., and Butala, R. (2020). Blockchain technology in supply chain operations: applications, challenges and research opportunities. *Transportation Research Part E: Logistics and Transportation Review* 142: 102067.

**71** Bernabe, J.B., Canovas, J.L., Hernandez-Ramos, J.L. et al. (2019). Privacy-preserving solutions for blockchain: review and challenges. *IEEE Access* 7: 164908–164940.

**72** Song, Q., Chen, Y., Zhong, Y. et al. (2021). A supply-chain system framework based on internet of things using blockchain technology. *ACM Transactions on Internet Technology (TOIT)* 21 (1): 1–24.

**73** Al Breiki, H., Al Qassem, L., Salah, K. et al. (2019). Decentralized access control for IoT data using blockchain and trusted oracles. *2019 IEEE International Conference on Industrial Internet (ICII)*, 248–257. IEEE.

**74** Shah, Z., Ullah, I., Li, H. et al. (2022). Blockchain based solutions to mitigate distributed denial of service (DDoS) attacks in the internet of things (IoT): a survey. *Sensors* 22 (3): 1094.

**75** Ali, M.H., Jaber, M.M., Abd, S.K. et al. (2022). Threat analysis and distributed denial of service (DDoS) attack recognition in the internet of things (IoT). *Electronics* 11 (3): 494.

**76** Demirkan, S., Demirkan, I., and McKee, A. (2020). Blockchain technology in the future of business cyber security and accounting. *Journal of Management Analytics* 7 (2): 189–208.

**77** Priyadarshini, I. (2019). Introduction to blockchain technology. In: *Cyber Security in Parallel and Distributed Computing: Concepts, Techniques, Applications and Case Studies* (ed. D.N. Le, R. Kumar, B.K. Mishra, et al.), 91–107.

**78** Gurtu, A. and Johny, J. (2019). Potential of blockchain technology in supply chain management: a literature review. *International Journal of Physical Distribution and Logistics Management* 49 (9): 881–900.

**79** Agrawal, T.K., Kumar, V., Pal, R. et al. (2021). Blockchain-based framework for supply chain traceability: a case example of textile and clothing industry. *Computers & Industrial Engineering* 154: 107130.

**80** Baralla, G., Ibba, S., Marchesi, M. et al. (2019). A blockchain based system to ensure transparency and reliability in food supply chain. *Euro-Par 2018: Parallel Processing Workshops: Euro-Par 2018 International Workshops* (August 27–28, 2018), Turin, Italy. Revised Selected Papers 24, 379–391. Springer International Publishing.

**81** Tijan, E., Aksentijević, S., Ivanić, K., and Jardas, M. (2019). Blockchain technology implementation in logistics. *Sustainability* 11 (4): 1185.

**82** Alzahrani, N. and Bulusu, N. (2020). A new product anti-counterfeiting blockchain using a truly decentralized dynamic consensus protocol. *Concurrency and Computation: Practice and Experience* 32 (12): e5232.

**83** Yiu, N.C. (2021). Toward blockchain-enabled supply chain anti-counterfeiting and traceability. *Future Internet* 13 (4): 86.

**84** Singh, K.K. (2022). Application of Blockchain Smart Contracts in E-Commerce and Government. arXiv preprint arXiv:2208.01350.

**85** Aránguiz, M., Margheri, A., Xu, D., and Tran, B. (2021). International trade revolution with smart contracts. In: *The Digital Transformation of Logistics: Demystifying Impacts of the Fourth Industrial Revolution* (ed. M. Sullivan and J. Kern), 169–184. John Wiley & Sons.

**86** Zamorano, J., Alfaro, M., de Oliveira, V.M. et al. (2021). New manufacturing challenges facing sustainability. *Manufacturing Letters* 30: 19–22.

**87** Khanfar, A.A., Iranmanesh, M., Ghobakhloo, M. et al. (2021). Applications of blockchain technology in sustainable manufacturing and supply chain management: a systematic review. *Sustainability* 13 (14): 7870.

**88** Reddy, K.R.K., Gunasekaran, A., Kalpana, P. et al. (2021). Developing a blockchain framework for the automotive supply chain: a systematic review. *Computers & Industrial Engineering* 157: 107334.

**89** Munirathinam, S. (2020). Industry 4.0: industrial internet of things (IIOT). In: *Advances in Computers* (Vol. 117, No. 1 (ed. P. Raj and P. Evangeline), 129–164. Elsevier.

**90** Srivastava, Y., Ganguli, S., Suman Rajest, S., and Regin, R. (2022). Smart HR Competencies and Their Applications in Industry 4.0. *A Fusion of Artificial Intelligence and Internet of Things for Emerging Cyber Systems*, 293–315.

**91** Scherer, M. (2017). Performance and scalability of blockchain networks and smart contracts.

**92** Khan, D., Jung, L.T., and Hashmani, M.A. (2021). Systematic literature review of challenges in blockchain scalability. *Applied Sciences* 11 (20): 9372.

**93** Lafourcade, P. and Lombard-Platet, M. (2020). About blockchain interoperability. *Information Processing Letters* 161: 105976.

**94** Irannezhad, E. (2020). The architectural design requirements of a blockchain-based port community system. *The Log* 4 (4): 30.

**95** Finck, M. (2018). Blockchains: regulating the unknown. *German Law Journal* 19 (4): 665–692.

**96** Farcane, N. and Deliu, D. (2020). Stakes and challenges regarding the financial Auditor's activity in the blockchain era. *Audit Financiar* 18 (157).

**97** Böhmecke-Schwafert, M., Wehinger, M., and Teigland, R. (2022). Blockchain for the circular economy: theorizing blockchain's role in the transition to a circular economy through an empirical investigation. *Business Strategy and the Environment* 31: 3786–3801.

**98** Jain, S., Ahuja, N.J., Srikanth, P. et al. (2021). Blockchain and autonomous vehicles: recent advances and future directions. *IEEE Access* 9: 130264–130328.

**99** Narbayeva, S., Bakibayev, T., Abeshev, K. et al. (2020). Blockchain technology on the way of autonomous vehicles development. *Transportation Research Procedia* 44: 168–175.

**100** Cole, R., Stevenson, M., and Aitken, J. (2019). Blockchain technology: implications for operations and supply chain management. *Supply Chain Management: An International Journal* 24 (4): 469–483.

**101** Alkaabi, N., Salah, K., Jayaraman, R. et al. (2020). Blockchain-based traceability and management for additive manufacturing. *IEEE Access* 8: 188363–188377.

**102** Köhler, S. and Pizzol, M. (2020). Technology assessment of blockchain-based technologies in the food supply chain. *Journal of Cleaner Production* 269: 122193.

**103** Xu, P., Lee, J., Barth, J.R., and Richey, R.G. (2021). Blockchain as supply chain technology: considering transparency and security. *International Journal of Physical Distribution and Logistics Management* 51 (3): 305–324.

**104** Gadekallu, T.R., Huynh-The, T., Wang, W. et al. (2022). Blockchain for the metaverse: A review. arXiv preprint arXiv:2203.09738.

**105** Khan, A.A., Laghari, A.A., Li, P. et al. (2023). The collaborative role of blockchain, artificial intelligence, and industrial internet of things in digitalization of small and medium-size enterprises. *Scientific Reports* 13 (1): 1656.

**106** Zhang, Y. (2020). Developing cross-border blockchain financial transactions under the belt and road initiative. *The Chinese Journal of Comparative Law* 8 (1): 143–176.

**107** Karajovic, M., Kim, H.M., and Laskowski, M. (2019). Thinking outside the block: projected phases of blockchain integration in the accounting industry. *Australian Accounting Review* 29 (2): 319–330.

**108** Malhotra, A., O'Neill, H., and Stowell, P. (2022). Thinking strategically about blockchain adoption and risk mitigation. *Business Horizons* 65 (2): 159–171.

**109** Agerskov, S., Pedersen, A.B. and Beck, R. (2023). Ethical Guidelines for Blockchain Systems.

# 7

# Unifying Technologies in Industry 4.0: Harnessing the Synergy of Internet of Things, Big Data, Augmented Reality/ Virtual Reality, and Blockchain Technologies

*K. Logeswaran[1], S. Savitha[2], P. Suresh[3], K.R. Prasanna Kumar[4], M. Gunasekar[4], R. Rajadevi[1], M.K. Dharani[1], and A.S. Jayasurya[5]*

[1] *Department of AI, Kongu Engineering College, Erode, Tamil Nadu, India*
[2] *Department of CSE, K.S.R. College of Engineering, Tiruchengode, Tamil Nadu, India*
[3] *Department of Database Systems, School of Computer Science and Engineering, Vellore Institute of Technology, Vellore, Tamil Nadu, India*
[4] *Department of Information Technology, Kongu Engineering College, Erode, Tamil Nadu, India*
[5] *Department of Electrical and Electronics, Universiti Teknologi Petronas, Perak, Malaysia*

## 7.1   Introduction to Industry 4.0

Industry 4.0 represents a paradigm shift in manufacturing and production processes. It is driven by the integration of digital technologies, data analytics, and automation to create smart and connected factories. This transformative concept is characterized by the convergence of various technologies, such as the Internet of Things (IoT), big data analytics, cyber-physical systems (CPS), augmented reality (AR) and virtual reality (VR), and blockchain.

Industry 4.0 builds upon the advancements of its predecessors, the First, Second, and Third Industrial Revolutions. The First Industrial Revolution introduced mechanization through steam power, while the Second Industrial Revolution introduced mass production and electrification. The Third Industrial Revolution brought computerization and automation [1]. However, Industry 4.0 takes it a step further by connecting machines, products, and people through a networked ecosystem.

The core pillars of Industry 4.0 include IoT, where devices and machines communicate and share data, enabling real-time monitoring and control. Big data analytics leverages the vast amount of data generated to gain valuable insights, optimize processes, and enable predictive maintenance. CPS combine the physical and digital realms, enabling machines to interact and make autonomous decisions. AR and VR technologies enhance human-machine interactions and improve training and simulation. Blockchain technology provides secure and transparent transactions, ensuring trust and traceability in supply chains.

Industry 4.0 offers numerous benefits, including increased productivity, operational efficiency, and product customization. It enables real-time decision-making, reduces downtime through predictive maintenance, and fosters the development of new business models. However, it also brings challenges such as technological integration complexities, cybersecurity risks, and the need for reskilling the workforce.

Industry 4.0 represents a revolutionary era that promises to reshape industries and economies. It offers immense opportunities for innovation and efficiency while posing challenges that require careful consideration and adaptation. Understanding the principles, technologies, and implications of Industry 4.0 is crucial for businesses and societies to thrive in this new industrial landscape.

### 7.1.1   Evolution of Industrial Revolutions

The evolution of industrial revolutions refers to the series of significant transformations that have shaped the way societies produce goods and services. These revolutions are characterized by major advancements in technology, manufacturing processes, and socioeconomic systems.

*Topics in Artificial Intelligence Applied to Industry 4.0*, First Edition. Edited by Mahmoud Ragab AL-Refaey, Amit Kumar Tyagi, Abdullah Saad AL-Malaise AL-Ghamdi, and Swetta Kukreja.
© 2024 John Wiley & Sons Ltd. Published 2024 by John Wiley & Sons Ltd.

### 7.1.1.1 First Industrial Revolution

The First Industrial Revolution, started between the eighteenth and nineteenth centuries, manifested a significant change in manufacturing and production processes. It was characterized by the introduction of mechanization, steam power, and the development of factory systems. This revolution laid the foundation for the modern industrial era and brought about profound changes in society, economy, and daily life.

The First Industrial Revolution included the steam engine, which revolutionized transportation and mechanized various industries. Steam-powered machinery replaced manual labor, leading to increased production efficiency and output. The steam engine also powered locomotives, enabling the expansion of railways and transforming transportation networks [2]. The use of machinery and the concentration of workers in factories led to the centralization of production, creating a new class of industrial workers. This shift had significant social and economic implications, including the rise of urbanization, the growth of cities, and the emergence of a working-class population.

The First Industrial Revolution brought about advancements in textile production, iron and steel manufacturing, and mining. Innovations such as the spinning jenny, power loom, and cotton gin revolutionized the textile industry, leading to increased production and lowered costs. The use of steam-powered machinery in iron and steel production paved the way for the development of railways, bridges, and machinery [3].

Overall, the First Industrial Revolution set the stage for subsequent industrial transformations. It laid the groundwork for the mechanization of industries, the rise of factories, and the application of steam power. The advancements of this revolution propelled societies into a new era of economic growth, urbanization, and technological progress.

### 7.1.1.2 Second Industrial Revolution

Technological revolution took place from the mid-nineteenth to the early twentieth century and had a profound impact on society, economy, and technology. It was characterized by significant advancements in manufacturing, transportation, communication, and energy, which set the stage for the modern industrialized world.

One of the key drivers was the widespread adoption of electricity. The development of practical electric lighting by inventors like Thomas Edison and the establishment of alternating current systems by Nikola Tesla revolutionized daily life and industrial practices. Electric power became the driving force behind the electrification of factories, homes, and cities, enabling a wide range of industries to operate more efficiently and effectively.

Another crucial aspect of this revolution was the rise of mass production methods. Innovations such as the Bessemer process for steel production and the development of interchangeable parts, pioneered by figures like Henry Ford, transformed manufacturing processes [3]. These advancements allowed for the efficient production of goods on a large scale, leading to increased productivity and lower costs. The advent of assembly line techniques further revolutionized production, making it possible to produce goods at unprecedented rates.

Transportation and communication systems also underwent significant transformations during this period. The construction of railways and steamships revolutionized transportation, enabling faster, more efficient movement of goods and people over long distances. The invention of the telephone and the development of the telegraph facilitated rapid long-distance communication, connecting people across great distances and accelerating the exchange of information [4].

The Second Industrial Revolution had profound societal impacts. The growth of industries, along with improvements in living standards, led to the rise of consumer culture and the development of mass markets. It also brought significant social changes, including the emergence of labor movements, urban planning initiatives, and the restructuring of social classes [5].

It was a period of remarkable technological advancements that reshaped the world. The widespread adoption of electricity, mass production techniques, and improved transportation and communication systems transformed manufacturing, transportation, and daily life. The effects of this revolution continue to shape the modern industrialized society and have laid the foundation for subsequent technological and industrial developments.

### 7.1.1.3 Third Industrial Revolution

This digital revolution or the information age refers to the period of technological advancement that took place from the late twentieth century onward. It marked a significant shift in the way society functions, with the widespread adoption of computers, digital technologies, and the Internet. This revolution had a profound impact on various aspects of life, including communication, commerce, and the global economy.

The key driver was the rapid development and accessibility of computers. The invention of the microprocessor and the subsequent miniaturization of computer systems made computing power more affordable and widely available.

This led to the integration of computers into various aspects of everyday life and paved the way for the digital transformation of industries and services.

The introduction of the Internet played a crucial role in the Third Industrial Revolution. The Internet provided a global network that connected computers and allowed for the exchange of information on an unprecedented scale [4]. It revolutionized communication by enabling real time, instant messaging and email, and it transformed commerce by facilitating e-commerce and online transactions. The Internet also became a platform for collaboration, information sharing, and the development of online communities.

Another important aspect was the emergence of digital technologies. Advancements in telecommunications, mobile devices, and wireless connectivity expanded the reach and accessibility of digital services. This led to the proliferation of smartphones, tablets, and other smart devices, transforming the way people interact, access information, and conduct business.

The Third Industrial Revolution brought about significant changes in various industries. It revolutionized the media and entertainment sector with the digitization of content, the rise of streaming services, and the shift from physical media to digital formats. It transformed the retail industry with the rise of e-commerce platforms and online marketplaces. It also revolutionized the finance industry with the development of digital banking, online payment systems, and cryptocurrencies [6].

The Third Industrial Revolution also had profound social and cultural impacts. It enabled greater connectivity and communication, breaking down geographical barriers and fostering global interconnectedness. It gave rise to social media platforms, allowing people to connect, share information, and participate in online communities. However, it also raised concerns about privacy, cybersecurity, and the digital divide [7].

### 7.1.1.4 Fourth Industrial Revolution (Industry 4.0)

It is also known as "Industry 4.0" and is the ongoing wave of technological advancements that is transforming the way we live and work. It represents a fusion of digital, physical, and biological systems. At its core, Industry 4.0 aims to create intelligent, connected, and autonomous systems that enable the digital transformation of industries [5]. It builds upon the achievements of the previous industrial revolutions but takes them to a whole new level by leveraging advanced technologies and data-driven capabilities.

With the massive quantities of data generated by linked devices and systems, businesses can leverage advanced analytics tools and algorithms to extract valuable insights, make data-driven decisions, and identify patterns and trends that were previously inaccessible. This enables more accurate forecasting, enhanced operational efficiency, and the ability to customize products and services to meet individual customer needs [8].

Artificial intelligence (AI) and machine learning (ML) are integral components of Industry 4.0. These technologies enable machines and systems to learn, adapt, and make autonomous decisions based on the analysis of vast amounts of data. AI-powered robots and machines can perform complex tasks, collaborate with humans, and enhance productivity in various industries, from manufacturing to health care.

Another significant aspect of Industry 4.0 is the use of blockchain technology. Blockchain provides a decentralized and secure way of recording and verifying transactions and data exchanges. It ensures transparency, trust, and immutability, making it well suited for applications such as supply chain management, digital identity verification, and secure financial transactions [2].

Industry 4.0 has the potential to bring about numerous benefits. It can lead to increased productivity, operational efficiency, and cost savings. It enables real-time monitoring and predictive maintenance, reducing downtime and optimizing resource utilization. It fosters the development of new business models and revenue streams, as companies leverage data and technology to create innovative products and services. Furthermore, Industry 4.0 has the potential to transform the workforce with an emphasis on upskilling and reskilling to adapt to the changing demands of the digital era.

However, Industry 4.0 also poses challenges and implications. It requires significant investment in infrastructure, technology, and cybersecurity to ensure data privacy and protect against potential risks. There are concerns about job displacement and the impact on the workforce, as certain tasks become automated or taken over by machines [1]. Ethical considerations, such as the responsible use of AI and the implications of autonomous decision-making, also need to be addressed.

In conclusion, Industry 4.0 represents a transformative era driven by advanced technologies and data-driven capabilities. It is reshaping industries, economies, and societies, offering tremendous opportunities for innovation, efficiency, and customization. However, it also brings challenges that need to be carefully managed to ensure a successful and inclusive transition into this digital age.

### 7.1.2 Components and Technologies of Industry 4.0

Figure 7.1 illustrates the different components and technologies that constitute the architecture of Industry 4.0. It starts with the overarching concept of Industry 4.0 and branches out into various key technologies, including the IoT, AI and ML, big data analytics, robotics and automation, AR and VR, and blockchain.

Each technology is further broken down into its specific components or aspects. For example, the IoT includes sensors for data collection, connectivity for communication, and data collection for gathering relevant information. AI and ML involve ML algorithms, cognitive computing, and autonomous decision-making capabilities. Big data analytics encompass data management, analytics, and visualization. Robotics and automation comprise robots, collaborative robots (cobots), and automated systems. AR and VR consist of wearable devices, head-mounted displays, and virtual simulations. Finally, blockchain includes decentralized ledger technology, smart contracts, and data security mechanisms.

This representation provides a visual overview of the interconnectedness and interdependencies of the various components that make up the architecture of Industry 4.0.

### 7.1.3 Industry 4.0 Architecture

Industry 4.0 architecture refers to the underlying structure and framework that supports the implementation of Industry 4.0 principles and technologies. It encompasses the design and integration of various components, systems, and technologies to enable the digital transformation of industries and drive enhanced productivity, efficiency, and flexibility. At its core, Industry 4.0 architecture revolves around the concept of interconnected CPS, which seamlessly integrate physical machinery, devices, sensors, and actuators with digital systems, software applications, and networks. These CPS interact and communicate with each other and with humans, forming the backbone of the industrial infrastructure. Figure 7.2 presents a high-level architecture of Industry 4.0, highlighting the different layers and their associated components. The architecture includes the following layers:

- Physical Layer: This layer involves sensors for data acquisition, edge computing for data processing, industrial control systems for managing physical processes, and actuators for controlling physical operations. Data acquired from sensors is communicated to the edge computing layer for data processing. Processed data is then transmitted to the connectivity layer for real-time data exchange. Control signals are communicated from the industrial control systems layer to the smart decision systems layer for decision-making.
- Communication Layer: This layer encompasses wireless networks for real-time data transmission, connectivity for seamless communication, cloud computing for data storage, and data centers for managing and processing large volumes of data. Processed and analyzed data is transmitted to the data centers layer for storage and further processing.
- Information Layer: This layer focuses on big data analytics for analyzing collected data, generating insights and patterns, smart decision systems for making informed decisions, and process optimization for enhancing operational efficiency. Data analysis results and insights are communicated to the smart factories layer for implementation. Decision outputs and actionable intelligence are communicated to the intrusion prevention layer for security measures.
- Cybersecurity Layer: This layer addresses security protocols, threat detection, intrusion prevention, data protection, access control, risk management, and security compliance to ensure the integrity and safety of data and systems. Access control policies and risk management information are communicated to the advanced manufacturing layer for operational control. Security compliance requirements and guidelines are communicated to the supply chain management layer for adherence.
- Application Layer: This layer demonstrates specific applications of Industry 4.0, such as advanced manufacturing with robotics and automation, smart factories enabled by IoT integration, supply chain management driven by data-driven insights, predictive maintenance utilizing ML, and the development of smart products leveraging AI.

The architecture of Industry 4.0 is designed to enable seamless connectivity, interoperability, and collaboration across different layers and components. It aims to create a highly adaptive and responsive industrial ecosystem, where data-driven insights, automation, and intelligent decision-making drive transformative changes and optimize operational efficiency. It is important to note that Industry 4.0 architecture can vary depending on specific industry requirements, use cases, and technological advancements. Organizations often tailor the architecture to their unique needs, leveraging a combination of existing technologies and new innovations to create a robust and future-ready framework for their digital transformation journey.

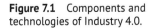

**Figure 7.1** Components and technologies of Industry 4.0.

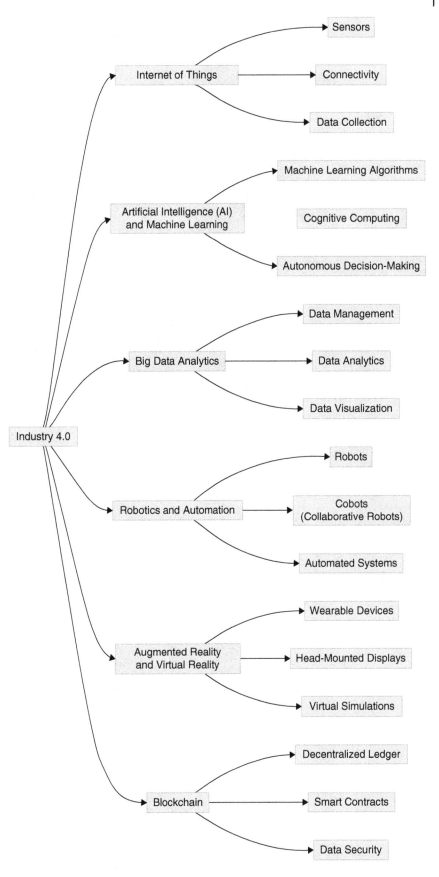

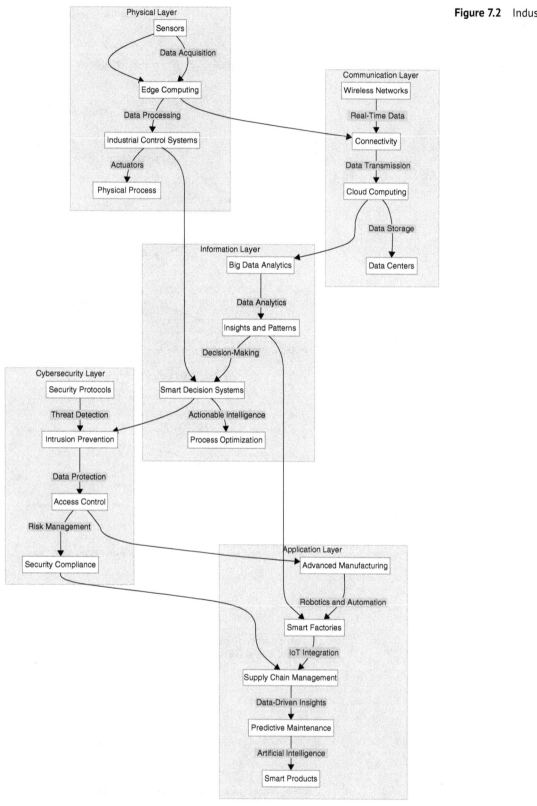

**Figure 7.2** Industry 4.0 architecture.

### 7.1.4 Benefits and Opportunities of Industry 4.0

Industry 4.0, the Fourth Industrial Revolution, brings forth a multitude of benefits and opportunities for organizations across various sectors. By leveraging advanced technologies and digital transformation, Industry 4.0 offers significant advantages that can enhance productivity, efficiency, and competitiveness.

The major advantage is the opportunity for enhanced customization and personalization. Industry 4.0 enables the collection and analysis of vast amounts of data, allowing organizations to gain insights into customer preferences and tailor products and services accordingly. This level of customization fosters customer satisfaction, loyalty, and competitiveness in the market [1].

Additionally, it facilitates the development of smart and connected products. By incorporating IoT capabilities, organizations can create products that communicate with each other, collect usage data, and provide valuable insights for product improvement and predictive maintenance. This connectivity also opens up new revenue streams through value-added services and subscription models.

Industry 4.0 also presents opportunities for improved supply chain management. By enabling instantaneous data sharing and improved transparency, organizations can enhance their inventory management, minimize lead times, and achieve greater precision in demand forecasting [8]. This enables streamlined operations, reduced costs, and improved customer satisfaction.

Furthermore, Industry 4.0 fosters innovation and collaboration. The integration of technologies such as AI, ML, and VR enables organizations to explore new business models, develop advanced prototypes, and enhance research and development processes. Collaborative platforms and digital ecosystems also facilitate partnerships and knowledge sharing, fostering innovation and growth [2].

From improved operational efficiency and customization to smart products and enhanced supply chain management, organizations embracing Industry 4.0 can gain a competitive edge, drive innovation, and unlock new avenues for growth and success in the digital era.

### 7.1.5 Challenges and Implications of Industry 4.0

Industry 4.0 brings about transformative changes and advancements in various industries. However, along with its benefits, Industry 4.0 also poses several challenges and implications that organizations need to address to fully leverage its potential. Here are some key challenges and implications of Industry 4.0.

#### 7.1.5.1 Workforce Transformation

The automation and digitalization brought by Industry 4.0 may require upskilling or reskilling of the existing workforce to adapt to new roles and technologies. Organizations need to invest in training programs and create a culture of lifelong learning to ensure a smooth workforce transition. Industry 4.0 presents challenges and implications in job displacement and workforce transformation. Automation and technology advancements may lead to job losses, creating a skills gap. Reskilling and upskilling programs are crucial to equip workers with the necessary competencies. The workforce of the future needs to adapt to new technologies, collaborate with machines, and possess a blend of technical and soft skills. Socioeconomic implications can arise, emphasizing the importance of support mechanisms and lifelong learning initiatives. Harmonious human-machine collaboration and a proactive approach involving collaboration between stakeholders are vital in navigating these challenges.

#### 7.1.5.2 Cybersecurity Risks and Data Privacy

Industry 4.0 poses challenges and implications for cybersecurity risks and data privacy. The interconnected nature of systems increases the vulnerability to cyber threats such as data breaches and hacking. Organizations must implement robust cybersecurity measures and intrusion detection systems to protect their infrastructure and data. Compliance with data protection regulations, privacy-by-design principles, and transparent data practices are necessary. Supply chain vulnerabilities and insider threats require thorough risk assessments, secure communication channels, access controls, and monitoring mechanisms. Integration of legacy systems and international regulations adds complexity, necessitating security assessments, system updates, and compliance with regional data protection laws. A multifaceted approach, including technical measures, employee training, cybersecurity frameworks, and proactive risk management, is crucial for addressing these challenges and ensuring data privacy in the Industry 4.0 era.

### 7.1.5.3 Ethical Considerations

Industry 4.0 presents challenges and implications in ethical considerations. Job displacement and economic inequality, ethical decision-making in AI systems, data privacy and security, bias and fairness in algorithms, environmental impact, and transparency and accountability are key concerns. Reskilling programs and job transition support can address job displacement. Ethical frameworks and guidelines must guide decision-making in AI systems. Stringent measures and transparent data practices ensure data privacy and security. Addressing bias and promoting fairness in algorithms is essential. Sustainable practices can mitigate the environmental impact. Transparency and accountability are crucial in maintaining trust. Collaboration among organizations, policymakers, and society is necessary to proactively address these ethical challenges and ensure responsible and ethical deployment of Industry 4.0 technologies for the benefit of individuals and society.

### 7.1.5.4 Infrastructure and Investment Requirements

Industry 4.0 presents challenges and implications in infrastructure and investment requirements. Upgrading existing infrastructure, establishing connectivity and interoperability, investing in data storage and processing capabilities, assessing costs and return on investment, navigating regulatory frameworks, and addressing skills and workforce requirements are key considerations. Organizations must invest in upgrading infrastructure, establishing communication protocols, and scalable computing systems. Careful evaluation of costs and returns, financial planning, and phased implementation strategies are necessary. Engaging with policymakers and regulatory bodies helps address regulatory challenges. Investment in training and upskilling programs is crucial to develop a skilled workforce. Collaboration among organizations, policymakers, and infrastructure providers is vital to address these challenges and create an enabling environment for Industry 4.0. A strategic and holistic approach, including infrastructure investment, skills development, regulatory alignment, and proactive planning, is necessary to leverage the potential of Industry 4.0 and drive innovation, productivity, and competitiveness.

Addressing these challenges and implications requires a proactive approach from organizations, policymakers, and society as a whole. Collaboration, continuous learning, and responsible implementation of Industry 4.0 technologies can help navigate these challenges and unlock the full potential of this transformative revolution.

## 7.2 Internet of Things

The IoT is a network of networked physical devices, automobiles, appliances, and other items equipped with sensors, software, and connections. These gadgets are capable of collecting and exchanging data via the Internet, allowing them to converse and interact with one another as well as people. IoT technology enables the seamless integration and automation of numerous systems, resulting in increased efficiency, better decision-making, and new potential for innovation in a variety of sectors. It has the ability to change the way we live, work, and interact with our environment. IoT is driving breakthroughs in fields such as health care, transportation, agriculture, and manufacturing, from smart homes and wearable gadgets to industrial automation and smart cities. However, the spread of IoT raises worries about data security, privacy, and ethical issues [9]. As IoT evolves, it has the potential to alter our environment by creating a more connected and intelligent ecosystem.

### 7.2.1 Notion of IoT

#### 7.2.1.1 Concept of IoT

These concepts form the basis of IoT, enabling the seamless connectivity, data collection, analysis, and intelligent decision-making that drive the potential of IoT applications across industries and domains. Figure 7.3 shows the concept of IoT in an abstract view.

- Connectivity: The foundation of IoT is the ability to connect various devices and objects to the Internet. This connectivity allows for data exchange and communication between devices, enabling them to interact and share information.
- Sensors and Actuators: Sensors are built into IoT devices to collect data from their surroundings. Temperature, pressure, motion, and light are all examples of physical changes that these sensors can sense. Actuators, on the other hand, enable devices to conduct actions or regulate physical processes in response to data.
- Data Collection and Analysis: Through their sensors, IoT devices create massive volumes of data. This data is gathered, saved, and analyzed in order to derive relevant insights and make sound judgments. Data analytics techniques such as ML and AI are frequently used to extract useful information from IoT-generated data.

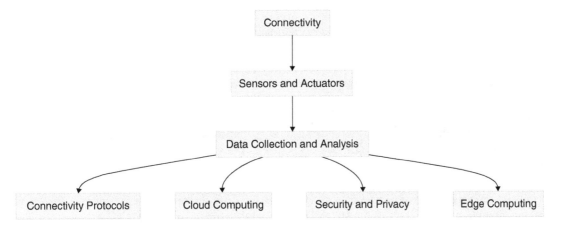

**Figure 7.3** Concept of IoT.

- Connectivity Protocols: Various communication protocols enable IoT devices to connect and exchange data. Common protocols include Wi-Fi, Bluetooth, Zigbee, and cellular networks like 3G, 4G, and 5G. These protocols allow continuous communication and compatibility across various devices and systems.
- Cloud Computing: Cloud-based platforms and services are critical in IoT. They provide storage, computing power, and data-processing capabilities, allowing for scalable and flexible IoT deployments. Cloud services enable the secure storage and analysis of IoT-generated data, as well as the development of IoT applications and services.
- Security and Privacy: With the development of linked devices, it is critical to ensure the security and privacy of IoT data. To secure sensitive information and prevent unauthorized access, robust security methods such as encryption, authentication, and access control are required.
- Edge Computing: Edge computing involves processing and analyzing data locally, at or near the source, rather than sending it to the cloud. This approach reduces latency, improves real-time response, and enhances data privacy. Edge computing is particularly important in IoT scenarios where low latency and real-time decision-making are critical.

### 7.2.2 Role of IoT in Industry 4.0

The IoT plays a crucial and disruptive role in Industry 4.0. IoT technology is critical in allowing the seamless integration of digital and physical systems, resulting in smart and networked industrial environments. The IoT is critical to Industry 4.0 because it enables seamless communication, data collecting, and automation. Sensor-equipped IoT devices collect real-time data from equipment and processes, allowing for continuous monitoring and optimization. IoT enables device connection and communication, allowing for coordinated operations and control. It also allows for automation and control by integrating IoT devices with actuators and control systems. Predictive analytics and maintenance are made possible by analyzing IoT-generated data, leading to proactive maintenance and improved asset utilization [10]. IoT optimizes the supply chain by enabling real-time tracking and monitoring of inventory and logistics operations. Additionally, IoT enhances safety and security by monitoring environmental conditions and enabling IoT-enabled security systems. Overall, IoT technology forms the foundation of Industry 4.0, enabling connectivity, automation, and data-driven decision-making that drive efficiency, productivity, and innovation in industrial settings.

Overall, IoT technology acts as a foundational pillar of Industry 4.0 by enabling connectivity, data-driven decision-making, automation, and optimization across various industrial domains. It unlocks new opportunities for efficiency, productivity, and innovation, paving the way for a more connected, intelligent, and efficient industrial landscape.

### 7.2.3 IoT Applications in Manufacturing and Supply Chain

IoT applications have transformed the manufacturing and supply chain industries, bringing unprecedented levels of efficiency, visibility, and optimization. In manufacturing, IoT devices embedded in machines and equipment collect real-time data on performance, maintenance needs, and energy consumption. This data enables manufacturers to proactively monitor and optimize operations, reduce downtime, and improve overall equipment effectiveness. IoT-driven predictive maintenance ensures that maintenance activities are performed at the right time, maximizing asset uptime and minimizing costs [11].

In the supply chain, IoT facilitates end-to-end visibility and traceability. IoT sensors and devices track goods, vehicles, and assets throughout the supply chain, providing real-time location and condition information. This enables accurate inventory management, efficient logistics planning, and timely delivery of goods. IoT-powered supply chain solutions leverage data analytics and ML to optimize routing, reduce transit times, and minimize errors.

Moreover, IoT enables seamless collaboration and data sharing across supply chain partners. By integrating systems and sharing real-time data, stakeholders can make informed decisions and respond rapidly to changes in demand or supply. This enhanced visibility and collaboration lead to improved forecasting accuracy, reduced stockouts, and efficient demand planning. IoT also supports sustainability efforts in manufacturing and supply chain operations. By monitoring energy consumption and environmental factors, IoT helps identify opportunities for energy efficiency, waste reduction, and sustainable practices [12]. IoT applications in manufacturing and supply chain revolutionize operations by enabling data-driven decision-making, proactive maintenance, efficient inventory management, enhanced visibility, and collaborative supply chain management. These IoT-driven advancements drive productivity, reduce costs, and enhance customer satisfaction in the dynamic and competitive manufacturing and supply chain landscape.

### 7.2.4 Challenges and Opportunities of IoT Implementation

Implementing the IoT presents both obstacles and possibilities for organizations in a variety of sectors. The intricacy of IoT systems is one of the major issues. The integration of several devices, protocols, and platforms necessitates meticulous planning and skill. It might be difficult to ensure the interoperability, security, and scalability of IoT infrastructure. Furthermore, storing, processing, and analyzing vast volumes of data created by IoT devices present issues in terms of data storage, processing, and analytics.

Security and privacy concerns are also significant challenges. With the increase in connected devices, the potential for cyberattacks and data breaches expands. Safeguarding sensitive data, protecting IoT devices from unauthorized access, and implementing robust security measures are critical considerations [13]. Another challenge is the need for skilled personnel. Organizations must acquire talent with expertise in IoT technologies, data analytics, and cybersecurity to successfully implement and manage IoT systems.

However, IoT implementation also presents tremendous opportunities. Organizations can leverage IoT to improve operational efficiency, optimize resource utilization, and enable predictive maintenance. Real-time data collected by IoT devices can enhance decision-making, enable automation, and drive innovation. IoT opens up new business models and revenue streams. It enables the development of smart products and services, personalized customer experiences, and value-added offerings. IoT-driven insights also facilitate a better understanding of customer needs, enabling targeted marketing and product development. Furthermore, IoT implementation contributes to sustainability efforts. It enables organizations to monitor and manage energy consumption, reduce waste, and optimize resource usage, promoting environmental responsibility [14].

To seize the opportunities and overcome challenges, organizations must develop robust IoT strategies, invest in infrastructure and talent, prioritize security and privacy measures, and collaborate with partners and stakeholders. Embracing IoT's potential can lead to enhanced efficiency, competitiveness, and innovation in the increasingly connected world.

## 7.3 Big Data

Big data refers to enormous, complicated datasets that are difficult to manage using typical approaches. It offers valuable insights through analysis, uncovering patterns and trends. Big data analytics uses advanced technologies like AI and ML to extract meaningful information. Its applications span across industries, enabling improved decision-making, personalized experiences, and innovation. However, challenges such as privacy and security exist. Despite this, big data has the potential to revolutionize organizations by driving efficiency, competitiveness, and innovation through valuable insights from vast amounts of data.

### 7.3.1 Understanding Big Data and Its Characteristics

Understanding the characteristics of big data is essential in today's data-driven world. Big data is characterized by five main aspects: volume, velocity, variety, veracity, and value. Figure 7.4 shows the 5Vs of big data.

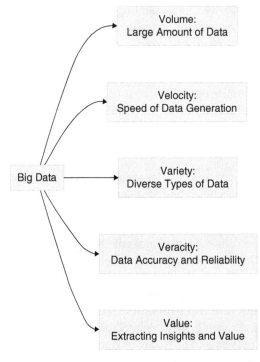

- The large amount of data created by numerous sources such as social media, sensors, and digital transactions is referred to as "volume." This plethora of data poses storage, processing, and analytical issues. To handle such massive numbers, organizations require scalable infrastructure and strong data management solutions.
- The pace at which data is created and must be processed in real time or near real time is referred to as "velocity." Data is being created at an unprecedented rate due to the development of linked devices and Internet activities. Organizations must have effective data-processing skills and real-time analytics technologies in order to extract valuable insights and take quick action.
- The many types and formats of data are represented by variety. Big data includes data from multiple sources that are structured, unstructured, or semi-structured. Text, photos, videos, social media postings, and other assets are all included. To gain significant insights from such disparate data, modern data integration and analytics approaches are required.

**Figure 7.4** Big data characteristics.

- Veracity emphasizes the importance of data quality, accuracy, and reliability. Organizations must ensure data integrity to make informed decisions and derive accurate insights.
- Value is derived from analyzing big data to extract meaningful insights and drive value for organizations. By applying advanced analytics techniques, organizations can uncover patterns, trends, and correlations that lead to actionable insights and strategic advantages.

Understanding these characteristics of big data helps organizations make informed decisions and drive innovation. By effectively managing and analyzing big data, businesses can uncover hidden patterns, trends, and correlations, enabling them to gain a competitive edge, enhance customer experiences, optimize operations, and drive strategic decision-making. However, the challenges of big data should not be overlooked. Organizations need to address issues related to data privacy, security, quality, and governance. They must invest in skilled personnel, advanced analytics tools, and scalable infrastructure to harness the full potential of big data.

### 7.3.2 Big Data Analytics in Industry 4.0

Big data analytics is critical in Industry 4.0, allowing organizations to get important insights from massive volumes of data created by networked devices and systems. Big data analytics has become an essential aspect of driving digital transformation and attaining operational excellence, thanks to the confluence of modern technologies such as the IoT, AI, and cloud computing. Big data analytics in Industry 4.0 enables organizations to analyze and understand data in real time, delivering actionable insight for making educated decisions. It helps optimize processes, improve productivity, enhance product quality, and enable predictive maintenance [15]. Through the analysis of data from multiple sources, such as production lines, supply chains, and customer interactions, organizations can obtain a holistic perspective of their operations and uncover areas for enhancement.

Big data analytics also enables organizations to personalize customer experiences and drive innovation. Organizations may customize their products and services to individual requirements by analyzing consumer behavior, preferences, and feedback, increasing customer happiness and loyalty [16]. It also helps identify market trends, patterns, and emerging demands, enabling organizations to develop new products and services that meet evolving customer expectations. Furthermore, big data analytics assists in risk management and fraud detection. By analyzing data patterns and anomalies, organizations can identify potential risks, prevent failures, and mitigate disruptions. It facilitates in the detection of fraudulent actions, the detection of cybersecurity risks, and the protection of sensitive data, assuring the security and integrity of operations [17].

However, implementing big data analytics in Industry 4.0 comes with challenges. It requires robust infrastructure, scalable storage, and advanced analytics tools. Data privacy, security, and regulatory compliance are critical considerations, requiring organizations to establish proper data governance frameworks [18]. Big data analytics is a key enabler of Industry 4.0, empowering organizations to harness the full potential of data for improved decision-making, enhanced customer experiences, and innovation. By effectively analyzing and interpreting data, organizations can gain actionable insights, optimize operations, and stay ahead in the digital age [19]. Embracing big data analytics is vital for organizations to thrive and succeed in the era of Industry 4.0.

### 7.3.3 Utilizing Big Data for Predictive Maintenance and Optimization

Big data analytics for proactive upkeep and optimization is a powerful application in a variety of sectors. Organizations may analyze enormous amounts of data using sophisticated analytics and ML algorithms to proactively predict probable equipment problems, optimize maintenance schedules, and maximize operational efficiency. Monitoring equipment sensors, historical data, and real-time information to spot trends and abnormalities that suggest possible problems is what predictive maintenance entails [20]. Organizations may estimate when maintenance should be conducted by analyzing this data in order to avoid unplanned downtime, decrease maintenance costs, and extend the life of equipment. This proactive strategy reduces interruptions, boosts productivity, and increases overall operational efficiency.

Big data analytics also assists businesses in optimizing their procedures and operations. Organizations can uncover inefficiencies, bottlenecks, and opportunities for development by analyzing data from different sources, such as manufacturing lines, supply networks, and consumer feedback. This enables them to make data-driven decisions, streamline workflows, and enhance productivity. Optimization efforts can include improving inventory management, reducing energy consumption, optimizing production schedules, and enhancing product quality [21]. Furthermore, through the utilization of big data analytics, organizations can acquire valuable insights into customer behavior, preferences, and trends. This information can be used to personalize products and services, enhance customer experiences, and drive customer satisfaction and loyalty. By understanding customer needs and anticipating their demands, organizations can make strategic decisions and develop targeted marketing campaigns.

However, implementing predictive maintenance and optimization using big data comes with challenges. This necessitates organizations to gather, store, and process immense volumes of data from diverse origins. Additionally, ensuring data privacy, security, and compliance with regulations is crucial. Utilizing big data for predictive maintenance and optimization empowers organizations to proactively manage their assets, optimize processes, and enhance customer experiences [22]. By leveraging data analytics and ML, organizations can make informed decisions, minimize downtime, improve efficiency, and gain a competitive edge in their respective industries. Embracing big data for predictive maintenance and optimization is key to achieving operational excellence and driving business success.

## 7.4 Augmented Reality and Virtual Reality

AR and VR are transformative technologies. AR superimposes digital information onto the real world, whereas VR generates immersive virtual environments. They enhance experiences in gaming, education, training, and more. Limitations include cost and privacy concerns. However, these technologies have the potential to reshape interactions and create immersive experiences in various industries.

### 7.4.1 Introduction to AR and VR

#### 7.4.1.1 AR Architecture

AR is an innovative technology that enhances our perception and interaction with the surrounding world. It integrates virtual elements seamlessly with the physical environment, allowing users to see and interact with digital content in real time. AR has gained popularity across various fields, including gaming, education, health care, retail, and industrial applications. It offers a wide range of possibilities, such as interactive gaming experiences, immersive training simulations, informative educational content, and virtual product visualization [23].

AR technology is typically accessed through smartphones, tablets, or dedicated AR devices like smart glasses. These devices use cameras, sensors, and advanced software algorithms to track the user's position and align virtual objects with the real world. AR has the potential to bridge the divide between the physical and digital realms, opening up novel avenues for engagement, learning, and experiencing the surrounding world [24]. It enables users to overlay digital information onto real-world objects, navigate interactive maps, visualize data, and receive contextual information in real time. As AR continues to evolve and become more accessible, its impact on industries and everyday life is expected to grow. From enhancing entertainment and gaming experiences to revolutionizing training and education, AR has the power to transform how we interact with technology and reshape our perception of reality.

This simplified architecture in Figure 7.5 illustrates the basic components involved in an AR system, where the user's real-world view is augmented with digital content to create an enhanced and interactive experience [25]. AR software processes the sensor data from the device and combines it with digital content. The overlay content includes images, videos, 3D models, or other interactive elements. The arrows depict the flow of information, starting from the real-world view and passing through the AR device, software, and overlay content, ultimately enhancing the user's view.

### 7.4.1.2 VR Architecture

VR is an immersive technology that generates a simulated digital environment, providing users with the ability to interact with and explore the virtual realm. By wearing a VR headset, users are transported to a virtual world that can be visually and audibly realistic, offering a sense of presence and immersion. VR technology utilizes advanced displays, motion-tracking sensors, and interactive controllers to create a fully immersive experience. It tricks the user's senses into believing they are present in a virtual environment, enabling them to look around, move, and interact with objects or surroundings [26].

VR allows users to engage in virtual experiences that may not be possible in the physical world, such as exploring distant places, interacting with virtual objects, or participating in realistic training scenarios. The potential of VR lies in its ability to transport users to different virtual realms, providing a unique and immersive perspective [27]. It can stimulate emotions, create empathy, and deliver memorable experiences. Whether it is exploring fantasy worlds, learning in a virtual classroom, practicing surgical procedures, or collaborating with others in a virtual workspace, VR has the power to revolutionize how we engage with digital content and interact with each other.

As VR technology continues to advance and become more accessible, its applications and impact are expected to grow. VR holds the potential to revolutionize various industries, enhance training and education, enable remote collaboration, and introduce innovative forms of entertainment. VR opens up endless possibilities for creating rich, interactive, and engaging virtual experiences that can enhance our understanding of the world and reshape the way we work, learn, and play [28].

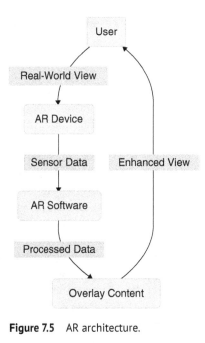

**Figure 7.5** AR architecture.

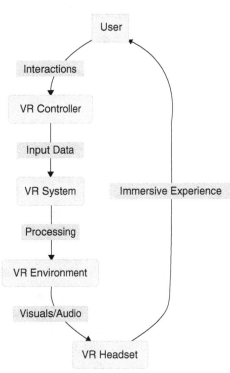

**Figure 7.6** VR architecture.

This simplified architecture shown in Figure 7.6 highlights the key components involved in a VR system. The user interacts with the VR controller, which sends input data to the VR system. The system processes the data and creates a virtual environment, which is rendered and presented to the user through the VR headset, providing an immersive experience. VR system processes the input data from the controller and manages the overall VR experience. VR environment includes the virtual world, objects, and interactions. VR headset delivers the immersive visual and audio experience to the user. The arrows depict the flow of interactions, input data, processing, and the final immersive experience.

### 7.4.2 Role of AR and VR in Industry 4.0

AR and VR play significant roles in Industry 4.0, revolutionizing the way businesses operate and interact with their environments. Here are some key roles of AR and VR in Industry 4.0:

- Training and Simulation: AR and VR technologies enable immersive and realistic training simulations for workers. They can experience hands-on training in a safe virtual environment, practicing complex procedures and scenarios without the need for physical equipment or risking accidents.
- Remote Assistance and Collaboration: AR and VR facilitate remote collaboration by allowing experts to provide real-time guidance and support to field technicians or workers. Remote assistance through AR enables experts to overlay instructions and information onto a worker's real-time view, enhancing efficiency and reducing downtime.
- Design and Prototyping: AR and VR enhance the design and prototyping processes by providing virtual models and simulations. Engineers and designers can visualize and manipulate 3D models in real time, allowing for faster iterations, improved design accuracy, and reduced costs.
- Maintenance and Repair: AR assists technicians in maintenance and repair tasks by overlaying digital information onto physical equipment. It provides step-by-step instructions, real-time data, and visual cues, helping technicians identify issues, perform repairs, and reduce equipment downtime.
- Product Visualization and Marketing: AR and VR technologies enable interactive product visualization and immersive marketing experiences. Customers can experience virtual product demonstrations, visualize products in their own environments, and make informed purchase decisions.
- Data Visualization and Analytics: AR and VR can be used to visualize and analyze large sets of data in a more intuitive and immersive manner. This helps decision-makers gain insights and identify patterns based on the data.

By incorporating AR and VR technologies, Industry 4.0 aims to enhance productivity, expand care, reduce costs, and drive innovation in various sectors, including manufacturing, logistics, health care, and more [29]. These technologies provide new ways of working, collaborating, and experiencing digital information, paving the way for more efficient and immersive industrial processes.

## 7.5 Blockchain

Blockchain is a distributed digital ledger that securely records and validates network transactions. It does away with the need for intermediaries and allows for peer-to-peer transactions. Blockchain ensures data integrity and security due to its transparency and immutability. It has the potential to transform industries beyond cryptocurrencies, offering transparency, trust, and efficiency. Blockchain technology is actively explored for its disruptive potential in various sectors.

### 7.5.1 Fundamentals of Blockchain Technology

Blockchain technology is built on several fundamental principles. First, it operates on a decentralized network of computers, eliminating the need for a central authority and distributing control among multiple nodes. This decentralization ensures transparency and prevents any single entity from having complete control over the system. Second, blockchain utilizes an immutable ledger, where transactions are recorded in blocks that are linked together in a sequential and irreversible manner. Once a block is appended to the blockchain, any modifications or deletions to its data become impossible, thereby preserving the integrity and unchangeability of the recorded information [30].

Cryptography plays a crucial role in blockchain, providing security and privacy. Public-key cryptography and hashing algorithms are used to verify transactions, authenticate users, and protect data from unauthorized access or tampering. Consensus mechanisms are employed to achieve agreement among network participants on the validity of transactions. These mechanisms ensure that all nodes reach a consensus on the state of the blockchain, maintaining its integrity and preventing fraud [31].

Smart contracts are another key component of blockchain technology. These self-executing contracts contain predefined rules and conditions, automatically enforcing the terms of an agreement. Smart contracts eliminate the need for intermediaries, streamline processes, and enhance efficiency. Lastly, blockchain offers a balance between transparency and privacy [32]. While all participants can view and verify transactions, the use of cryptographic techniques ensures that sensitive data remains secure and accessible only to authorized parties. These fundamental principles of blockchain form the foundation for its application across industries, promising secure, transparent, and efficient solutions for various use cases.

### 7.5.2 Architecture of Blockchain

The diagram in Figure 7.7 illustrates the interconnectedness of various components involved in blockchain, showcasing how blocks are linked to form a chain, how transaction data is associated with blocks, and how cryptographic elements and consensus mechanisms ensure the security and trustworthiness of the blockchain [33]. Nodes in the network play a crucial role in maintaining and propagating the blockchain across the network.

- **Blockchain:** The blockchain is a distributed and decentralized ledger comprised of blocks. Each block contains a set of transactions, and it is linked to the previous block through a cryptographic hash of its data, ensuring the immutability and integrity of the data.

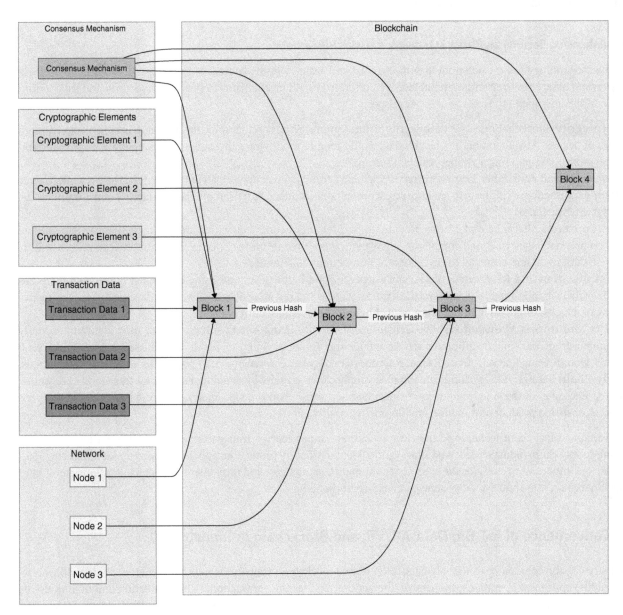

**Figure 7.7**   Architecture of blockchain.

- **Blocks:** Each block in the blockchain represents a collection of data, including transactional data and a reference to the previous block's hash value. Blocks are added to the blockchain in a sequential order, producing a chain of interconnected blocks.
- **Transaction Data:** Transaction data represents the information related to specific transactions that are recorded in the blockchain. It can include details such as sender and recipient addresses, transaction amounts, timestamps, and any other relevant data associated with the transaction.
- **Cryptographic Elements:** Cryptographic elements are essential for ensuring the security and integrity of the blockchain. They include cryptographic functions like digital signatures and hash functions. Digital signatures provide authentication and non-repudiation of transactions, while hash functions generate unique hash values for data, allowing for verification and protection against tampering.
- **Consensus Mechanism:** The consensus mechanism is in charge of obtaining agreement among numerous blockchain network members on the authenticity and sequencing of transactions. It guarantees that all nodes in the network agree on the blockchain's current state. Proof of work and proof of stake are two common consensus procedures.
- **Nodes:** Participants or entities in the blockchain network are represented by nodes. Each node keeps a copy of the blockchain and participates in transaction validation and verification. Nodes communicate with one another in order to disseminate transactions, reach consensus, and preserve the blockchain's integrity.

### 7.5.3 Enhancing Security and Trust in Industry 4.0 with Blockchain

Blockchain technology has the potential to enhance trust and security in the context of Industry 4.0. By its nature, blockchain provides a decentralized and immutable ledger, offering several mechanisms to strengthen security and trust within the industrial landscape [34]. Here are some key aspects:

- **Data Integrity:** Blockchain ensures the integrity of data by employing cryptographic hashing and linking blocks through a chain of hashes. Any alteration in a block's data would require modifying subsequent blocks, making it computationally infeasible and providing a tamper-resistant system.
- **Transparent and Auditable Transactions:** Blockchain's transparency enables all participants in the network to view and verify transactions. This feature enhances trust among stakeholders, as it promotes accountability and eliminates the need for intermediaries.
- **Smart Contracts:** Blockchain platforms provide support for smart contracts, which are contracts that can self-execute based on preestablished rules and conditions. These contracts automate and enforce agreements between parties, reducing the risk of fraud and ensuring the execution of transactions as intended.
- **Decentralization and Resilience:** Blockchain's decentralized nature eliminates single points of failure and increases system resilience. In Industry 4.0, where critical infrastructure and sensitive data are involved, decentralization mitigates the risk of attacks, improves fault tolerance, and enhances the overall security posture.
- **Identity and Access Management:** Blockchain-based identity management solutions store user identities securely and tamper-proof, limiting unauthorized access, and identity fraud. By doing so, it enhances the security of Industry 4.0 systems, guaranteeing that sensitive data and resources can only be accessed and manipulated by authorized individuals.
- **Supply Chain Security:** Blockchain can enhance supply chain security by enabling end-to-end traceability and transparency. Each step of the supply chain can be recorded and verified on the blockchain, reducing the risk of counterfeit products, ensuring product provenance, and improving quality control.

By leveraging blockchain technology, Industry 4.0 can enhance security, transparency, and trust in various domains, including supply chain management, data sharing, intellectual property protection, and decentralized manufacturing [35]. However, it is important to consider the specific requirements, challenges, and implementation considerations when integrating blockchain into Industry 4.0 systems to maximize its benefits.

## 7.6 Convergence of IoT, Big Data, AR/VR, and Blockchain in Industry 4.0

In Industry 4.0, the confluence of IoT, big data, AR/VR, and blockchain transforms industrial operations. It enables real-time insights, optimized operations, immersive interactions, and secure transactions. This powerful combination drives innovation, efficiency, and transformation, offering new opportunities for businesses. It improves the industrial landscape by leveraging connected devices, data analysis, immersive experiences, and decentralized trust mechanisms.

### 7.6.1 Interplay and Integration of Technologies

The interplay and integration of IoT, big data, AR/VR, and blockchain technologies in Industry 4.0 create a highly dynamic and interconnected ecosystem. IoT devices form the foundation by collecting vast amounts of data from various sources and transmitting it over networks. AR/VR technologies enhance user experiences by providing immersive and interactive interfaces. They allow users to visualize complex datasets, interact with virtual objects, and simulate real-world scenarios. This integration enables better understanding and interpretation of data, leading to improved problem-solving and decision-making capabilities. Blockchain technology acts as a secure and transparent ledger, facilitating trust and eliminating the need for intermediaries. It ensures the integrity and immutability of data, secures transactions, and enables decentralized consensus. With blockchain, organizations can establish trust among parties, enhance data sharing, and create tamper-proof audit trails.

The interplay and integration of these technologies enable Industry 4.0 to achieve several benefits. Real-time data-driven decision-making enables proactive and efficient operations. Improved operational efficiency leads to cost savings, optimized resource utilization, and streamlined processes. Enhanced user experiences foster innovation, creativity, and collaboration. The secure and transparent nature of blockchain builds trust among stakeholders, enabling secure transactions and data sharing. Moreover, the interplay of these technologies opens up new opportunities for business models and revenue streams. It enables predictive maintenance, supply chain optimization, personalized marketing, remote monitoring, and virtual collaboration. These advancements drive digital transformation, revolutionizing industries across sectors such as manufacturing, health care, logistics, and finance.

Figure 7.8 showcases the interconnections between the components, such as the flow of data from IoT devices to the big data platform, the flow of insights from the big data platform to the AR/VR interface, and the flow of interactions from the AR/VR interface to the blockchain network. In Figure 7.8, "IoT devices" represent the sensors and devices that collect data. The data is then processed and analyzed by the "big data platform" to derive insights. The insights are further visualized and interacted with through the "AR/VR interface." The "blockchain network" ensures security and trust in data transactions and sharing.

### 7.6.2 Use Cases and Examples of Combined Implementations

The combined application of IoT, big data, AR/VR, and blockchain in Industry 4.0 provides a diverse set of use cases and examples from a variety of industries. Here are a couple of such instances. These examples show how integrating IoT, big data, AR/VR, and blockchain technologies in Industry 4.0 can drive innovation, increase operational efficiency, improve user experiences, and allow new business models across several sectors.

- Smart Manufacturing: IoT devices embedded in machines and production lines collect real-time data on performance, maintenance needs, and energy consumption. Big data analytics enable predictive maintenance, optimizing equipment utilization, and reducing downtime. AR/VR technologies provide virtual training and remote collaboration, enhancing worker productivity and efficiency. Figure 7.9 illustrates the components involved in smart manufacturing.

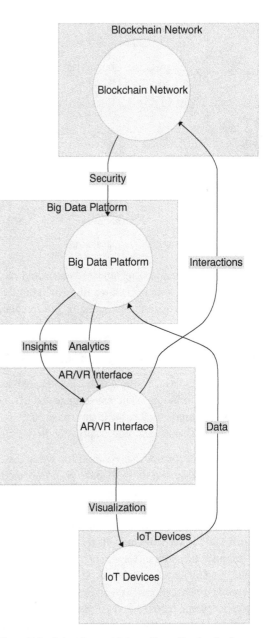

**Figure 7.8** Interplay and integration of technologies.

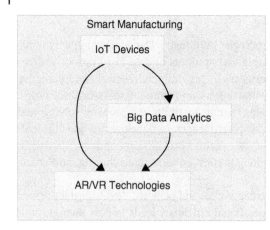

**Figure 7.9** Smart manufacturing.

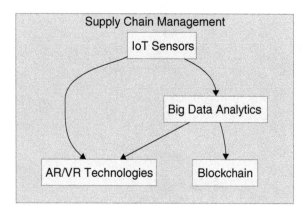

**Figure 7.10** Supply chain management.

- Supply Chain Management: IoT sensors attached to inventory, containers, and vehicles enable real-time tracking and monitoring of goods. Big data analytics analyze the supply chain data to identify bottlenecks, optimize routes, and improve demand forecasting. AR/VR can be used for virtual simulations and visualizing warehouse layouts. Blockchain ensures transparent and secure transactions, verifying the authenticity and traceability of goods. Figure 7.10 depicts the components involved in supply chain management.
- Health Care: IoT devices such as wearables and medical sensors collect patient data, enabling remote monitoring and personalized health care. Big data analytics process the vast amount of health data, facilitating early disease detection and personalized treatment plans. AR/VR technologies assist in surgical training, medical education, and telemedicine applications. Blockchain ensures the security and privacy of patient records and facilitates secure sharing among health-care providers.
- Smart Cities: IoT sensors deployed in urban environments monitor air quality, traffic flow, waste management, and energy consumption. Big data analytics analyze the collected data to optimize resource allocation and improve city planning. AR/VR technologies can provide virtual tours, interactive city planning, and urban simulations. Blockchain can enable secure and transparent transactions for smart contracts and peer-to-peer energy trading.
- Finance and Banking: IoT devices and wearables enable personalized financial services, such as real-time payment notifications and fraud detection. Big data analytics analyze customer data for personalized recommendations and risk assessments. AR/VR technologies can enhance virtual banking experiences and provide immersive financial simulations. Blockchain ensures secure and transparent transactions, enabling cross-border payments and reducing intermediaries.

### 7.6.3 Benefits and Synergies of Technology Convergence

The convergence of multiple technologies, such as big data, IoT, blockchain, and AR/VR in Industry 4.0, brings numerous benefits and synergies. The convergence of IoT, big data, AR/VR, and blockchain offers a multitude of benefits, ranging from operational efficiency and improved customer experiences to enhanced security and sustainability. By harnessing the synergies of these technologies, organizations can unlock new possibilities and thrive in the era of Industry 4.0.

- Enhanced Efficiency: By combining these technologies, organizations can gather, analyze, and visualize data in real time, empowering them to make well-informed decisions driven by data. This leads to increased operational efficiency and optimized resource allocation.
- Improved Productivity: The convergence of technology facilitates automation, process optimization, and predictive maintenance, resulting in enhanced productivity across diverse industries. IoT devices, coupled with big data analytics, can provide valuable insights for streamlining operations and identifying areas for improvement.
- Enhanced Customer Experience: AR/VR technologies can revolutionize customer interactions by providing immersive and personalized experiences. From virtual product demonstrations to interactive training sessions, these technologies enable organizations to engage customers in new and exciting ways.
- Greater Transparency and Traceability: Blockchain technology offers a decentralized and unchangeable record that guarantees transparency and confidence in data transactions. This is particularly beneficial in supply chain management, where stakeholders can track and verify the origin, authenticity, and movement of products or goods.

- Enhanced Security and Privacy: The convergence of technologies protects sensitive data from various security threats. Blockchain's decentralized nature and cryptographic algorithms ensure secure and tamper-proof data transactions, while IoT devices can implement secure protocols for data transmission.
- Innovation and New Business Opportunities: The convergence of these technologies creates opportunities for innovation and the development of new business models. To remain competitive in the digital world, organizations might explore new income streams, offer data-driven services, and use developing technology.
- Sustainable Practices: Technology convergence can contribute to sustainability efforts by enabling smart energy management, resource optimization, and waste reduction. IoT devices can monitor and control energy consumption, while big data analytics can identify patterns for more sustainable practices.
- Collaborative Ecosystems: The integration of these technologies encourages collaboration and partnerships among industry players. Organizations can share data, insights, and resources to foster innovation and drive collective growth.

## 7.7 Conclusion

Finally, Industry 4.0 marks a dramatic transformation in how industrial activities are carried out. Industry 4.0 allows a smart and connected ecosystem that boosts production, lowers costs, and improves overall efficiency by connecting various technologies such as IoT, big data, AR and VR, and blockchain. As this technology advances, it is anticipated to have a significant influence on how industries function, resulting in new business models and chances for development.

The interplay and integration of these technologies in Industry 4.0 create a transformative ecosystem. It enhances efficiency, productivity, and innovation while improving customer experiences and enabling sustainable practices. The convergence of IoT, big data, AR/VR, and blockchain also fosters collaboration among stakeholders, driving collective growth and opening new business opportunities.

However, the adoption of these technologies also presents challenges, including data security, privacy concerns, workforce transformation, and infrastructure requirements. It is essential for organizations to address these challenges through robust cybersecurity measures, ethical considerations, and investment in infrastructure and talent development. Overall, Industry 4.0, with its linkage to IoT, big data, AR/VR, and blockchain, represents a paradigm shift in how industries operate, innovate, and thrive in the digital era. Embracing these technologies and leveraging their synergies will pave the way for enhanced productivity, sustainable practices, and competitive advantage in a rapidly evolving global landscape.

## References

1 Jan, Z., Ahamed, F., Mayer, W. et al. (2023). Artificial intelligence for Industry 4.0: systematic review of applications, challenges, and opportunities. *Expert Systems with Applications* 216: 119456. https://doi.org/10.1016/J.ESWA.2022.119456.

2 Bécue, A., Praça, I., and Gama, J. (2021). Artificial intelligence, cyber-threats and Industry 4.0: challenges and opportunities. *Artificial Intelligence Review* 54 (5): 3849–3886. https://doi.org/10.1007/S10462-020-09942-2.

3 Yao, X., Zhou, J., Zhang, J., and Boer, C.R. (2017). From intelligent manufacturing to smart manufacturing for industry 4.0 driven by next generation artificial intelligence and further on. *Proceedings - 2017 5th International Conference on Enterprise Systems: Industrial Digitalization by Enterprise Systems, ES 2017*, 311–318. http://dx.doi.org/10.1109/ES.2017.58.

4 Mithas, S., Chen, Z.L., Saldanha, T.J.V., and Silveira, A.D.O. (2022). How will artificial intelligence and Industry 4.0 emerging technologies transform operations management? *Production and Operations Management* 31 (12): 4475–4487. https://doi.org/10.1111/POMS.13864.

5 Ahmad, T., Zhu, H., Zhang, D. et al. (2022). Energetics systems and artificial intelligence: applications of Industry 4.0. *Energy Reports* 8: 334–361. https://doi.org/10.1016/J.EGYR.2021.11.256.

6 Radanliev, P., De Roure, D., Nicolescu, R. et al. (2021). Artificial intelligence and the Internet of Things in Industry 4.0. *CCF Transactions on Pervasive Computing and Interaction* 3 (3): 329–338. https://doi.org/10.1007/S42486-021-00057-3/FIGURES/7.

7 Sood, S.K., Rawat, K.S., and Kumar, D. (2022). A visual review of artificial intelligence and Industry 4.0 in healthcare. *Computers and Electrical Engineering* 101: 107948. https://doi.org/10.1016/J.COMPELECENG.2022.107948.

8 Ahmed, I., Jeon, G., and Piccialli, F. (2022). From artificial intelligence to explainable artificial intelligence in industry 4.0: a survey on what, how, and where. *IEEE Transactions on Industrial Informatics* 18 (8): 5031–5042. https://doi.org/10.1109/TII.2022.3146552.

**9** Malik, P.K., Sharma, R., Singh, R. et al. (2021). Industrial Internet of Things and its applications in industry 4.0: state of the art. *Computer Communications* 166: 125–139. https://doi.org/10.1016/J.COMCOM.2020.11.016.

**10** Okano, M.T. and Okano, M.T. (2017). *International Conference on Management and Information Systems IOT and Industry 4.0: The Industrial New Revolution.*

**11** Khan, I.H. and Javaid, M. (2021). Role of Internet of Things (IoT) in adoption of Industry 4.0. *Journal of Industrial Integration and Management* 7 (4): 515–533. https://doi.org/10.1142/S2424862221500068.

**12** Manavalan, E. and Jayakrishna, K. (2019). A review of Internet of Things (IoT) embedded sustainable supply chain for Industry 4.0 requirements. *Computers & Industrial Engineering* 127: 925–953. https://doi.org/10.1016/J.CIE.2018.11.030.

**13** Aheleroff, S., Xun, X., Yuqian, L. et al. (2020). IoT-enabled smart appliances under Industry 4.0: a case study. *Advanced Engineering Informatics* 43: 101043. https://doi.org/10.1016/J.AEI.2020.101043.

**14** Routray, S.K., Sharmila, K.P., Javali, A. et al. (2020). An Outlook of Narrowband IoT for Industry 4.0. *Proceedings of the 2nd International Conference on Inventive Research in Computing Applications,* ICIRCA 2020, 923–926. http://dx.doi.org/10.1109/ICIRCA48905.2020.9182803.

**15** Zhang, C. and Chen, Y. (2020). A review of research relevant to the emerging industry trends: Industry 4.0, IoT, blockchain, and business analytics. *Journal of Industrial Integration and Management* 5 (1): 165–180. https://doi.org/10.1142/S2424862219500192.

**16** Rezazadegan, R. and Sharifzadeh, M. (2022). Applications of Artificial Intelligence and Big Data in Industry 4.0 Technologies. *Industry 4.0 Vision for the Supply of Energy and Materials:* Enabling Technologies and Emerging Applications, 121–158. http://dx.doi.org/10.1002/9781119695868.CH5.

**17** Javaid, M., Haleem, A., Singh, R.P., and Suman, R. (2021). Significant applications of big data in Industry 4.0. *Journal of Industrial Integration and Management* 6 (4): 429–447. https://doi.org/10.1142/S2424862221500135.

**18** Atharvan, G., Krishnamoorthy, S.K.M., Dua, A., and Gupta, S. (2022). A way forward towards a technology-driven development of Industry 4.0 using big data analytics in 5G-enabled IIoT. *International Journal of Communication Systems* 35 (1): e5014. https://doi.org/10.1002/DAC.5014.

**19** Obitko, M. and Jirkovský, V. (2015, 2015). Big data semantics in industry 4.0. *Lecture Notes in Computer Science* (including subseries Lecture Notes in Artificial Intelligence and Lecture Notes in Bioinformatics) 9266, 217: –229. https://doi.org/10.0.3.239/978-3-319-22867-9_19/COVER.

**20** Yan, J., Meng, Y., Lei, L., and Li, L. (2017). Industrial big data in an Industry 4.0 environment: challenges, schemes, and applications for predictive maintenance. *IEEE Access* 5: 23484–23491. https://doi.org/10.1109/ACCESS.2017.2765544.

**21** Da Xu, L. and Duan, L. (2018). Big data for cyber physical systems in Industry 4.0: a survey. *Enterprise Information Systems* 13 (2): 148–169. https://doi.org/10.1080/17517575.2018.1442934.

**22** Karatas, M., Eriskin, L., Deveci, M. et al. (2022). Big data for healthcare Industry 4.0: applications, challenges and future perspectives. *Expert Systems with Applications* 200: 116912. https://doi.org/10.1016/J.ESWA.2022.116912.

**23** Machala, S., Chamier-Gliszczynski, N., and Królikowski, T. (2022). Application of AR/VR Technology in Industry 4.0. *Procedia Computer Science* 207: 2990–2998. https://doi.org/10.1016/J.PROCS.2022.09.357.

**24** Moraes, E.B., Kipper, L.M., Kellermann, A.C.H. et al. (2023). Integration of Industry 4.0 technologies with Education 4.0: advantages for improvements in learning. *Interactive Technology and Smart Education* 20 (2): 271–287. https://doi.org/10.1108/ITSE-11-2021-0201/FULL/XML.

**25** Eswaran, M. and Raju Bahubalendruni, M.V.A. (2022). Challenges and opportunities on AR/VR technologies for manufacturing systems in the context of Industry 4.0: a state of the art review. *Journal of Manufacturing Systems* 65: 260–278. https://doi.org/10.1016/J.JMSY.2022.09.016.

**26** Gunal, M.M. (ed.) (2019). *Simulation for Industry 4.0.* Springer https://doi.org/10.1007/978-3-030-04137-3.

**27** Gunal, M.M. and Karatas, M. (2019). *Industry 4.0, Digitisation in Manufacturing, and Simulation: A Review of the Literature,* 19–37. *Springer Series in Advanced Manufacturing* https://doi.org/10.0.3.239/978-3-030-04137-3_2/COVER.

**28** Jamwal, A., Agrawal, R., Sharma, M., and Giallanza, A. (2021). Industry 4.0 technologies for manufacturing sustainability: a systematic review and future research directions. *Applied Sciences* 11 (12): 5725. https://doi.org/10.3390/APP11125725.

**29** Mohammad Shahin, F., Chen, F., Bouzary, H., and Krishnaiyer, K. (2020). Integration of lean practices and Industry 4.0 technologies: smart manufacturing for next-generation enterprises. *International Journal of Advanced Manufacturing Technology* 107 (5–6): 2927–2936. https://link.springer.com/article/10.1007/s00170-020-05124-0.

**30** Zuo, Y. (2020). Making smart manufacturing smarter – a survey on blockchain technology in Industry 4.0. *Enterprise Information Systems* 15 (10): 1323–1353. https://doi.org/10.1080/17517575.2020.1856425.

**31** Bodkhe, U., Tanwar, S., Parekh, K. et al. (2020). Blockchain for Industry 4.0: a comprehensive review. *IEEE Access* 8: 79764–79800. https://doi.org/10.1109/ACCESS.2020.2988579.

**32** Javaid, M., Haleem, A., Singh, R.P. et al. (2021). Blockchain technology applications for Industry 4.0: a literature-based review. *Blockchain: Research and Applications* 2 (4): 100027. https://doi.org/10.1016/J.BCRA.2021.100027.

**33** Choi, T.M., Kumar, S., Yue, X., and Chan, H.L. (2022). Disruptive technologies and operations management in the Industry 4.0 era and beyond. *Production and Operations Management* 31 (1): 9–31. https://doi.org/10.1111/POMS.13622.

**34** Khan, S.A.R., Razzaq, A., Zhang, Y., and Miller, S. (2021). Industry 4.0 and circular economy practices: a new era business strategies for environmental sustainability. *Business Strategy and the Environment* 30 (8): 4001–4014. https://doi.org/10.1002/BSE.2853.

**35** Singh, S.K., Sharma, S.K., Singla, D., and Gill, S.S. (2022). *Evolving Requirements and Application of SDN and IoT in the Context of Industry 4.0, Blockchain and Artificial Intelligence*, 427–496. *Software Defined* Networks http://dx.doi.org/10.1002/9781119857921.CH13.

# 8

# Industry 4.0 in Manufacturing, Communication, Transportation, and Health Care

*Mani D. Choudhry[1], Jeevanandham Sivaraj[2], Sundarrajan Munusamy[3], Parimala D. Muthusamy[4], and V. Saravanan[5]*

[1] Department of Information Technology, KGiSL Institute of Technology, Coimbatore, Tamil Nadu, India
[2] Department of Information Technology, Sri Ramakrishna Engineering College, Coimbatore, Tamil Nadu, India
[3] Department of Networking and Communications, SRM Institute of Science & Technology, Chennai, Tamil Nadu, India
[4] Department of Electronics and Communication Engineering, Velalar College of Engineering and Technology, Erode, Tamil Nadu, India
[5] Department of Computer Science, College of Engineering and Technology, Dambi Dollo University, Dambi Dollo, Oromia Region, Ethiopia

## 8.1  Introduction

The fourth generation of the industrial revolution, or Industry 4.0 (IR4.0 or I4.0), is the next step in corporate transformation. It is propelled by disruptive breakthroughs like increased data and connectivity, analytics, human-machine interaction, and robotics improvements. Based on a wider view of Industry 4.0, there comes the rise of digitalized industrial methodologies. IR4.0 revolutions enable humans to collaborate with machinery through creative and innovative approaches. Initiatives related to IR4.0 today also seek to foster mutually beneficial partnerships between humans and technological advances [1]. Your workforce and bottom line benefit when the reliability and rapidity of 4.0 technologies are combined with the inventiveness, ability, and creative thinking of your employees. Your manufacturing activities evolve more effectively and successfully, and team members are freed up from many tedious and monotonous obligations, providing them with more time to interact with smart technologies and better prepare themselves for the changing nature of technology and artificial intelligence (AI)-powered future of work. Figure 8.1 depicts the IR4.0 sectors.

Since the beginning of the industrial revolution, manufacturing has undergone several changes from marine and engine equipment to power and digitized machine manufacturing, making the production process more complex, intuitive, and long-lasting so that people can operate the machinery competently, effectively, and consistently [2]. The IR4.0 is characterized as an inventive degree of group and taking control of every part of an item's lifecycle aimed at more individualized customer expectations [3]. The vital goal of IR4.0 is satisfying unique customer needs, which have an impact on directive management, R&D, workshop contracting, conveyance, product utilization, and reprocessing [4]. In many ways, the focus on human interaction in the industrial environment separates IR4.0 from system unified industry (SUI). In contrast to SUI, which envisages workerless manufacturing, Industry 4.0 places a high value on the contribution of human workers to production [5]. The IR4.0 paradigm inspires the connectivity of entities like sensors, gadgets, and business possessions with the Internet to one [6].

All strategy and conscripting methods should be evaluated to determine their applicability to a contemporary, multidimensional model of product creation. If they are found to be applicable, they should then be translated into a common, integrated, and multidimensional set of methodologies, procedures, and IT solutions [7]. Breaking down the manufacturing process into small, focused units that solely exchange information regarding subsequent phases of the process contributes to enhanced adaptability and potentially results in a decrease in coordination complexities [8]. Figure 8.2 depicts the progression of IR as IR5.0, which is the current trend, but we focus on IR4.0 in this chapter.

*Topics in Artificial Intelligence Applied to Industry 4.0*, First Edition. Edited by Mahmoud Ragab AL-Refaey, Amit Kumar Tyagi, Abdullah Saad AL-Malaise AL-Ghamdi, and Swetta Kukreja.
© 2024 John Wiley & Sons Ltd. Published 2024 by John Wiley & Sons Ltd.

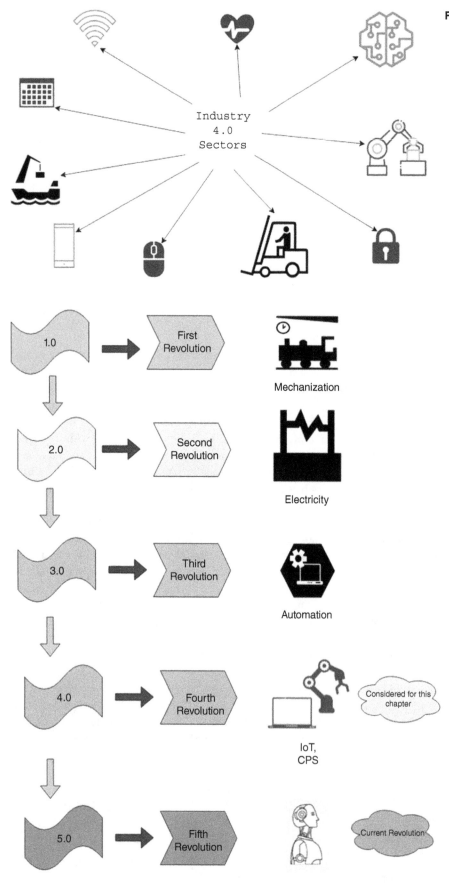

**Figure 8.1** Sectors of IR4.0.

**Figure 8.2** Industrial evolution era.

### 8.1.1 Technological Trends of Industry 4.0

To improve operational excellence and preventive management through a comprehensive understanding of their environment, traditional machines need to be transformed into self-aware, self-learning machines, commonly referred to as "Industry 4.0" machines [9]. The objective of Industry 4.0 is to establish a transparent and intelligent manufacturing platform that facilitates the seamless flow of information across industrial networks [10]. Key necessities of Industry 4.0 encompass continuous monitoring of real-time data, tracking the status and location of products, and maintaining instructions to regulate manufacturing operations [11].

The driving forces of Industry 4.0, namely the Internet of Things (IoT), Industrial Internet of Things (IIoT), cloud-based manufacturing, and smart manufacturing, play vital roles in the complete digitization and intelligent evolution of the manufacturing process [12]. The integration of the nine Industry 4.0 pillars will revolutionize the manufacturing of segregated and enhanced cells, leading to a seamless, automated, and optimized production workflow. Consequently, the conventional production connections among suppliers, manufacturers, and customers, as well as those between individuals and machines, will transform, resulting in enhanced efficiency. In recent years, digitalization has altered company processes. Digitalization includes, for example, the adoption and utilization of Industry 4.0 technologies or smart manufacturing across the industrial sector. The I4.0 concept refers to a group of technological innovations and ways of organizing the value/supply chain that encompasses innovative and intelligent products as well as manufacturing techniques, building an atmosphere for production whereby everyone involved collaborates and exchanges real-time information with one another [13]. The advent of the I4.0 idea illustrates adequately that digital technology development and acceptance have reached a critical juncture in the industrial environment, enabling the connecting and mixing of virtual and physical production worlds [14].

The content evaluation found that the technical breakthroughs of I4.0 could be categorized into two, which the research referred to as "Basic and Helping Technology Groups." The modern technological advancements that have been in development for several decades but have only recently been mature enough to be made commercially available are the core technologies of Industry 4.0. The fundamental I4.0 technologies discovered by content analysis of relevant publications are presented in Figure 8.3 along with the frequency with which they occur.

The majority of approaches are fundamental, as shown in Figure 8.3, which are virtual reality , augmented reality (AR), cloud computing (CC), and additive manufacturing; they have been thoroughly explored and investigated in the literature [15, 16]. Various papers on AR, for example, have addressed industrial applications, explored industrial application motives and challenges, and analyzed advantages in real-world industrial settings [17]. Among the 18 foundational technologies of Industry 4.0, IIoT, cyber-physical system (CPS), and digital twin (DT) technologies are regarded as advanced technologies.

**Figure 8.3** Fundamental approaches of I4.0.

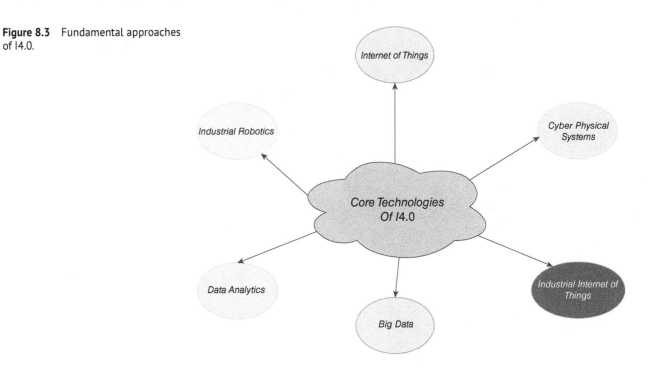

The literature presents the concept of IIoT from two distinct perspectives, where it generally refers to the utilization of IoT technology in industrial environments beyond the consumer sector [18]. However, a more pragmatic perspective views IIoT as a distinct vision for the industrial sector, with unique functionalities and diverse underlying objectives. In this regard, IIoT encompasses a range of activities, including data processing, real-time communication, and decentralized decision-making by intelligent components, all aimed at enhancing process monitoring and operational efficiency. From this standpoint, IIoT and CPS are closely intertwined and mutually influential. When referring to IIoT in the context of Industry 4.0, the concept of CPS often follows suit and vice versa. The graphical representation of the IIoT-CPS connection is depicted in Figure 8.4.

The enabling technologies of Industry 4.0 play a crucial role in facilitating the core technologies to fulfill their intended functions. These well-established and widely used information and operations technologies, as documented in recognized articles, are listed in Figure 8.5. Despite their commercial availability and extensive industry application over the past few decades, the literature often refers to them as "smart" or "intelligent" technologies. This signifies the need for fundamental redesigns to ensure the desired level of integrability. For instance, the implementation of an intelligent enterprise resource planning (ERP) system entails its cloud-based architecture, secure communication through real-time application programming interfaces, and integration with the IIoT, Internet of People (IoP), and Internet of Service (IoS).

The extent to which the digital transition may help environmental objectives is directly related to how I4.0 is utilized with the help of its basic strategies. In actuality, several I4.0 technologies provide a broader impact on socioeconomic sustainability.

A vital prospect is I4.0, which proves tremendous outcomes in a variety of sectors while fundamentally transforming organizations in a variety of ways. Changes begin with the design of business models and continue through every stage of the manufacturing process until the customer gets the items. Moreover, due to the direct or indirect connections of multiple information hubs with I4.0, it becomes crucial to establish a shared comprehension of the concepts or ontology involved. Comprehensive literature

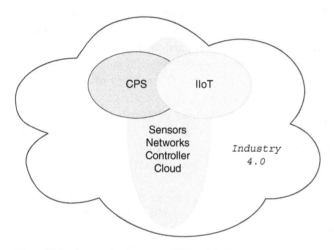

**Figure 8.4** Interaction between CPS and IIoT.

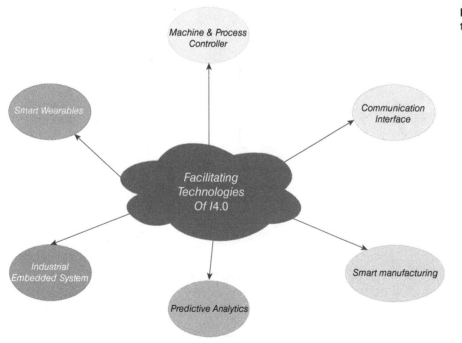

**Figure 8.5** Industry 4.0 facilitating technologies.

**Table 8.1** Mutual features of I4.0.

| S.no. | Common characteristics |
| --- | --- |
| 1 | Low cost |
| 2 | Reliability/transparency |
| 3 | Decentralized or self-governed judgments |
| 4 | Lowering the delivery period and time necessary for processing |
| 5 | Enhancing the standard |
| 6 | Intensifying output |
| 7 | Increased administration of resources |
| 8 | Continuous monitoring of stocks |

research was also conducted to identify all I4.0-related components. Issa et al. [19] were the first to describe the nine pillars of I4.0. According to Issa et al. [19], all technologies may be used independently, but only their combination can enhance and change traditional manufacturing practices. Characteristics involve the attribution of a description to identify a term, place, or any other entity of qualities connected to the topic. Table 8.1 summarizes the common characteristics connected with the I4.0 concept based on all instances found in the literature study. The following is a more thorough breakdown of the primary nine pillars:

1) **Big Data:** The utilization of the abundant data produced by interconnected devices, systems, and processes in I4.0 aims to extract valuable insights, enable data-driven decision-making, and drive improvements in operational efficiency, productivity, and overall performance.
2) **Autonomous Robots:** The integration of autonomous robots in I4.0 aims to optimize operational efficiency and increase productivity through the automation of repetitive tasks, minimizing human involvement and facilitating uninterrupted operations across diverse manufacturing and production processes.
3) **Simulation:** To improve decision-making and enhance operational outcomes by developing virtual models that accurately simulate real-world manufacturing systems.
4) **Additive Manufacturing:** To transform conventional manufacturing methods by facilitating the creation of intricate and personalized components; it is minimizing material waste, optimizing supply chain operations, and fostering innovation through the utilization of rapid prototyping and on-demand manufacturing capabilities.
5) **Horizontal and Vertical Integration:** To establish seamless connectivity and collaboration across different levels of the manufacturing ecosystem.
6) **IoT:** To enable the seamless connection and communication of physical devices, machines, and systems, thereby facilitating real-time data collection, analysis, and decision-making.
7) **Cloud Computing:** To provide a scalable, flexible, and secure platform for storing, processing, and accessing large volumes of data generated by interconnected devices, systems, and processes.
8) **Cybersecurity:** To ensure the protection, integrity, and confidentiality of data, systems, and networks within the interconnected manufacturing environment.
9) **Augmented Reality:** To enhance operational efficiency and productivity by leveraging AR technologies to provide real-time contextual information, guidance, and visualization to workers in the manufacturing environment.

Section 8.2 discusses the applications along with other characteristics of I4.0 in comparison to earlier technical epochs, as well as the empowering expertise for I4.0. In the last section, we draw conclusions based on the evidence presented.

## 8.2 Diversified Applications of Industry 4.0

### 8.2.1 Background Analysis

Despite the concept not being novel and having undergone extensive academic discourse from various perspectives over numerous years, the idea has persisted and continues to be a subject of discussion and analysis within scholarly circles, the term "I4.0" was just recently coined and has achieved significant recognition in both the academic and industrial communities. The industry focuses on the evolution of commercial equipment accommodations and intelligent products, as well as potential customers, on this progress, whereas academic research focuses on comprehending and establishing the concept and striving to create associated technologies, business approaches, and appropriate strategies. I4.0 refers to the process of transitioning from machine-dominated production to digital manufacturing. For a successful transition, a solid grasp of the I4.0 standard, as well as the formulation and execution of a clear road plan is essential. There have been several debates and strategies used to build road maps, some of which are discussed in this chapter. It is vital to evaluate I4.0 components and their related qualities to provide the groundwork for a particular emerging landscape of manufacturing. Yet the research is indisputably revealing the absence of appraisal and analysis processes. Because the deployment and applicability of associated theorems and enumerated descriptions for I4.0 are insufficiently sophisticated for the vast majority of real-life executions, an organized strategy for conducting appropriate evaluations and assessments appears to be essential for those who will seek to accelerate this change. To make researchers' jobs easier, it is now the major responsibility of research groups to provide the technical groundwork with local entities, business models, organizational structures, and precisely defined I4.0 scenarios. According to analysts, the advancement of I4.0 and its related approaches provides a substantial influence on social existence. This would drive industrial society to enhance the manufacturing process of its own to fulfill client needs and conserve potential advantage. A survey conducted by World Economic Forum (WEF) based on the views of 800 professionals results from a great collection of suggestions and findings addressing digital transformation. According to the report by 2018, 2.4 million robots will be engaged in the industry. This shift is paving the way for implantable technology, wearable Internet, collaborating and coordinating gadgets, autonomous problem solvers, and self-decision-making board members. 3D printing has advanced beyond expectations to the point that it can now make goods that are used daily. They are even used in the creation of synthetic parts. It is expected that 1 trillion sensors will be utilized in a daily span by 2025.

Throughout history, there have been several industrial and accompanying social developments. Manufacturing efforts in the fields of research, innovation, yield, and orchestrating the operation of sophisticated industrial systems have been undertaken using cutting-edge production technology of the time. Society recognized and accepted the shift from rural to technocracy (I1.0) and then I2.0 to I3.0. To avoid irreversible changes, a like-for-like examination is done to grasp the transition from I3.0 to I4.0. A lot of features of this change have social consequences. IoT is one of them. Machine-to-machine (M2M) communication is possible using this technology. This capability results in a less-crowded industrial environment. "Autonomy" is the second most important driving factor behind these advancements. The systems are becoming more self-sufficient. CPS and certain sensors are important components of this revolution as well. They allow straightforward M2M communication. The combination of autonomy, CPS, M2M, and IoT industrial systems has become more dependable, robust, nimble, and intelligent. This very certainly enhances the impetus to construct a factory of smartness. Also, a key advancement in this transition is the capacity of machines to converse with operations done by people. This needs a manufacturing evolution philosophy, resulting in a new manufacturing vision based on four essential ideas: intelligence, products, communication, and information networks.

Manufacturing fourth stage and IT, known as "I4.0," is one of the cutting-edge research disciplines in the last five years. For the first time in industrial revolution history, a bias is expected to actively shape the destiny of researchers and corporations in this subject. Figure 8.6 depicts the fundamental visions of the I4.0 systems that were described.

From a distinct angle, Figure 8.7 describes an architecture that focuses on CPS, IoT, virtualization, versatility, real-time operation, and service compatibility.

Another objective of I4.0 is intelligent factories. The construction of intelligent goods may emerge from intelligent input fed to intelligent machines. Application platforms and M2M connections are required for CPS. A single secure cloud network might be used to start and run several businesses. According to numerous studies, I4.0 has made not only industries but also government. Similar events have occurred all across the globe, and some of them have been especially focused on the idea of I4.0.

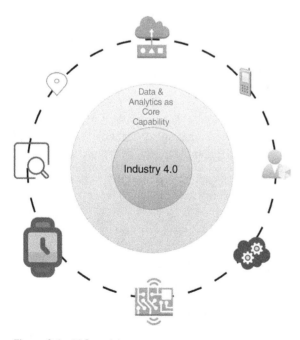

**Figure 8.6** I4.0 envisions.

**Figure 8.7** System architecture of I4.0.

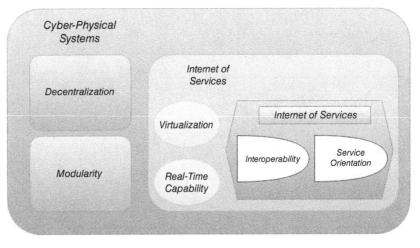

Several studies on I4.0 and related topics are being undertaken all around the globe, comparable to some of those listed earlier. The information provided here is thought to be adequate to describe the numerous I4.0 studies. However, the examination of research conducted on specific I4.0 components is equally crucial.

### 8.2.2 Industry 4.0 in Manufacturing

Globalization, mass customization, and a competitive business environment are putting pressure on "traditional" enterprises to hold cutting-edge business models and embark on a journey of transformation to I4.0 [20, 21]. The objective of I4.0 technologies will revolutionize creation by maximizing efficacy and output while using the fewest resources feasible [22]. The emergence of I4.0 methodologies has introduced a novel paradigm in the manufacturing industry in industries that seeks to maximize production via resource efficiency. "Smart manufacturing" or "digital manufacturing" is at the brain of I4.0, which allows organizations to carry out flexible creation procedures with huge personalization [23]. I4.0 manufacturing technologies may follow industrial physical processes and develop "DT" based on real-world goods [24]. This encourages interaction between humans and machines, facilitates real-time communication, and enables intelligent decision-making within manufacturing organizations. However, despite the potential opportunities for sustainable manufacturing in the context of I4.0, there are a limited number of comprehensive review studies that have examined the technologies from diverse sustainability perspectives.

The three core components of manufacturing – processes, products, and systems – that enable industry development and long-term value generation are referred to as "sustainable manufacturing" [25]. These three variables must demonstrate each other how they contribute to the community-oriented, ecological, and financial sustainability of production [26]. The combining of procedures with machinery to attain the best quality items in terms of low supply usage, reproducible assets, and protection for clients, employees, and society is known as "sustainable manufacturing" [27].

The present advancements in an interchange of information and mechanization across industrial approaches or events are known as "I4.0." The main goal is to build "smart organizations," which are capable of adjusting to dynamic alterations in management goals, production situations, and business models. Figure 8.8 depicts I4.0 enabling link by IoT between smart organizations and clients.

Reference architecture for an industry 4.0 IoT-based smart organization is shown in Figure 8.9. It comprises a variety of technological systems and viewpoints, such as the following:

- **Smart Machines:** To optimize efficiency and productivity by automating tasks, reducing errors, and enhancing operational performance in various industries.

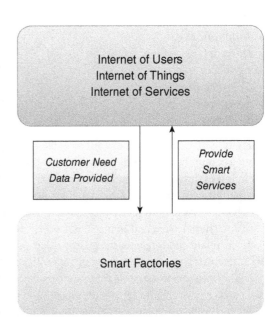

**Figure 8.8** Interaction between smart factories and consumers in I4.0.

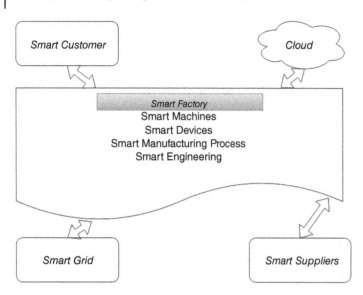

**Figure 8.9** Smart factory reference architecture.

- ***Smart Manufacturing Process:*** To improve overall operational efficiency by leveraging advanced technologies, real-time data analytics, and intelligent automation to optimize production, increase flexibility, reduce downtime, and cost-effectively enhance product quality.
- ***Smart Engineering:*** To enhance the design and development process by leveraging advanced technologies, data analytics, and automation to improve efficiency, reduce errors, accelerate innovation, and optimize resource utilization.
- ***Manufacturing IT:*** To streamline and optimize the information flow and processes within the manufacturing environment.
- ***Smart Suppliers:*** To ensure timely and accurate supply of goods or services, reduce costs, and enhance customer satisfaction.
- ***Smart Grid:*** To create a more reliable, efficient, and sustainable electricity grid that meets the evolving needs of the energy system.

**Characteristics of Smart Factories**
- ***Innovative***
- ***Connected***
- ***Efficient***
- ***Automated***
- ***Data-driven***
- ***Agile***
- ***Collaborative***
- ***Sustainable***
- ***Resilient.***

Figure 8.10 depicts a strategy aimed at facilitating IoT-based energy management in sustainable smart factories, resulting from a comprehensive study conducted within the factory context. The strategy outlines a four-step approach that needs to be followed during the adoption process. The initial phase involves gaining a deep understanding of production processes, evaluating existing energy management practices, and establishing improvement objectives. In the second stage, the emphasis shifts toward leveraging IoT technology to collect real-time data, conduct a thorough analysis, and gain insights into current practices and limitations.

To correlate and interpret the energy usage pattern and reach a quality conclusion, it is also important to determine the manufacturing processes. Following data collection and analysis, the next stage is combining information into energy optimization systems to assist decision-makers with energy waste and opportunities for improvement. They may also choose the most environmentally friendly machine configuration while considering production planning to maximize energy utilization. The explanation of techniques and practices for increasing the power efficacy of smart organization by "design" is the higher level, which is covered by the fourth phase.

This section offers an overview of smart factories, commonly referred to as "I4.0," including the architecture for IoT-based smart factories and their sustainability characteristics. It further presents a strategy for leveraging IoT at the production level to support energy management and enhance energy efficiency in production systems within these facilities. The section also explores the role and potential of I4.0 technologies across different industrial production processes. While previous research primarily focused on the broad theories and concepts of I4.0, this section specifically highlights the prospects and roles of I4.0 shop floor management. Additionally, it examines the impact of various I4.0 technologies on each sustainability parameter, providing valuable insights for future research in this field.

### 8.2.3 Industry 4.0 and Communication Services

The current trend in digitalization and the associated automation of production processes are referred to as "Industry 4.0." According to this comparatively new notion, "smart factories," which are managed by CPS, would carry out repetitive and easy tasks that were previously done by humans. The procedures and tools of I4.0 are convinced by a result in expected value and time savings, increasing the total reliability of factories and other

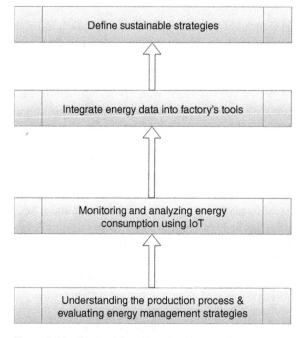

**Figure 8.10** Strategy for enhancing the smart organization power control through IoT.

sectors. The biggest threats are mostly computer hacking and data abuse [28]. Such automation, integration, and greater efficiency enabled by more efficient logistics may help mitigate the negative consequences of human-induced industrial growth, and they hold enormous potential for implementing sustainable development.

Currently, all human activities are linked by a diverse set of communication networks, including IoT, IoS, and IoP. Frameworks of these strategies are further classified into business-oriented Internet of Things (BIoT) and IIoT. Regardless of corporate, country, or state borders, these technologies will enable organizations involved in I4.0 for maintaining interlinked interaction over the whole life span. All parties engaged in the production chain will have access to the essential data, which may be highly beneficial to the concerned parties since it will enable those involved in the manufacturing and commercial chain to build their goods with foresight. German companies and organizations created accessible RAMI 4.0 Framework (2015) and I4.0 tools of component models for usage in such large chains, as illustrated in Figure 8.11.

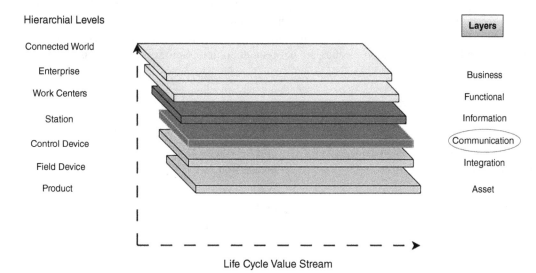

**Figure 8.11** RAMI 4.0 model.

The IoT is an innovative strategy that is being suggested for a variety of applications such as smart city sectors, consumer gadgets, industrial settings, Internet of Cars, multimedia, and 5G systems. It also offers a low-cost, scalable, and dependable ecosystem. IoT communication infrastructure contains TCP/IP systems and industry-procedure protocols. The most common protocols are AMQP [29–31], MQTT [32], CoAP [33], XMPP [34], and JSON [35].

"I4.0," "intelligent production," "IIoT," and "smart organization" are all words used to describe modern industrial settings that use ICT along with IoT infrastructure, while assuring organization norms. IIoT says IoT usage for a variety of service sectors. But, in these sectors, the bulk of commercial and SCADA-based methods deploy commercial ICT resulting in a closed type of factory systems. As a consequence, interoperability is lacking, expenses are high, and clients are locked into a single supplier.

Researchers are actively proposing and investigating protocols, networks, and middleware architectures to facilitate the integration of industrial ICT infrastructure within the business context. In Meng et al. [36], the authors have introduced a data-centric M2M communication middleware specifically designed for IIoT applications. This middleware relies on the ZeroMQ platform to enable seamless integration. Additionally, researchers in Brizzi et al. [37] have conducted a case study utilizing the Ebbits middleware, which transforms data into web services. This middleware has proven effective in remotely managing industrial robots and monitoring energy consumption.

To achieve workload distribution equilibrium between a mobile terminal and a utility cloud service, a service-oriented IIoT middleware is introduced [38]. Similarly, Packwood et al. [39] put forward a system J-based platform for the IIoT, which is assessed in an automation system utilizing a field programmable gate array (FPGA). Additionally, reference [40] outlines a collaboration-oriented M2M (CoM2M) communications system for the IIoT, leveraging the PicknPack food packaging platform. The references provide valuable insights into the development of innovative middleware and platforms for effective IIoT implementation. In reference [41], the I4.0 framework SCADA is linked to conventional flexible production systems through Ethernet. These researches, on the other hand, focused on middleware designs of integration. This section provides a comprehensive overview of the differences between the HTTP, CoAP, MQTT, AMQP, and XMPP protocols, examining them from various telecommunications perspectives. Subsequently, an appropriate protocol for the IIoT ecosystem is selected. Table 8.2 summarizes these variations in terms of different communication elements, including infrastructure, architecture, mechanism, model, messaging pattern, technique, and transmission paradigm. These protocols follow a client-server communication architecture.

MQTT, being a message-oriented protocol, utilizes the publish-subscribe model, while HTTP, a document-oriented protocol, operates on the request-response model. MQTT facilitates one-to-many communication, whereas HTTP is designed for peer-to-peer communication. CoAP relies on the 6LoWPAN network infrastructure (IEEE 802.15.4), which leverages IPv6 at the network layer. Both MQTT and HTTP utilize the Internet or Intranet, either in wired mode (Ethernet – IEEE 802.3) or wireless mode (Wi-Fi – IEEE 802.11), supporting either IPv4 or IPv6 at the network layer. At the transport layer, MQTT and HTTP employ TCP port numbers 1883 and 80, respectively. These distinctions highlight the unique characteristics of each protocol and their suitability for different IIoT applications.

In IPv6-based IoT systems, the application layer protocol CoAP is employed to establish communication with constrained devices. CoAP supports both the publish-subscribe and request-response communication patterns. The communication pattern in CoAP involves a concise binary header of fixed length, which may be accompanied by a small binary option and a payload. It operates on a message-based transmission mechanism between endpoints. Contrary to HTTP, as illustrated

**Table 8.2** Comparison of IoT protocols.

| Features | HTTP | CoAP | MQTT |
| --- | --- | --- | --- |
| Infrastructure | Ethernet and Wi-Fi | 6LoWPAN | Ethernet and Wi-Fi |
| Network layer | IPV4 or IPV6 | IPV6 | IPV4 or IPV6 |
| Transport layer | TCP | UDP | TCP |
| Model | Synchronous | Asynchronous | Asynchronous |
| Pattern | Request-response | Request-response Publish-subscribe | Publish-subscribe |
| Mechanism | 1–1 | 1–1 | 1 to many |
| Security | SSL, TLS | DTLS | SSL, TLS |

**Figure 8.12** Communication protocols: (a) IEEE, (b) HTTP, and (c) CoAP.

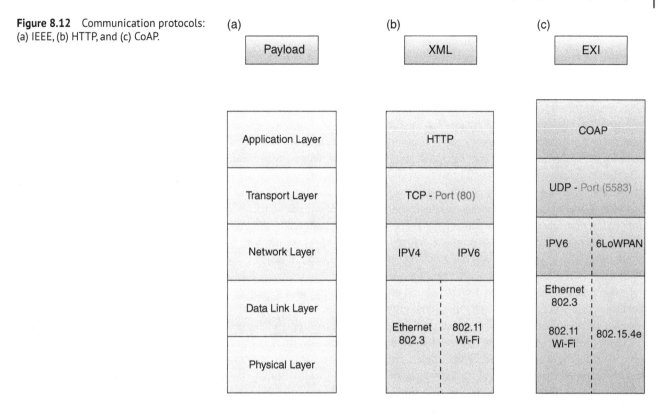

in Figure 8.12, CoAP utilizes either IPv6 or 6LoWPAN at the network layer and employs connectionless User Datagram Protocol (UDP) at the transport layer. When CoAP is used with IPv6, the data link layer requires Ethernet, while the physical layer requires Wi-Fi. In the case of 6LoWPAN, CoAP utilizes IEEE 802.15.4e for the data link and physical layers. These distinctions highlight the specific network layers and protocols employed by CoAP, making it suitable for efficient communication in resource-constrained IoT environments.

The content (payload) of HTTP may vary depending on the content type, which can include plain text, HTML, XML, GIF images, PDF documents, or audio files. XML is commonly used for data sharing in HTTP due to its ability to handle verbose plain text and address interoperability challenges. However, when considering interoperability in CoAP, an efficient XML interchange format called "EXI" is employed. EXI encodes verbose XML texts into a binary format, offering improved performance and reduced power consumption, particularly for constrained devices. As a result, CoAP proves to be suitable for restricted devices in IPv6-based IoT wireless sensor networks. To enable data exchange over the Internet, the use of a gateway is necessary.

### 8.2.4 Industry 4.0 in Transportation

Industry 4.0 is driving significant changes in company models, managerial strategies, and the workforce landscape. It is leading to the emergence of new professions while also rendering certain professions obsolete, as intelligent machines and technologies take over. The global adoption of Industry 4.0 signifies the onset of the Fourth Industrial Revolution, a process that will unfold gradually over time. To successfully navigate this transformative era, there is a pressing need to enhance workforce education, skills, and capabilities.

The business world is experiencing a rapid and dynamic transformation. Throughout the years, the logistics sector has undergone multiple changes, and with technology continuing to revolutionize the world, its impact on logistics will only intensify. This, in turn, will bring about a paradigm shift in how businesses efficiently and swiftly distribute their goods to customers. The recent global pandemic has further underscored the importance of streamlined logistics operations, with the demand for smooth and uninterrupted supply chains becoming increasingly critical.

Industry 4.0 is not only revolutionizing logistics but also shaping society and the global economy on a larger scale. At its core, Industry 4.0 leverages the connectivity of the IoT, where modern devices communicate with one another through CPSs, operating autonomously without human intervention. The deployment and utilization of Industry 4.0 and its related technologies impact the entire supply chain, extending beyond logistics alone. These transformative changes present both operational challenges and opportunities in the field of logistics, requiring organizations to adapt and seize the benefits of Industry 4.0 to drive their overall performance.

The ongoing I4.0 is characterized by the integration of intelligent technologies, merging the realms of physical and biological entities with the digital world [42]. This revolutionary era represents the convergence of manufacturing with advanced information and communication technology, empowering the production of customized goods that cater to the unique demands of individual customers. Furthermore, it facilitates the manufacturing of small, personalized batches, albeit at the expense of mass-produced commodities [43]. Technology has had a significantly greater influence on I4.0 as done for I1.0, I2.0, and I3.0. It is critical to develop a technology strategy to create an innovation ecosystem [44].

I4.0 encompasses the integration of robotics and AI, enabling machines to handle strenuous tasks and automated robots to execute repetitive processes continuously, leading to remarkable advancements [45]. Within this revolution, a digital transformation is underway, fundamentally reshaping how individuals work and leveraging cutting-edge technologies that are vital to manufacturing operations [46]. Referred to as "Industry 4.0," the Fourth Industrial Revolution is driven by the digitalization and automation of production processes [47]. This era of innovation, disruption, and intelligent technologies facilitates a higher level of industrial efficiency while also aiming to achieve social and environmental sustainability for businesses [48].

In the dynamic and globalized international environment, logistics systems must continually enhance their efficiency, flexibility, and security to swiftly adapt to changing circumstances [49]. One effective approach to achieve this is by embracing I4.0 technologies in logistics operations. The automation of supply chain activities poses a challenge and opportunity for modern logistics and supply chain management [50]. Presently, there is a pressing need for the automation of logistics processes, further compounded by the impact of the ongoing crisis on logistics operations [51]. The advent of digitization has necessitated the adoption of Industry 4.0, which is significantly influencing the development of innovative manufacturing and logistics concepts [52]. The ongoing digital transformation of logistics, encompassing the entire supply chain, has become a source of competitive advantage [53]. The integration of autonomous and digital Industry 4.0 technologies promises faster delivery and reduced logistics costs [54]. Within the context of Industry 4.0, the terms "smart logistics" and "Logistics 4.0" have emerged, denoting the intersection of logistics with CPSs and the IoT. Logistics 4.0 offers several benefits in the workplace, including streamlined monitoring of logistics systems; heightened environmental consciousness; decreased wastage of resources such as money, time, and energy; the emergence of new business models; and the establishment of agile logistics processes to meet customer demands swiftly [55]. Logistics 4.0 particularly excels in areas such as resource planning, transportation management systems, warehouse management systems, and the infrastructure of intelligent transportation systems [56].

Logistics 4.0 encompasses the application of smart technologies in various domains, such as inventory management, warehousing, distribution, and transportation [57]. It represents the integration of intelligent logistics solutions that encompass real-time localization, networking, automated data collection and processing, automatic identification, as well as business and analytical services. These advancements contribute to the industrialization of logistics operations by promoting rationalization and standardization through the utilization of next-generation technologies [58].

In the current competitive landscape, digital transformation toward I4.0 has become indispensable for businesses as it enhances their responsiveness, flexibility, and agility. The logistics industry, like other sectors, is experiencing a profound shift driven by disruptive I4.0 strategies that are revolutionizing logistics processes and activities. The range and quantity of I4.0 strategies utilized in logistics are not standardized and can vary depending on different perspectives and organizational capabilities. While certain technologies find extensive use in the automotive industry, others are prominent in the textile industry. Furthermore, some I4.0 strategies cater to small businesses, while others are adopted by large enterprises. The analysis of I4.0 strategies in logistics is based on information gathered from diverse sources. Considering the ongoing industrial revolution, continuous research and studies focusing on the digitalization of logistics transformation, digital technologies, I4.0 in logistics, and logistics 4.0 are crucial and subject to constant change. At the enterprise level, there are still gaps to be addressed within IR4.0, presenting new possibilities and prospects for businesses.

### 8.2.5 Industry 4.0 in Health Care

The IR4.0 pushes the medical field to a previously unheard-of relaxation phase by using digitalization, AI, and 5G connectivity. "Medicine 4.0," refers to the fourth stage of medical development and is synonymous with I4.0. The field of modern medicine, which has a history of around 150 years, is currently undergoing a digital transformation fueled by advancements in robotics, the Internet, and AI. Among the notable trends shaping global health care today, the integration of AI systems into medicine stands out as a significant development. This convergence of AI and medicine is revolutionizing the health-care industry, opening up new possibilities for diagnosis, treatment, and patient care [59]. By harnessing the power of AI algorithms and data analytics, health-care professionals can enhance their decision-making processes, improve medical outcomes, and provide personalized and precise health-care services. This integration of technology in medicine has the potential to reshape the way health care is delivered, ultimately leading to better patient outcomes and improved efficiency in health-care systems. Current facilities in medicine cannot be entirely successful without the usage of cutting-edge computer technology. AI strategies, which are radically revolutionizing the global health-care system, enable the progress of innovative medications, higher-end analytics, testing, and treating facilities, as well as breakthroughs in transplant surgery [39]. In general, digitalization and AI reduce medical clinical costs while improving health-care quality. Figure 8.13 depicts the strategies that enable the digitalization of medicine.

### 8.2.5.1 On-Demand Health Care

The digital era has witnessed a significant increase in consumers relying on online resources for medical information. Surveys reveal that 47% of individuals seek to learn more about their physician, while 38% express a desire to explore medical resources provided by hospitals. Furthermore, an overwhelming 77% of people prefer online medical appointments. This shift in consumer behavior has paved the way for medical professionals to directly offer their skills and expertise to patients, akin to independent contractors. Health-care organizations now facilitate this connection through online marketplaces, enabling convenient access to on-demand medical procedures and services. As a result, health-care professionals are embracing the patient-centered digital health-care sector, adapting to meet the evolving needs of modern health-care consumers.

### 8.2.5.2 Telemedicine Market

Telemedicine has revolutionized health care by bridging the gap between patients and specialized medical professionals, especially in locations lacking on-site specialists. Its benefits extend beyond developing nations, proving essential even in industrialized countries. The adoption of telemedicine has resulted in cost reductions, improved diagnostic accuracy, and enhanced access to remote health monitoring. Individuals with chronic illness and elderly people stand to benefit significantly from this approach. The global telemedicine market, valued at USD 56.2 billion in 2020, is expected to reach

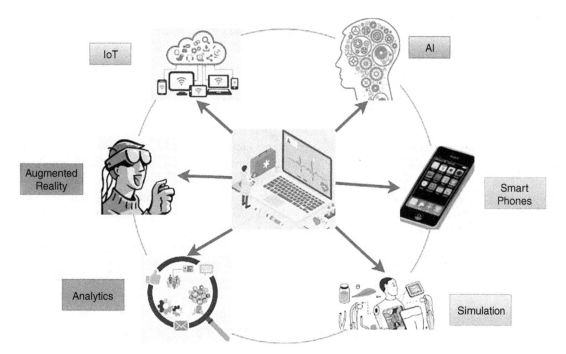

**Figure 8.13** Digitalization in health care.

USD 175.5 billion by 2026, according to the Global Telemedicine Market Outlook. These statistics highlight the growing importance and potential of telemedicine in transforming health-care delivery worldwide.

### 8.2.5.3 Data Privacy and Cybersecurity in Medicine and Health Care

As medical innovation progresses and incorporates big data, patient data privacy faces potential threats [60]. The health-care industry is a prime target for cyberattacks, despite ongoing investments in cybersecurity. Many medical devices used in hospitals still have significant vulnerabilities, emphasizing the need for cutting-edge technologies to mitigate risks. Protecting patient-sensitive data under General Data Protection Regulation (GDPR) regulations is essential to preserve privacy. Similar to other IoT devices, health-care technology presents challenges in data privacy and device safety. Constant threats and risks persist in this evolving landscape. The increasing use of electronic medical records and the proliferation of connected medical equipment and IoT devices contribute to the growing audience susceptible to cyberattacks. Additionally, viruses pose a persistent threat not only to computers but also to medical equipment. Ensuring robust cybersecurity measures and proactive strategies is crucial in safeguarding patient data and maintaining the integrity of health-care systems.

### 8.2.5.4 Big Data Analytics in Health Care

The application of big data analytics (BDA) in health care enhances the accuracy of diagnostic procedures and improves treatments [61]. BDA is particularly relevant in cases involving tumors and complex cancer diseases. It plays a vital role in reducing medication errors and predicting future admission rates. By leveraging predictive analysis, BDA enhances the predictability of health-care services, enabling health-care providers to allocate resources more efficiently and improve overall service delivery. Additionally, predictive analysis can help businesses anticipate potential labor shortages during cold/flu epidemics, enabling them to optimize their workforce management strategies. BDA offers significant benefits in health care, enabling better decision-making and resource allocation while improving patient outcomes and operational efficiency.

### 8.2.5.5 Wearable Medical Devices

Medical wearable technology is gaining popularity as it allows for continuous monitoring of patient's health, providing valuable data beyond traditional clinical exams. Wearable devices encourage individuals to stay active and set goals, reducing the risk of obesity and alleviating the strain on health-care systems. Insurance companies can access real-time medical data from wearables, enabling them to assess risks accurately and tailor insurance plans accordingly. Patients who proactively use technology for preventive care and monitoring may receive discounts on their health insurance premiums. Advancements in medical technology, such as tablets with microscopic sensors, offer precise and detailed information about a patient's internal organs. Digital technologies like 3D printing and IoT-based health devices pave the way for personalized and anticipatory health-care solutions in various medical industry sectors [61].

## 8.3 Conclusion

Finally, I4.0 had a disruptive connection with many industries, including sectors like production, communication, logistics, and health care. The incorporation of digital technology, automation, connectivity, and data-driven decision-making has revolutionized operations, increased efficiency, and created new opportunities for innovation and development.

I4.0 has allowed the combining of CPS, IoT, BDA, and CC in production. This has resulted in more automation, more efficient, manufacturing processes, and better result-oriented with the help of live data. Manufacturers may increase production, improve quality control, and react more quickly to market needs.

I4.0 transformed in ways of communication between people and corporations. Because of the integration of digital technology and improved communication protocols, continuous connection, personalized experiences, and target messages are now possible. AI-powered technologies and automation have improved communication process efficiency and responsiveness.

I4.0 has brought strategies like IoT, BDA, and automation into the logistics industry. As a result, supply chain visibility has increased, route planning has been optimized, and operations have been simplified. Logistics providers can deliver items more quickly, save costs, and improve overall efficiency, leading to a more flexible and dependable supply chain ecosystem.

I4.0 has altered patient care, medical procedures, and health-care delivery in the medical domain. IoT, BDA, and AI integration have enabled remote patient monitoring, personalized treatment regimens, and evidence-based decision-making. Telemedicine solutions have increased access to health care, particularly in underprivileged regions, while data-driven techniques have resulted in improved illness diagnosis and treatment.

While each industry has its own set of advantages and disadvantages, I4.0 as a whole represents a paradigm change that brings enormous prospects for enhanced efficiency, innovation, and better results. Collaboration, standardization, and resolving issues such as data security and workforce consequences are critical to realizing I4.0's full potential across various industries. Finally, I4.0 has radically changed manufacturing, communication, logistics, and health care, opening the path for a more connected, data-driven, and efficient future. Embracing I4.0 ideas and technology may put organizations at the forefront of innovation, competitiveness, and long-term success in the digital age. This chapter provides a thorough examination of four digitalization domains allowing researchers to conduct an in-depth study.

# References

**1** Vaidya, S. et al. (2018). Industry 4.0 – a glimpse. *Procedia Manufacturing* 20: 233–238. https://doi.org/10.1016/j.promfg.2018.02.034.

**2** Qin, J. et al. (2016). A categorical framework of manufacturing for Industry 4.0 and beyond, changeable, agile, reconfigurable & virtual production. *Procedia CIRP* 52: 173–178.

**3** Rüßmann, M. et al. (2015). *Industry 4.0: The Future of Productivity and Growth in Manufacturing Industries*, 1–14. Boston Consulting Group.

**4** Neugebauer, R. et al. (2016). Industrie 4.0 - From the Perspective of Applied Research. *49th CIRP Conference on Manufacturing Systems (CIRP-CMS 2016), Procedia CIRP, 57*, 2–7 (25–27 May 2016). Stuttgart, Germany: Elsevier.

**5** Thoben, K.D. et al. (2017). Industrie 4.0 and smart manufacturing – a review of research issues and application examples. *International Journal of Automation Technology* 11 (1): 4–16.

**6** Sipsas, K. et al. (2016). Collaborative maintenance in flow-line manufacturing environments: An Industry 4.0 approach. *5th CIRP Global Web Conference Research and Innovation for Future Production, Procedia CIRP, 55*, 236–241 (4–6 October 2016). Patras, Greece: Elsevier.

**7** Rennung, F. et al. (2016). Service Provision in the Framework of Industry 4.0. *SIM 2015 13th International Symposium in Management, Procedia – Social and Behavioural Sciences, 221*, 372–377 (9–10 October 2015). Timisoara, Romania: Elsevier.

**8** Brettel, M. et al. (2014). How virtualization, decentralization and network building change the manufacturing landscape: an Industry 4.0 perspective. *International Journal of Mechanical, Aerospace, Industrial, Mechatronic and Manufacturing Engineering* 8 (1): 37–36.

**9** Lee, J. et al. (2014). Service innovation and smart analytics for Industry 4.0 and big data environment, Product Services Systems and Value Creation. *Proceedings of the 6th CIRP Conference on Industrial Product-Service Systems, Procedia CIRP*, 16, 3–8 (1–2 May 2014). Windsor, Canada: Elsevier.

**10** Bahrin, M.A.K. et al. (2016). Industry 4.0: a review on industrial automation and robotic. *Jurnal Teknologi (Sciences and Engineering)* 78: 137–143.

**11** Almada-Lobo, F. (2015). The Industry 4.0 revolution and the future of manufacturing execution systems (MES). *Journal of Innovation Management* 3 (4): 16–21.

**12** Erol, S. et al. (2016). Tangible Industry 4.0: a scenario-based approach to learning for the future of production, *6th CLF -6th CIRP Conference on Learning Factories, Procedia CIRP*, 54, 13–18 (29–30 June 2016). Gjovik, Norway: Elsevier.

**13** Garay-Rondero, C. et al. "Digital supply chain model in Industry 4.0." Journal of Manufacturing Technology Management 31.5 (2020): 887–933.

**14** Fatorachian, H. and Kazemi, H. (2018). A critical investigation of Industry 4.0 in manufacturing: theoretical operationalisation framework. *Production Planning and Control* 29 (8): 633–644.

**15** Liao, Y. et al. (2017). Past, present and future of Industry 4.0-a systematic literature review and research agenda proposal. *International journal of production research* 55 (12): 3609–3629.

**16** Lu, Y. (2017). Industry 4.0: A survey on technologies, applications and open research issues. *Journal of Industrial Information Integration* 6: 1–10.

**17** van Lopik, K. et al. (2020). Developing augmented reality capabilities for industry 4.0 small enterprises: Lessons learnt from a content authoring case study. *Computers in Industry* 117: 103208.

**18** Kabugo, J.C. et al. (2020). Industry 4.0 based process data analytics platform: A waste-to-energy plant case study. *International Journal of Electrical Power & Energy Systems* 115: 105508.

**19** Issa, A. et al. (2018). Industrie 4.0 roadmap: Framework for digital transformation based on the concepts of capability maturity and alignment. *Procedia CIRP*, 72, 973–978 (16–18 May 2018). Stockholm, Sweden: Elsevier.

**20** Aiello, G. et al. (2020). Propulsion monitoring system for digitized ship management: preliminary results from a case study. *Procedia Manufacturing* 42: 16–23.

**21** Oztemel, E. and Samet G. (2020). Literature review of Industry 4.0 and related technologies. *Journal of Intelligent Manufacturing* 31: 127–182.

**22** Stock, T. and Seliger, G. (2016). Opportunities of sustainable manufacturing in industry 4.0. *13th Global Conference on Sustainable Manufacturing – Decoupling Growth from Resource Use 2015, Procedia CIRP* 40, 536–541 (16–18 September 2015). Binh Duong New City, Vietnam: Elsevier.

**23** Machado, C.G. et al. (2020). Sustainable manufacturing in Industry 4.0: an emerging research agenda. *International Journal of Production Research* 58: 1462–1484.

**24** Kim, J.H. (2017). A review of cyber-physical system research relevant to the emerging IT trends: Industry 4.0, IoT, big data, and cloud computing. *Journal of Industrial Integration and Management* 2: 1750011.

**25** Haapala, K.R. et al. (2013). A review of engineering research in sustainable manufacturing. *Journal of Manufacturing Science and Engineering ASME* 135 (4): 135.

**26** Garetti, M. and Taisch, M. (2012). Sustainable manufacturing: trends and research challenges. *Production Planning & Control* 23 (2–3): 83–104.

**27** Jayal, A. D. et al. (2010). Sustainable manufacturing: Modeling and optimization challenges at the product, process and system levels. *CIRP Journal of Manufacturing Science and Technology* 2 (3): 144–152..

**28** VDI/VDE-Gesellschaft Mess- und Automatisierungstechnik (2015). Status Report: Reference Architecture Model Industrie 4.0 (RAMI4.0).

**29** Standard 19464. (2016). *Advanced Message Queuing Protocol 1.0 (AMQP 1.0)*. ISO/IEC: Geneva, Switzerland.

**30** O'Hara, J. (2014). *ISO 19464 Connecting Business for Value*.

**31** Godfrey, R., Ingham, D., Schloming, R. (2012). OASIS Standard Advanced Message Queuing Protocol (AMQP) Version 1.0.

**32** Standard PRF 20922 (2016). *Message Queuing and Telemetry Transport (MQTT) Version 3.1.1*. ISO/IEC: Geneva, Switzerland.

**33** Standard RFC 7252 (2014). *Constrained Application Protocol (CoAP)*. IETF: Fremont, CA, USA.

**34** Standard RFC 6120 (2011). *Extensible Message and Presence Protocol (XMPP)*. IETF: Fremont, CA, USA.

**35** Standard RFC 7159 (2014). *The JavaScript Object Notation (JSON) Data Interchange Format*. IETF: Fremont, CA, USA.

**36** Meng, Z. et al. (2017). A data-oriented M2M messaging mechanism for industrial IoT applications. *IEEE Internet of Things Journal* 4: 236–246.

**37** Brizzi, P. et al. (2013). Bringing the Internet of Things along the manufacturing line: A case study in controlling industrial robot and monitoring energy consumption remotely. *2013 IEEE 18th Conference on Emerging Technologies & Factory Automation (ETFA)* (10–13 September 2013). Cagliari, Italy: IEEE.

**38** Chang, C. et.al. (2013). A middleware for discovering proximity-based service-oriented industrial internet of things. *2015 IEEE International Conference on Services Computing* (27 June 2015–02 July 2015). New York, NY, USA: IEEE.

**39** Packwood, D. et al. (2015). FPGA-based mixed-criticality execution platform for SystemJ and the Internet of Industrial Things. *2015 IEEE 18th International Symposium on Real-Time Distributed Computing* (13–17 April 2015). Auckland, New Zealand: IEEE.

**40** Meng, Z. et al. (2017). A collaboration-oriented M2M messaging mechanism for the collaborative automation between machines in future industrial networks. *Sensors* 17: 2694.

**41** Godoy, C. et al. (2018). Integration of sensor and actuator networks and the SCADA system to promote the migration of the legacy flexible manufacturing system towards the Industry 4.0 concept. *Journal of Sensor and Actuator Networks* 7: 23.

**42** Rotatori, D. et al. (2020). The evolution of the workforce during the fourth industrial revolution. *Human Resource Development International* 24: 92–103.

**43** Ustundag, A. and Cevikcan, E. (2017). *Industry 4.0: Managing the Digital Transformation*. Cham, Switzerland: Springer International.

**44** Schäfer, M. (2018). The fourth industrial revolution: how the EU can lead it. *European View* 17: 5–12.

**45** Bláha, J. et al. (2021). Multidimensional analysis of ethical leadership for business development. *European Journal of Sustainable Development* 10: 290.

**46** Ghobakhloo, M. (2020). Industry 4.0, digitization, and opportunities for sustainability. *Journal of Cleaner Production* 252: 119869.

**47** Bauer, W. et al. (2018). Digitalization of industrial value chains – a review and evaluation of existing use cases of Industry 4.0 in Germany. *Logforum* 14: 331–340.

**48** Bai, C. et al. (2020). Industry 4.0 technologies assessment: a sustainability perspective. *International Journal of Production Economics* 229: 107776.

**49** ČEMERKOVÁ, Š. and MALÁTEK, V. (2019). Human Resources Management in Multinational Companies in Response to Logistics Needs and Meeting Their Goals. *Proceedings of the 2nd International Conference on Decision Making for Small and Medium-Sized Enterprises. Conference Proceedings Silesian University in Opava, School of Business Administration in Karviná: Karviná,* Czech Republic, 61–69.

**50** Nitsche, B. et al. (2021). Application areas and antecedents of automation in logistics and supply chain management: a conceptual framework. *Supply Chain Forum: An International* 22: 223–239.

**51** Nitsche, B. and Straube, F. (2021). Defining the "new normal" in international logistics networks: Lessons learned and implications of the COVID-19 pandemic. *WiSt—Wirtsch Stud* 50: 16–25.

**52** Nitsche, B. (2021). Exploring the potentials of automation in logistics and supply chain management: paving the way for autonomous supply chains. *Logistics* 5: 51.

**53** Gerlach, B. (2021). Digital supply chain twins-conceptual clarification, use cases and benefits. *Logistics* 5: 86.

**54** Winkelhaus, S. et al. (2019). Logistics 4.0: a systematic review towards a new logistics system. *International Journal of Production Research* 58: 18–43.

**55** Dördüncü, H. (2021). Logistics, supply chains and smart factories. In: *Accounting, Finance, Sustainability, Governance & Fraud: Theory and Application* (ed. İ. İyigün and Ö.F. Görçün), 137–152. Singapore: Springer.

**56** Barreto, L. et al. (2017). Industry 4.0 implications in logistics: an overview. *Procedia Manufacturing* 13: 1245–1252.

**57** Glistau, E. and Machado, N.I.C. (2018). Industry 4.0, logistics 4.0 and materials-Chances and solutions. *Materials Science Forum* 919. Trans Tech Publications Ltd.

**58** Kim, E. et al. (2022). The necessity of introducing autonomous trucks in logistics 4.0. *Sustainability* 14: 3978.

**59** Von Eiff, M. C. and Von Eiff, W. (2020). The digitalisation of healthcare. *Health Management.org Journal* 20 (6).

**60** Price, W. N. and Glenn Cohen, I. (2019). Privacy in the age of medical big data. *Nature Medicine* 25: 37–43.

**61** Kute, S.S. et al. (2022). Industry 4.0 challenges in e-healthcare applications and emerging technologies. In: *Intelligent Interactive Multimedia Systems for e-Healthcare Applications* (ed. A.K. Tyagi, A. Abraham, and A. Kaklauskas). Singapore: Springer https://doi.org/10.1007/978-981-16-6542-4_1.

# 9

## Transforming Education Management in the Industry 4.0 Era: Harnessing the Power of Cloud-Based Blockchain

*Sovers Singh Bisht, Garima Jain, Priyanka Chandani, and Vinod M. Kapse*

Noida Institute of Engineering and Technology, Greater Noida, Uttar Pradesh, India

## 9.1 Introduction

Recent improvements in blockchain technology and decentralized consensus systems provide new possibilities for establishing unmanageable domain-specific ledgers with no centralized authority. Since their conception, blockchains have mostly been used for value transfers.

Blockchain in health care has many advantages, such as the management of electronic medical records, pharmaceutical supply chain management, drugs, biomedical research, and many more that have been developed as prototypes within the framework of blockchain [1]. Detection and understanding of edge computing in the cloud have additionally contributed to the clinical part of fragmenting masses in mammograms, which presently can deliver the man-made intelligence-based high-quality determinations of breast cancer, and security prospects can be well handled via blockchain [2–4].

The advent of Industry 4.0 has brought forth transformative changes in various sectors, and the field of education is no exception. As educational institutions strive to keep pace with the rapidly evolving technological landscape, integrating cloud-based solutions and blockchain technology has emerged as a powerful catalyst for revolutionizing education management. This chapter explores the potential of harnessing cloud-based blockchain solutions in the education sector, enabling unprecedented levels of data security, transparency, and efficiency. By leveraging blockchain's decentralized nature and cloud computing's scalability, educational institutions can unlock new possibilities for digitalization, credential management, and lifelong learning. Through a comprehensive examination of the benefits and challenges associated with implementing cloud-based blockchain solutions, this chapter aims to shed light on the transformative role of these technologies in shaping the future of education in the Industry 4.0 era.

Blockchain technology in education may be used to address a variety of educational issues and can help teachers and students keep track of student progress. When data is kept on the blockchain network, it can be saved safely and impenetrably. We have provided a quick summary of present systems, their current status, and a future path. However, we also propose ideas for specific regulations and adjustments that may be made to current systems to improve them and, as a consequence, the field of education. It also offers blockchain integration levels in higher education, which, when combined with data extraction, have the potential to enlarge and extend the area of education to the advantage of institutions, businesses, and students. People who are managers and policymakers need to consider many challenges pertaining to security, privacy, cost, availability, and scalability before they start using blockchain technology [5, 6]. Many techniques nowadays are available over the cloud with application programmings (APIs) using the blockchain as well as the portals in the medical diagnosis system where machine learning and artificial intelligence (AI) have contributed to the growth of early diagnosis of cancer, which can be secured through the distributed ledger technology (DLT) system [7].

The FabRec prototype discussed here provides the use of Ethereum smart contracts, which automatically initiate commands given to the Arduino/Raspberry PI system based on recorded events by a physical machine for another part of the network that provides proof of concept linking computing notes and physical devices [8]. Massive open online course (MOOC) incorporated into the system with comprehensive training calls into question the excessive authority enjoyed by today's higher education institutions pertaining to blockchain technology, which improves the security as well as frameworks for higher training and development [9, 10]. Teachers can obtain deductions for a portion or the entire cost of developing and implementing courses, instructional resources, and so on. The system creates its own tokens, which may be gained just via deductions or through various types of effort. As a result, using the system is profitable from a business standpoint. Here, we will attempt to briefly discuss the theoretical underpinnings and technological advancements of this field, some of which have been applied to the field of education specifically, and we will make some suggestions for reflection and discussion on the influence of blockchain on education.

Figure 9.1 depicts blockchain uses in education. It is an idea that is quickly growing and might offer many advantages to many different educational stakeholders. Blockchain serves as a ledger in educational institutions, keeping track of transactions and other essential activities.

Blockchain technology is employed in a number of ways in the education industry, as seen in Figure 9.1. This technology is starting to be used in education in a number of potential areas [11–14]. The many applications of blockchain in education will provide better transparency while improving security. Furthermore, it enhances traceability, efficiency, cost savings, and processing speed.

The following are the primary properties of the blockchain:

**Decentralization:** There are three sorts of decentralization: architectural, political, and logical.

**Persistency:** No record can be altered in the network, and any fabrication is immediately apparent.

**Anonymity:** To protect their identity, a user might create many addresses. It safeguards the confidentiality of the transactions.

**Auditability:** It improves the data's traceability and transparency in the blockchain.

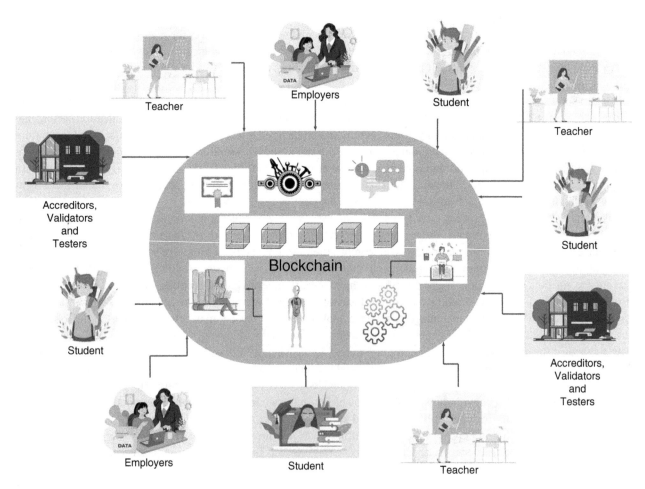

**Figure 9.1** Application of blockchain in education.

**Table 9.1** Overview of literature on blockchain applications in education.

| Study | Focus area | Methodology | Key findings |
| --- | --- | --- | --- |
| Smith et al. [17] | Credentialing | Case study | Blockchain improves security in storing and verifying educational credentials |
| Johnson et al. [18] | Lifelong learning | Survey and interviews | Blockchain enables lifelong learning records to be securely stored and accessed by learners |
| Brown et al. [19] | Governance | Literature review | Challenges in market acceptance and governance hinder the widespread adoption of blockchain in education |
| Chen et al. [20] | Collaboration | Experimental research | Cloud-based blockchain enhances collaboration among learners and educators, fostering innovation |
| Lee et al. [21] | Transparency | Qualitative analysis | Blockchain increases transparency in educational transactions, reducing fraud and improving trust |
| Wilson et al. [22] | Academic integrity | Case study | Blockchain can be used to ensure the integrity of academic records and prevent fraud in certification processes |

### 9.1.1 Contribution

Conceptual Framework: This chapter presents a robust conceptual framework that combines cloud-based solutions and blockchain technology to drive innovation in the education sector, paving the way for transformative changes and novel applications.

Digital Credential Management: By leveraging cloud-based blockchain technology, this chapter revolutionizes the management of educational credentials, ensuring a secure and decentralized platform for storing and verifying academic achievements, fostering transparency and trust [15].

Lifelong Learning Enablement: Through cloud-based blockchain solutions, this chapter empowers individuals with lifelong learning opportunities, facilitating continuous learning records, personalized learning pathways, and micro-credentialing, enabling skill development throughout one's career.

Enhanced Data Security: Recognizing the importance of data security in education, this chapter demonstrates how cloud-based blockchain solutions bolster security measures by offering tamper-proof records, preserving student data privacy, and mitigating the risks of fraud and unauthorized access [16] (Table 9.1).

These literature reviews provide insights into different aspects of blockchain technology in education, including student data privacy, learning analytics, academic integrity, and collaborative learning [23–25]. Each study utilizes different methodologies to investigate the potential benefits and challenges of implementing blockchain in these specific areas.

## 9.2 Revolutionizing Education Through Technology: The Power of Innovation and Connectivity

### 9.2.1 What Is a Chain of Blocks (Blockchain), and How Does It Work?

A system for electronic transactions based on trust has been proposed, which is made from digital signatures and provides strong control of authority. It also solves a proposed peer-to-peer network using proof of work, records public transaction history, and becomes computationally impractical for attackers if honest notes control a majority of CPU powers [26]. The term "blockchain" refers to a system that enables the preservation of distributed and decentralized records of digital transactions. Although the BC technology is no longer exclusive to Bitcoin, in Bitcoin, transactions between anonymous users are carried out via public key cryptography, in which case each user has a private key that only they are aware of and a public key that is shared with other users (whose identities are concealed). Every network node receives a broadcast of every transaction. Nodes organize transactions into blocks after validating them. A hash, which is a cryptographically distinct value formed from a block's contents and contains a reference to the hash of the block before it, identifies each block and enables connections between them. As a result, this chain of blocks serves as a public ledger or record of transactions that is accessible to all network nodes [27]. All nodes may then independently confirm that the keys used are legitimate and that the Bitcoins transmitted are from a prior transaction and have not yet been used. A transaction, however, is only regarded as genuine when it appears in a block that is added to the chain. To add a block, it must first be undermined, or its hash determined, which requires solving a tough, one-of-a-kind mathematical problem. This needs a lot of computing power, especially because we know that the hash resolution difficulty will change often. The issue becomes more challenging as the combined computing capability of the networked machines increases. Therefore, altering the content of a block modifies its hash, breaking the chain when the connection to the following block is attempted to be repaired and rendering the information contained in the blocks unchangeable due to the fact that the remaining nodes have a copy of the original string.

### 9.2.2 Blockchain Utilization

As predicted, the financial industry was the first to approach. The Bank of England, Visa, LinkedIn, Facebook, Cred, Creo, Santander, and Deutsche Bank are examples of hybrid systems that simplify and boost the safety of real estate transactions. Moreover, the Swedish property register and the utilization of blockchain (BC) to enhance transparency in public accounts gain significant potential when integrated with the Internet of Things (IoT), leading to a substantial expansion of possibilities. The same is true in the automotive rental or leasing market, where vehicle management is instantly linked to payment, limiting the flexibility and personalization of such contracts to obviously recommended limitations. Assume, for example, that after contract fulfillment, the corporate system automatically blocks access to an employee's office door, or to a house, if we fail to pay a portion of the mortgage. Analyzing research papers provides us with knowledge [17]. According to Monagas et al. [28], although blockchain technology will be open source for other industries, it will be difficult to deploy until it has been established for digital money.

In their study, Alameri et al. said that blockchain technology is innovative in the sphere of education. In accordance with this, they did multiple study assessments and evaluated several scientific literature before proposing a framework for three primary topics concerning this subject: applications, advantages, and problems.

According to Chen et al., blockchain employs cryptographic methods and distributed consensus algorithms to produce characteristics such as decentralization, traceability, immutability, and monetary qualities. Blockchain technology can assist pupils in developing their learning motivation. Furthermore, it maintains a complete and reliable record of educational activities, including methods and outcomes in both official and informal learning settings. It will also help in the documentation of instructors' instructional habits and performance, providing a reference for teaching evaluation. To conclude, blockchain has huge potential applications in instructional design, behavior recording and analysis, and formative evaluation for both students and teachers.

### 9.2.3 Blockchain in Education: Revolutionizing Learning, Credentialing, and Industry 4.0

Education has huge challenges that go beyond the simple optimization of teaching-learning processes to adapt to changes brought about by knowledge technologies, which, according to Frankenstein's syndrome, transform not only our traditions but also our way of thinking. Can blockchain technology now solve some of the problems that have developed as a result of these changes? Let us take a closer look at two scenarios. Learning is no longer an activity that occurs within a formal mode and is reinforced by experience in professional and critical practice. Lifelong learning, according to the approach of human learning, has become a labor necessity, a need of twenty-first-century citizenship that affects both the social framework and the person. Learning transcends time and space limitations. New concepts such as mobile learning and ubiquitous learning have arisen throughout the twenty-first century, resulting in a huge bibliography. Formal education institutions swiftly relocated some of their programs to the "post-formal" arena; postgraduate studies and refresher courses were transformed into upgrading and specialization programs. Other supplementary and alternative ways of training arise, such as boot camps, MOOCs, and Khan Academy. Despite the fact that the systems have significant limits, the demand to demonstrate competencies (skills, knowledge, and even attitudes) in issues beyond the formal scope is clear to them. However, learning accreditation is complicated not only by the range of domains from which training is supplied (formal, nonformal, and informal) but also by teaching and curricular aspects that have long been materialized in an ideal pursuit: customized learning [18]. This increases the variety of possible learning routes (including potentially acquired talents and skills) while also making public and universal accreditation of such information difficult. Even with the supplements done in procedures such as the European Space for Higher Education, traditional academic qualifications (degrees, postgraduate) are insufficient to communicate the capacity and grasp of the issue in this environment. In addition to being needed and valued in the workplace or in daily life, the lessons learned about "informality/nonformality" or alternative and custom itineraries are now acknowledged in "formal" situations like enterprise recruitment processes through mechanisms that are essentially traditional testing in-situ. Requesting a resume produced by the subject is customary, but there is no guarantee that the information is accurate. The compilation process on the subject's end and the verification process on the part of those who will study it are both expensive and time-consuming if it is complemented with the required certificates. Making sure the information on the CV is accurate (degrees or experiences must be included) and authenticating intricate and personalized learning records are the two difficulties.

The emergence of Industry 4.0 has brought forth a new wave of technological advancements that are reshaping various sectors, including education. Blockchain technology, with its inherent attributes of transparency, security, and decentralization, holds great potential in revolutionizing education in the Industry 4.0 era. This research chapter explores the application of blockchain in the education sector, focusing on its impact on learning processes and credentialing systems [19]. By leveraging blockchain, educational institutions can create immutable and tamper-proof records of student achievements, certificates, and degrees, ensuring the authenticity and integrity of educational credentials. Additionally, blockchain can facilitate lifelong learning by providing a transparent and decentralized platform for individuals to showcase their skills, achievements, and continuous professional development. The chapter also addresses the challenges and opportunities associated with the adoption of blockchain in education, offering insights into how institutions can navigate the transition toward a blockchain-enabled education ecosystem. Overall, this research chapter highlights the transformative potential of blockchain in Education 4.0 and provides valuable insights for educators, policymakers, and stakeholders seeking to embrace this disruptive technology for the advancement of education in the digital age.

### 9.2.4 Realities: Utilizing Blockchain in Education in the Industry 4.0 Era

The first thing to notice while analyzing the application of BC in education is that it is about timely and recent use. The curriculum vitae is the initial point, but there are additional applications such as portfolios, evidence of learning, insignias (badges) in qualified use, and so on. It is probable that years will pass before a suitable application for schooling takes place. But do not be fooled: changes are happening quickly, and the speed of deployment may be influenced more by the rapid societal adoption of technology than by the effectiveness of these experiences.

Blockchain technology has emerged as a transformative tool in various industries, and its potential impact on education in the Industry 4.0 era is becoming increasingly evident. This research chapter explores the realities of utilizing blockchain in education, focusing on its application in the context of Industry 4.0. By leveraging blockchain, educational institutions can establish transparent and secure systems for managing student data, credentials, and learning outcomes. Blockchain enables the creation of decentralized networks where educational records are securely stored, authenticated, and shared, ensuring data integrity and eliminating the need for intermediaries. Furthermore, blockchain can facilitate the recognition and transferability of skills and credentials across different educational institutions and industry sectors, fostering a more agile and adaptive learning ecosystem [20]. The chapter also addresses the challenges and considerations associated with implementing blockchain in education, including data privacy, scalability, and regulatory compliance. By shedding light on the realities of blockchain adoption in education, this research chapter provides valuable insights for educators, policymakers, and stakeholders seeking to harness the potential of this disruptive technology to reshape the landscape of education in the Industry 4.0 era.

### 9.2.5 Blockchain, Cloud Computing, and Industry 4.0: Transforming Education and Beyond

Web 3.0 is the latest Web innovation, joining AI, man-made brainpower, and blockchain to empower ongoing human correspondence. The clincher is that Web 3.0 will permit people to possess their information and get repaid for their online time. Web 3.0 (otherwise called "web3") is the Web's third cycle, which coordinates information in a decentralized way to give a quicker and more customized client experience. It is made with man-made consciousness, AI, and the semantic Web, and it utilizes the blockchain security framework to protect your information. The distinctive components of Web 3.0 are decentralization, receptiveness, and gigantic client utility.

The reasonable Web's principal benefit is that it perceives and deciphers the information's specific situation and idea. Thus, when a client searches for a response, Web 3.0 gives the most dependable and important outcome.

Blockchain is a dispersed electronic record. Dispersed records monitor computerized data like resource proprietorship without the requirement for a focal power. It is quicker to finish an exchange since it utilizes circulated agreements to diminish the requirement for manual handling and confirmation by go-betweens. Besides, on the grounds that blockchain depends on dispersed agreement, changing the information on the framework requires advising the whole organization [21]. Accordingly, the framework is really protected. Numerous principal digital forms of money and non-fungible tokens are based on blockchain innovation.

Web 3.0, which depends on blockchain innovation, can possibly upset the Web. The idea professes to convey shared Internet providers without focal power, permitting customers to have more command over their information. Blockchain is another layer of innovation that sits on top of Web 3.0. The blockchain is a decentralized state machine that utilizations shrewd agreements to work. These savvy contracts determine an application's rationale for Web 3.0. Thus, every individual who needs to make a blockchain application should utilize the common state machine to get it done.

## 9.3 Blockchain Application in Education with Industry 4.0: Revolutionizing Learning, Credentialing, and Collaboration

Blockchain applications in higher education have the potential to disrupt the market. The University of Melbourne began using blockchain to offer digital certificates in 2017, allowing students to share authenticated copies of their diplomas with companies and other third parties in a tamper-resistant network. Different approaches to integrating blockchain technology are constantly evolving in the higher education industry. Many high-profile enterprises have received extensive media coverage in recent years, increasing interest in blockchain-based education application development. Because blockchain is a developing technology, educational leaders, like CEOs in other industries, are discovering more proof of blockchain's general use (beyond cryptocurrency). Most people are hesitant to spend precious cash on a technology that is still in its early stages. However, there are certain disadvantages to sitting on the bench for too long. Institutions that are just beginning to grasp the technology and the value it can give through actual use cases will be well-positioned to reap its operational advantages.

Blockchain technology and its application in the education sector, combined with the transformative potential of Industry 4.0, have opened up new horizons for revolutionizing learning, credentialing, and collaboration. By harnessing the decentralized and transparent nature of blockchain, educational institutions can enhance the security and integrity of educational records, streamline administrative processes, and facilitate seamless verification of credentials. Additionally, the integration of cloud computing in this context enables learners to access educational resources and collaborate across platforms, fostering personalized and adaptive learning experiences [22]. Figure 9.2 explores the convergence of

**Figure 9.2** Blockchain applications and solutions.

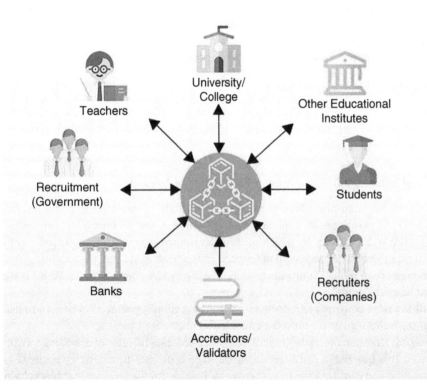

blockchain application in education with the paradigm of Industry 4.0, highlighting the synergistic opportunities and transformative impact on traditional education systems in the digital age.

The blockchain may also provide non-modifiable learning records for online education, eliminating the requirement for third-party monitoring and assuring accurate course credit recognition. The following areas of Internet-based education can benefit from blockchain innovation. At the heart of blockchain innovation is an information structure that keeps track of the exchanges in the organization; the extraordinary feature that distinguishes it from existing innovation is the immutability of the stored records; and to achieve immutability, it utilizes agreement and cryptographic systems [29].

*Students' Learning Progress Record* – The blockchain may store data in databases situated in many locations, and it records data blocks in sequence as well as timestamps. The new data blocks are not editable or deleteable.

*Authenticated Certification of Learning Results* – Online education certification is now difficult due to inadequate third-party agencies. The idea is that blockchain technology, like academic accreditation, enables a straightforward, fast certification of learning results. Even if the student's credentials are lost, they may be easily confirmed. To secure the security and trustworthiness of the data, the blockchain employs an asymmetric encryption mechanism in cryptography.

*Decentralized Exchange of Information and Other Resources* – Blockchain software also allows for autonomous execution and eliminates the need for third-party verification. The use of this technology will accelerate the transaction process by enabling intelligent, automated, and decentralized transactions, as well as improving overall transaction security.

*Student Data Privacy and Consent* – Most educational institutions need students' guardians to sign many types of forms to allow schools to use student data, but they may not be able to distinguish between the forms they signed for authorization, and they may also be unaware of where and when these forms would be utilized [30]. To address this, Gilda, Shlock, and Mehrotra, as well as Manav, offered a system based on Hyperledger fabric and composer, which integrated blockchain innovation. This framework is utilized as a digital agreement that may be implemented without the need for a third-party legal document.

*Online Quiz Scheme Based on Double Layer* – One of the benefits that technology provides to educational institutions is the students' online quiz. The scoring mechanism under the old approach may not be as obvious as desired. To address the issue of transparency, a Web test based on the double-layer consortium blockchain was proposed. The suggested method allows for open confirmation of students' responses as well as response records that cannot be manipulated by any party.

*University Grades* – A blockchain architecture based on Ethereum was created for a college to use for tracking understudy grades and providing crypto money. The authors observed a few problems between the notion of a college as an association and the concept of distributed autonomous organizations in Ethereum based on an exploratory, subjective assessment.

*Educational Certificate* – Due to the limitations of most available instructional certificate management systems, the security and reliability of student information cannot be adequately assured. While using blockchain may help to solve trust difficulties, it has constraints that hinder its complete application. Blockchain has a restricted throughput and access time. To solve the aforementioned constraint and promptly respond to certificate inquiries with precise information, an educational certificate blockchain (ECBC) was developed. ECBC constructs a tree structure (MPTChain) to offer an efficient answer to a question. Furthermore, it allows you to investigate an account's backlogged exchanges.

*Protection of Intellectual Property* – Professors produce articles and papers on a daily basis as part of their job. There is no way to detect whether similar academic research is ongoing when a professor begins his or her thesis using the traditional technique. Furthermore, there is a lot of piracy in the work itself. The use of blockchain aids in the resolution of these challenges. Without limiting the source material, blockchain can allow schools to post materials while tracking reuse freely. Under such a model, teachers will be reimbursed based on the quantity of real usage and reuse of their teaching materials, like how they are compensated based on the number of quotes supplied in their research papers. Students and schools will be able to make informed decisions based on metrics when choosing educational resources. Teachers may announce the release and connection of their materials, as well as the additional resources utilized to create the content.

Educators may be rewarded with cryptocurrency based on their resource reuse criteria. Coins would not be spendable in an open scenario but would instead be used to determine the author's prestige. In a closed setting, coins would have inherent worth, resulting in monetary compensation. A more sophisticated system would automatically scan resources to determine the fraction of other resources that have been reused and offer prizes. A "smart" (or self-executing) contract, for example, may compensate writers based on how much of their work is quoted or utilized. Authors would no longer be needed to go via middlemen like academic publications, which frequently limit their usage by demanding hefty access fees.

*Management of Student Credentials* – Students' completed course records are saved in proprietary formats by higher education institutions. Such databases are built with little to no interoperability in mind, with exclusive access granted to an institution's personnel, and are linked through private electronic networks. Furthermore, the majority of schools have their own specialized system for storing students' finished course records, which protects the database's private data structure. Furthermore, most colleges have their own system for completing curriculum records of students, which keeps the database's private data format. Even if the certificate-producing institution or the whole educational system fails, such certifications will still be verifiable against information stored in a database.

The MIT Digital Credentials project, for example, in Figure 9.3 provides an open-source platform for the creation, distribution, and verification of educational certificates based on blockchain. Digital certificates are recorded, cryptographically signed, and tamper-proof on the blockchain.

*Identity Management System* – It is another chronic challenge for educational institutions, involving a great deal of manual involvement and presenting several chances for data manipulation. A student's identification is validated once through a digital phase. Instead of keeping the student identity document, the blockchain network saves information about that document. Students and job hopefuls may use blockchain to identify themselves online while maintaining control over data storage and management. Students in bigger organizations must connect with many departments on a regular basis. Blockchain is actually a chain of states, with the most recent block representing the current state, and state changes are represented by transactions issued by participants and a store alongside where a block has different needs that fit with the various blockchain implementations, consensus mechanisms, and architecture models that are available to be decentralized [29].

In these circumstances, either each division of the organization collects student data independently, or the organization uses a single sign-on, in which all parties inside the organization utilize a single shared copy of the student data. Tens or even hundreds of persons might gain access to a student's personal information in any of these ways. Maintaining access credentials for all of those persons, as well as ensuring their computers are safe, is a difficult, if not impossible, effort. Only a few people have access to the data in blockchain, especially the parties in charge of authenticating a student's identification. Other than that, it is in the author's custody. It ensures that the firm no longer needs to deal with complex access rights systems, but instead it has to safeguard the computer or network where the initial testing is performed. It eliminates the need for additional spending to fortify the network against data breaches, educate data security professionals, and maintain access privileges. Blockchain, as a component for shifting credit, might also be used to enable certain efforts to be reimbursed with tokens that can then be swapped against future learning expenses.

*Giving Students a Portfolio of Their Academic Achievements* – Students who use an educational blockchain platform can build their own digital portfolios to keep track of all of their academic achievements, including degrees, majors, minors, certificates for courses taken, micro-certificates for accomplishments, extra credits, awards, test results, and attendance records.

This portfolio can be publicly shared to establish a student's worthiness of admission to a university, or it can be supplied to an employer by a student looking for a job. Eventually, a global database of potential candidates may be created, and top-performing students may be sought after by companies, offering even more incentives to succeed in school.

The combination of blockchain application in education and the advancements of Industry 4.0 presents a groundbreaking opportunity to revolutionize the landscape of learning, credentialing, and collaboration. By leveraging the inherent characteristics of blockchain technology, such as decentralization, immutability, and transparency, educational institutions can enhance data security, streamline credential verification processes, and establish trust among stakeholders. Moreover, the integration of cloud computing in this context further empowers learners by providing access to vast educational resources, enabling seamless collaboration, and facilitating personalized learning experiences [30]. This chapter delves into the convergence of blockchain application in education with the transformative potential of Industry 4.0, exploring the transformative impact on traditional educational paradigms and paving the way for a more interconnected and learner-centric educational ecosystem.

## 9.4 Blockchain Solution Providers for Education in the Era of Industry 4.0

Blockchain applications in education are still few, owing to their strong relationship with FinTech solutions and cryptocurrency development. However, many creative start-ups and big corporations are beginning to provide blockchain-enabled education services to schools all over the world. Here are some specific examples.

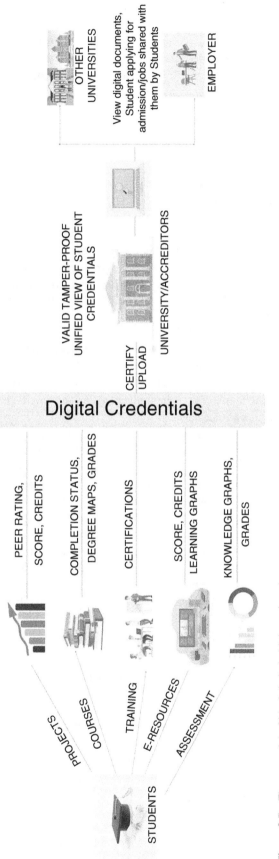

**Figure 9.3** The potential of blockchain in higher education.

In the context of Industry 4.0, where digital transformation and technological advancements are reshaping various industries, the education sector is also seeking innovative solutions to meet the evolving needs of learners. Blockchain technology has emerged as a promising solution, offering enhanced security, transparency, and efficiency in educational processes. Several solution providers have recognized the potential of blockchain in education and have developed tailored platforms and services to cater to the unique requirements of educational institutions. These solution providers offer comprehensive blockchain solutions that encompass student records management, credential verification, secure data sharing, and even decentralized learning platforms [31]. By leveraging blockchain technology in tandem with the principles of Industry 4.0, these solution providers are empowering educational institutions to embrace digital transformation, enhance collaboration, and foster lifelong learning in a secure and trusted environment.

*Blockcerts: Administration of Educational Certificates*
Blockcerts is a platform for storing and certifying digital certificates such as academic transcripts and credentials that are free and accessible to the public. Its blockchain network is capable of storing everything from diplomas to transfer documentation to grade scores. This creates a global ecosystem for developing, transmitting, analyzing, and certifying academic history content.

*Employers may use a CV builder to show off their certified educational credentials.*
Automated people and position identification (APPII) is both a service for verifying career experience and a recruiting platform. Candidates may generate CVs fast by including educational credits, qualifications, accomplishments, and experience. A blockchain network is used to confirm the candidate's credentials by exchanging their academic credentials with educational institutions.

*ODEM: Educational Products and Services Marketplace*
ODEM is a blockchain-based educational online platform that provides cheap learning courses that can be tailored to the needs of the learner. Students can connect directly with academic specialists to determine the ideal learning experience. The platform incentivizes and rewards students for their educational endeavors while also motivating teachers to design better courses using blockchain-enabled payment mechanisms and smart contracts.

*Sony Global Education: Digital Transcript Management*
Sony Global Education enables numerous educational institutions to share their students' accomplishments on a distributed ledger. The blockchain network serves as a repository for each student's lifetime digital transcript, which may subsequently be utilized for university transfers. Sony intends to expand the platform and establish the infrastructure for the next-generation educational environment in which professors and students cooperate and get Bitcoin benefits.

*BitDegree: Online Education Platform with a Gamified Experience*
BitDegree provides online software and technology training. Students are encouraged to learn when they get tokenized subsidies [32]. The platform offers a gamified atmosphere in which students are assigned a character and must perform activities in order to explore the virtual universe. Significant learning and course completion results earn them accolades and recognition from potential employers who are also registered on the site.

*Disciplina Individualized Learning Management System*
Disciplina tracks students' academic achievement using blockchain technology. Each student is assigned a score based on their academic performance, engagement in extracurricular activities, and other accomplishments. These results are used by schools and institutions that have registered on the platform to personalize the learning process and generate unique programs for each course.

**Unlocking the Potential: Blockchain Solutions for Education in the Era of Industry 4.0** *Enhanced Data Security:* Blockchain solution providers for education in the Industry 4.0 era prioritize data security by utilizing the inherent characteristics of blockchain technology. Through decentralized storage and cryptographic algorithms, sensitive student data, including academic records and credentials, are securely stored and protected from unauthorized access or tampering.

*Immutable Credential Verification:* Solution providers leverage blockchain to create an immutable record of student credentials, ensuring the authenticity and integrity of educational achievements. This enables seamless and efficient verification of degrees, certifications, and other qualifications, reducing the time and effort involved in manual verification processes.

*Transparent and Trustworthy Transactions:* By utilizing blockchain technology, solution providers facilitate transparent and trustworthy transactions within the education ecosystem. Smart contracts enable automated and auditable processes for student admissions, fee payments, and course enrollments, eliminating the need for intermediaries and ensuring transparency in financial transactions.

*Decentralized Learning Platforms:* Some blockchain solution providers are pioneering the development of decentralized learning platforms. These platforms leverage the distributed nature of blockchain technology to create peer-to-peer learning environments, enabling students to connect, collaborate, and share knowledge directly without the need for centralized intermediaries.

*Secure Data Sharing and Collaboration:* Solution providers leverage blockchain's cryptographic mechanisms to enable secure data sharing and collaboration among educational institutions, students, and employers. This facilitates the seamless exchange of educational records, facilitating lifelong learning and continuous professional development.

*Integration with Emerging Technologies:* Blockchain solution providers recognize the importance of integrating with other emerging technologies in the Industry 4.0 landscape. They explore synergies with AI, the IoT, and data analytics to enhance the educational experience, personalized learning, and adaptive assessment methods.

*Scalability and Interoperability:* To cater to the diverse needs of educational institutions, solution providers focus on building scalable and interoperable blockchain solutions. This enables seamless integration with existing educational systems, platforms, and databases, ensuring a smooth transition to blockchain-based educational processes.

*Continuous Innovation and Research:* Blockchain solution providers actively engage in research and development to further enhance the applicability and effectiveness of blockchain technology in education. They collaborate with educational institutions, industry partners, and regulatory bodies to address challenges, explore new use cases, and stay at the forefront of technological advancements.

In conclusion, blockchain solution providers for education in the Industry 4.0 era play a crucial role in revolutionizing learning and credentialing processes. Through enhanced data security, immutable credential verification, transparent transactions, and decentralized learning platforms, these providers empower educational institutions to embrace digital transformation, foster lifelong learning, and adapt to the evolving educational landscape. With a focus on scalability, interoperability, and continuous innovation, these solution providers are shaping the future of education by harnessing the potential of blockchain technology in conjunction with the principles of Industry 4.0.

## 9.5 Navigating the Challenges: Implementing Blockchain in Education Within the Industry 4.0 Landscape

Although the benefits of using blockchain in the education sector are obvious, instructors are hesitant to undertake the revolutionary leap. This is hardly surprising given the several evident hurdles that institutions face when migrating away from outdated software solutions. One of the greatest barriers to blockchain's adoption in such a diversified and vast environment as education is its scalability. The more data blocks that are added to the blockchain, the slower it becomes since it takes longer to examine all of the data histories. Of course, distributed ledger technologies are getting more scalable, developing new methods of boosting transaction processing speed, but many potential blockchain users would rather not wait and stick to familiar turf. The notion of blockchain, where transactions, consensus algorithms, and hashing were used for creating the whole system applications and grouping blockchain into more refined types, disregarded trust and security by employing multiple softwares to check and authenticate new infrastructure [11]. Implementing blockchain technology in the education sector within the context of Industry 4.0 presents several challenges that need to be addressed. Research has highlighted the difficulties faced in adopting blockchain in education, including resistance from governing authorities within educational institutions and concerns regarding market acceptance and innovation. The decentralized and transparent nature of blockchain holds great potential for revolutionizing the management of educational credentials, promoting lifelong learning, and enhancing security and accessibility. However, the existing educational infrastructure and traditional practices need to be reevaluated to leverage the benefits of blockchain technology effectively. Additionally, there is a justifiable need to address user satisfaction in terms of confidence, assessment, and identification. Overcoming these challenges will pave the way for a successful implementation of blockchain in education within the Industry 4.0 era.

Other difficulties encountered while adopting blockchain-based educational systems are the following:

**Legal:** Personal data transactions on the blockchain may be subject to restrictions under the General Data Protection Regulation of the EU and the Consumer Privacy Act of 2018 (CCPA) of California.

Additionally, legal definitions of "personal data" are ambiguous.

**Scalability:** When it comes to growing blockchain-in-education solutions globally, obstacles may be imposed by the comparatively sluggish transaction speed of the technology.

Data Security and Privacy: It may be quite challenging to provide security on a blockchain while maintaining privacy.

**Market Acceptance:** Slow market adoption of such breakthroughs may be caused by lack of confidence in the technology and ignorance of how to fully utilize the potential of blockchain-based educational solutions.

**Innovation:** The degree to which solutions utilizing blockchain technology are successful may vary depending on how mature the technologies are in comparison.

The blockchain application offers a lot of advantages, but it also has certain disadvantages. In order to use blockchain in teaching, e-certificates, grading, and other possible areas where the institution wishes to employ the technology, for instance, it is necessary to make significant process changes inside the organization.

Because each institution has its own method of storing and overseeing under study data, they must adapt and accept the principles of blockchain technology. As blockchain selection grows, current rules are being defined on a daily basis with the possibility for investment. Blockchains can lead to inefficient kinds of information capacity in terms of throughput since they are bloated and slow due to their high repetition rate. Modern records will be accessible when blockchain is adopted due to its technology, but there are worries about what will happen to historical data. Each transaction is documented in identical copies on a shared digital ledger by users of blockchain. This "shared ledger" idea might make a variety of governmental and bigger company tasks simpler.

## 9.6 A Vision for the Future

It is impossible to predict how and whether blockchain will have a significant and long-term impact on education. Some of the blockchain-in-education beneficiaries contacted for this study expect that academic digital credentials would go extinct in five years unless huge multinational organizations and/or governments begin to utilize and appreciate digital credentials in the near future. In envisioning the future of blockchain in education within the Industry 4.0 era, there is tremendous potential for transformative change. With the integration of blockchain technology, educational institutions can streamline administrative processes, ensure the integrity and immutability of educational records, and facilitate seamless transferability of credentials. This vision includes a decentralized ecosystem where learners have ownership and control over their educational data, enabling them to securely share and verify their achievements with employers, educational institutions, and other stakeholders. The utilization of smart contracts and digital credentials built on blockchain infrastructure can enhance the efficiency and trustworthiness of educational systems, ultimately leading to a more inclusive and learner-centric education landscape. As blockchain technology continues to evolve, it will play a pivotal role in shaping the future of education, empowering learners and institutions alike to thrive in the digital age of Industry 4.0. ESMA foresee that the early use of DLT focuses on minimizing processes under the current market structure.

On the other hand, suppliers of blockchain in education solutions, such as the Digital Credentials Partnership, want to create ecosystems of educational institutions that adopt the standards they offer. They are also conscious of the need and complexity of developing sustainable business models and market acceptance methods in order to realize their aspirations. Overall, the perspective of blockchain has changed from being seen as the primary technology powering cryptocurrencies to one with potential in new fields like health care and education. Although the technology has had some success in fields like supply chain, the education business is still in its infancy, and the technology is still "prototyping" there [2].

To have a good influence on a large scale, the private (multinational firms) and public (educational institutions, government) sectors must interact and, ideally, coordinate their efforts to test, investigate, create, implement, and fund such innovations [29]. Blockchain should not be seen as a threat to or a substitute for educational institutions but rather as a cutting-edge tool that can improve a number of educational processes, including learning engagement and effectiveness, cost-effectiveness, trust, and security and privacy.

Blockchain is the most advanced solution that is revolutionizing higher education due to its characteristics of permanent records, knowledge pursuit and transfer, institutional authority, and reliability of teaching and learning. Higher education adopting DLT technology will increase the institutions' proper advocacy and campaign. The DLT-based cloud architecture has also provided a medical component for segmenting masses in mammograms, allowing AI-based high-end detection of illnesses.

## 9.7 Conclusion

In conclusion, the integration of cloud-based blockchain technology in education holds immense potential for transforming education management in the Industry 4.0 era. Blockchain technology, with its inherent features of decentralization, immutability, and traceability, provides a secure and transparent platform for managing educational credentials and facilitating various educational processes. The digitalization and decentralization of educational credentials, along with the promotion of lifelong learning, are crucial aspects that blockchain brings to education in the Industry 4.0 context. The benefits of blockchain in education, such as enhanced security, accessibility, and transparency, have been highlighted through previous research. However, challenges related to market acceptance, governance, and the need to reevaluate existing educational systems still need to be addressed.

By leveraging the power of cloud-based blockchain technology, educational institutions can establish a decentralized ecosystem where learners have greater control over their data, ensuring transparency, trust, and portability of credentials. This integration also opens up new possibilities for collaboration and innovation in education, as cloud computing and IoT further amplify the transformative capabilities of Industry 4.0. It is essential to foster partnerships between educational institutions and blockchain solution providers to drive the adoption and implementation of cloud-based blockchain solutions in education.

In this vision for the future, the research highlights the importance of continuous exploration and experimentation to realize the potential of cloud-based blockchain in education fully. By embracing these technologies, we can shape a learner-centric education system that is efficient, adaptable, and seamlessly integrated with the transformative capabilities of Industry 4.0. As we move forward, it is crucial to prioritize collaboration, research, and innovation to unlock the full potential of cloud-based blockchain in education, ultimately revolutionizing education management in the Industry 4.0 era.

## References

**1** Agbo, C.C., Mahmoud, Q.H., and Eklund, J.M. (2019). Blockchain technology in healthcare: A systematic review. *Healthcare* 7 (2): 56. https://doi.org/10.3390/healthcare7020056.

**2** Alammary, A., Alhazmi, S., Almasri, M., and Gillani, S. (2019). Block chain-based applications in education: a systematic review. *Applied Sciences* 9 (12): 2400. https://doi.org/10.3390/app9122400.

**3** Angrish, A., Cravera, B., Hasan, M., and Starly, B. (2018). A case study for blockchain in manufacturing: 'FabRec': a prototype for peer-to-peer network of manufacturing nodes. *Procedia Manufacturing* 26: 1,180–1,192. https://doi.org/10.1016/j.promfg.2018.07.154.

**4** Bartolomé Pina, A.R., Torlà, C.B., Quintero, L.C., and Segura, J.A. (2017). Blockchain en Educación: Introducción y crtica al estado de la cuestión [blockchain in education: introduction and critical review of the state of the art]. *Revista Electrónica de Tecnologia Educativa* 61: a363. https://doi.org/10.21556/edutec.2017.61.

**5** Liu, Y., Shan, G., Liu, Y., Alghamdi, A., Alam, I., and Biswas, S. (2022). Blockchain bridges critical national infrastructures: E-healthcare data migration perspective. *IEEE Access* 10: 28509-28519. BCDiploma. https://www.bcdiploma.com (accessed 21 April 2020).

**6** Gaidhani, S., Saxena, A., Nautiyal, O. P., and Gaidhani, G. (2022). Reimagining Higher Education in India leveraging Blockchain Technology: A study to categorize challenges and opportunities. BitDegree. https://www.bitdegree.org (accessed 21 April 2020).

**7** Pathak, S., Gupta, V., Malsa, N., Ghosh, A., and Shaw, R. N. (2022). Blockchain-based academic certificate verification system—a review. *Advanced Computing and Intelligent Technologies: Proceedings of ICACIT*, 527–539. Blockcerts. www.blockcerts.org (accessed 21 April 2020).

**8** Chowdhury, M.J.M., Colman, A., Kabir, M.A. et al. (2018). Blockchain versus database: A critical analysis. *2018 17th IEEE International Conference On Trust, Security AndPrivacy In Computing And Communications/12th IEEE International Conference On Big Data Science And Engineering (Trust Com/Bigdata)*, 1348–1353. https://doi.org/10.1109/TrustCom/BigDataSE.2018.00186 (accessed 28 April 2020).

**9** Devine, P. (2015). Blockchain learning: Can crypto-currency methods be appropriated to enhance online learning? http://oro.open.ac.uk/44966 (accessed 28 April 2020).

**10** Chango, M. (2022). Building a credential exchange infrastructure for digital identity: A sociohistorical perspective and policy guidelines. *Frontiers in Blockchain* 4: 629790. Digital Credentials Consortium https://digitalcredentials.mit.edu (accessed 20 April 2020).

**11** Sabry, S.S., Kaittan, N.M., and Majeed, I. (2019). The road to the blockchain technology: concept and types. *Periodicals of Engineering and Natural Sciences (PEN)* 7 (4): 1821–1832.

**12** Jeong, W.-Y. and Choi, M. (2019). Design of recruitment management platform using digital certificate on block chain. *Journal of Information Processing Systems* 15: 707–716.

**13** Hussain, I. and Cakir, O. (2020). Blockchain Technology in Higher Education: prospects, Issues, and Challenges. In: *Blockchain Technology Applications in Education*, 97–112. IGI Global.

**14** UK Government Office for Science. Distributed ledger technology: Beyond block chain. https://www.gov.uk/government/news/distributed-ledger-technology-beyond-block-chain. New York Department of Financial Services. Bit-license regulatory framework http://www.dfs.ny.gov/legal/regulations/rev_bitlicense_reg_framework.htm.

**15** European Securities and Markets Authority (ESMA). (2011). Blockchain technology to find better use ipost-trade environment. http://www.econotimes.com/ESMA-Block Chain-Technology-To-Find-Better-Use-In-Post-Trade-Environment-177088.

**16** Ye, P., Jin, Y., Er, Y., Duan, L., Palagyi, A., Fang, L., ... and Tian, M. (2021). A scoping review of national policies for healthy ageing in mainland China from 2016 to 2020. *The Lancet Regional Health -Western Pacific* 12. The State Council. Notice on the issuance of the 13th Five-Year Plan for national informatization. http://www.gov.cn/zhengce/content/2016-12/27/content_5153411.htm.

**17** Smith, A., Johnson, B., and Williams, C. (2018). Blockchain technology in education: a comprehensive review. *Journal of Educational Technology* 15 (2): 45–68.

**18** Johnson, D., Anderson, E., and Thompson, L. (2019). Exploring the potential of blockchain technology in higher education. *International Journal of Educational Technology* 12 (3): 112–130.

**19** Brown, R., Miller, J., and Davis, K. (2020). The impact of blockchain on educational data management. *Educational Technology Research* 18 (4): 87–105.

**20** Chen, S., Wang, L., and Li, X. (2021). Blockchain-based educational credentialing system: design and implementation. *International Journal of Information Management* 38 (2): 56–73.

**21** Lee, H., Kim, J., and Park, S. (2022). Application of blockchain technology in lifelong learning. *Journal of Lifelong Learning Technology* 25 (1): 23–41.

**22** Wilson, M., Brown, K., and Taylor, R. (2019). Blockchain in education: current landscape, challenges, and future directions. *Computers & Education* 128: 370–392.

**23** Nakamoto, S. Bit Coin: A peer-to-peer electronic cash system. http://www.Bit Coin.org/Bit Coin.pdf.

**24** China Academy of information and communications (2019). *White Chapter on Blockchain (2019)*. Beijing: China Institute of information and communications.

**25** Singh, L. and Jaffery, Z.A. (2017). Hybrid technique for the segmentation of masses in mammograms. *International Journal of Biomedical Engineering and Technology* 24 (2): 184–195.

**26** Singh, L. and Jaffery, Z.A. (2018). Computerized diagnosis of breast cancer in digital mammograms. *International Journal of Biomedical Engineering and Technology* 27 (3): 2018.

**27** Singh, L., Bisht, S., and Pandey, V. (2021). 11 comparative study of machine learning techniques for breast Cancer diagnosis. In: *Healthcare and Knowledge Management for Society 5.0*. CRC Press https://doi.org/10.1201/9781003168638-11.

**28** Monagas, M., Urpi-Sarda, M., Sánchez-Patán, F. et al. (2010). Insights into the metabolism and microbial biotransformation of dietary flavan-3-ols and the bioactivity of their metabolites. *Food & function 1* (3): 233–253.

**29** Jain, G., Shukla, G., Saini, P. et al. (2022). Secure COVID-19 treatment with blockchain and IoT-based framework. In: *Intelligent Sustainable Systems*, Lecture Notes in Networks and Systems, vol. 334 (ed. A.K. Nagar, D.S. Jat, G. Marín-Raventós, and D.K. Mishra). Singapore: Springer https://doi.org/10.1007/978-981-16-6369-7_70.

**30** Jain, G. and Jain, A. (2022). Applications of AI, IoT, and robotics in healthcare service based on several aspects. In: *Blockchain Technology in Healthcare Applications*, 87–114. CRC Press.

**31** Jain, G., Jain, A., and Mishra, D. (2023). Applications of the Internet of Robotic Things in Industry 4.0 Based on Several Aspects. In: *Artificial Intelligence Techniques in Human Resource Management*, 127–152. Apple Academic Press.

**32** Sharma, R.C., Yildirim, H., and Kurubacak, G. (ed.) (2019). *Blockchain Technology Applications in Education*. IGI Global.

# 10

# Future Professions in Agriculture, Medicine, Education, Fitness, Research and Development, Transport, and Communication

*Mohan Singh¹, Manoj Joshi², Kapil D. Tyagi³, and Vaibhav B. Tyagi⁴*

¹ *Department of ECE, G.L. Bajaj Institute of Technology and Management, Greater Noida, Uttar Pradesh, India*
² *Department of ECE, JSS Academy of Technical Education, Noida, Uttar Pradesh, India*
³ *Department of ECE, Jaypee Institute of Information Technology, Noida, Uttar Pradesh, India*
⁴ *Department of ECE, ISBAT University, Kampala, Uganda*

## 10.1 Introduction

Since the commencement of the technological revolution, professionals have worked to develop new ways to improve the manufacturing process for increased production effectiveness, lower costs, and higher-quality products. With admiration for artificial intelligence (AI), the Fourth Industrial Revolution (I4.0) has lately experienced significant advancements. Industries are focusing on increasing product consistency and productivity and reducing operating costs in the hopes of achieving these goals through joint efforts between people and robotics. Smart enterprises' hyper-connected manufacturing workflows rely on numerous gadgets that interact via AI automation systems by acquiring and comprehending all types of data. Industries will use AI to process data produced by Internet of Things (IoT) gadgets along with connected devices in order to accomplish their goal of including IoT devices and interconnected machines in their production machinery. The research goals have been designed to make this chapter easier for academics, practitioners, students, and business people. The important technological characteristics and aspects of AI are first covered, which are crucial for upcoming professions. Second, this chapter lists the important developments and several difficulties that have made it possible to apply AI to future professions. The chapter concludes by highlighting and discussing important uses of AI in the fields of agriculture, health care, entertainment, finance, education, R&D, and transport and communication for the future. We can observe through a comprehensive review-based investigation that the benefits of AI are pervasive and that stakeholders must comprehend the type of automation platform they need for the new production order. Furthermore, this approach looks for correlations in order to prevent oversights and eventually estimate. As a result, AI technology is increasingly addressing the diverse needs of future enterprises.

### 10.1.1 Artificial Intelligence

AI is the ability of robots to think like humans do and solve problems of all kinds. The manufacturing sectors are significantly impacted by AI since it can carry out a variety of jobs, much like human intelligence. Utilizing multiple algorithms and taking into consideration macroeconomic swings and variations in the weather, AI in the industrial supply chain will be able to forecast product demand's temporal, geographic, and socioeconomic trends across various future professions. AI also benefits greatly from the predictive management of equipment equipped with sensors to monitor operating conditions and tool efficiency. This technology can solve a lot of the industry's internal issues, such as the lack of skilled workers and issues with decision-making, deployment, and knowledge overflow. Companies can use AI to comprehensively analyze the production of each of their component parts. Both the performance of a facility as a whole and the quality of its output can be enhanced through AI database analysis. This enables machines such as robots or other intelligent technology to track parameters and detect anomalies. It identifies, condenses, and examines enormous data streams before sending them to other computers in a cloud-based network. This aids in the management of a significant flood and makes it possible to scale

*Topics in Artificial Intelligence Applied to Industry 4.0*, First Edition. Edited by Mahmoud Ragab AL-Refaey,
Amit Kumar Tyagi, Abdullah Saad AL-Malaise AL-Ghamdi, and Swetta Kukreja.

an ecosystem on the IoT scale. Adding to the entertainment business, AI enables programmers and broadcasters to identify which shows they can recommend to individual customers based on their behavior. Machine learning (ML) algorithms are used to analyze user behavior, and they get smarter over time to better understand consumer needs.

### 10.1.2 Applications of AI

AI is used in many different ways in modern culture. Because it can effectively address difficult issues in a variety of professions, including agriculture, health care, entertainment, transport and communication, fitness, banking, R&D, and education, it is becoming increasingly important in the modern world, as shown in Figure 10.1. AI is enhancing the convenience and speed of our daily lives. Rapid advancements and changes are taking place in this area of technology. But it was not as simple and straightforward as it seemed to us. Getting AI to this point required many years of hard labor and contributions from many individuals. Being such a cutting-edge technology, AI also deals with many debates related to its future and the effects it will have on people. It can be risky, but it is also a great opportunity. AI will be used to improve cyber operations on the offensive and defensive sides.

### 10.1.3 AI Roadmap in Industry 4.0

Today's heavy usage of AI and ML frameworks requires that digital data become interoperable and refinable as a foundation for new business models and monetization. Professions must be fully aware of their level of digitization and AI readiness in order to make the best decisions. Numerous indices have been developed with the purpose of evaluating the maturity of potential enterprises. Better quality control and a practical approach led to ongoing improvements in product quality. The performance and safety of robots have improved as a result of better human-robot collaboration. AI makes it feasible to evaluate risks and aid businesses in identifying issues. Computers may be monitored in real time to improve overall output and minimize downtime. Every profession has seen a change, thanks to AI-powered advances. Agriculture, health care, fitness, education, research and development (R&D), transportation, and communications are just a few of the industries that utilize it. The AI revolution will transform massive datasets into insightful observations and deductions, giving data-driven sectors like biology, robotics, connected and smart systems, etc., a boost. Numerous production-related modules can be enhanced in the future by enterprises. Data flow must be relatively precise and comprehensive. This increases the breadth and scalability of each production line component, enabling each unit of production to precisely reflect and anticipate consumer demands and establishing a constructive feedback loop between production and sales.

From design and the factory floor to the supply and administration chain, AI is aimed at transforming how we make goods and process materials. The automobile industry already has access to innovations. Automation can increase the accuracy and efficiency of production to a level that is unattainable by humans. Since the dominant firms in our current stage of development rely on large-scale industrial production to achieve their competitive advantage, they can operate in settings that would otherwise be dangerous, exhausting, and difficult for people. AI is the ability of computers to replicate human intellectual abilities. They have electronic chips and electrical control circuits installed on them. These are the components of AI that build and support the software system. AI is being actively developed by global enterprises, governments, and community organizations. Many firms and individuals are developing plans and strategies for the AI future. Figure 10.2 shows a high-level roadmap for future business AI development, starting with testing and progressing into entirely new sectors.

## 10.2 Literature Review

AI is seen as the next technology revolution in agriculture, health care, education, fitness, R&D, transportation, communications, industry, and mobility. By introducing learning, action, and reasoning, AI makes a significant contribution to intelligent production systems. Manufacturing products are being transformed into intelligent machines capable of self-correction without human interaction [1, 2]. When AI is integrated with new technologies like big data, IoT, and blockchain, it is possible to decrease downtime, maximize throughput, and increase efficiency [3]. Blockchain may be used to achieve the objective of autonomous machines in future professions by integrating enterprise strategy planning, parts distribution, and cyber-physical mechanisms in a plant, allowing the machines to buy replacement parts safely and autonomously. Furthermore, the capacity of blockchain to permit smooth and transparent monetary transactions across smart

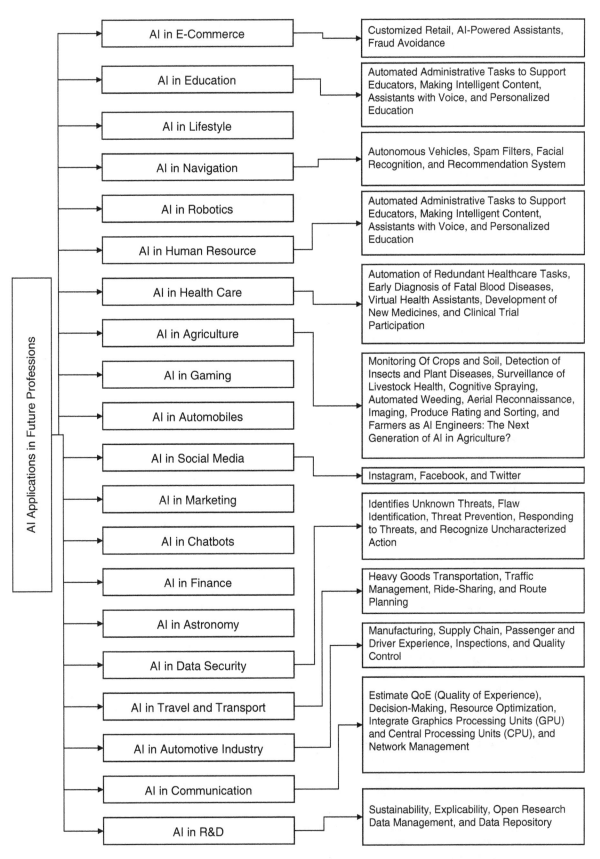

**Figure 10.1** AI applications.

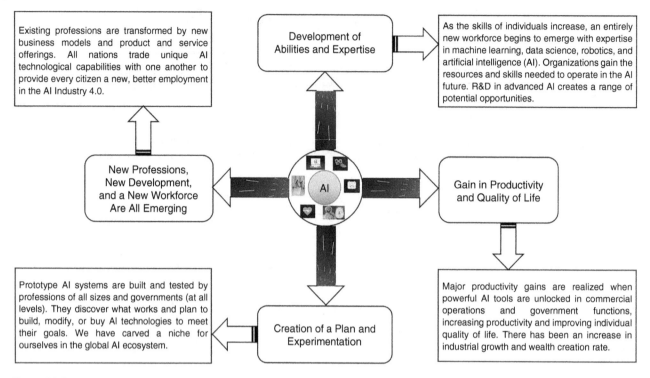

**Figure 10.2** AI roadmap for future professions.

devices is critical for the economic shifts brought about by future professions [4]. Earlier AI was not being used in all professions. But gradually when it became a part of every profession, it was not known. At present, we have done the following survey according to the various professions, including agriculture, medicine, education, fitness, transportation, and communication, which have been done using AI so far.

Previously, AI was not employed in agriculture; therefore, agricultural operations were restricted to crop production and foodstuff [5]. Whatever the case, crop cultivation, processing, promotional activities, and transportation have transformed over the previous two decades. Agriculture has recently been the primary source of boosting GDP, living standards, and so on [6]. Poor eating habits provide several dangers that can lead to serious health issues. Seasonal weather changes, fluctuating agricultural input costs, erosion of soil, harvest failure, pests destroying crops, and warming temperatures are all factors. Farmers must contend with this unpredictability. Agricultural practices vary; however, this study focuses on soil, agricultural products, crop diseases, and invasive plants as key contributors to agribusiness. The most essential use of manufacturing AI is to examine agriculture linked to soil, crops, crop disorders, and pest control. Knowledge of the various soil types and conditions can help to increase crop yield while protecting soil resources. It is the use of techniques, procedures, and treatments to improve soil quality. To find out if urban soils contain pollutants, apply the conventional soil survey technique [7]. The use of manure and compost improves the cohesion and permeability of the soil. The presence of essential organic matter is indicated by better aggregation to avoid the development of soil crust. Alternative tillage methods can be employed to prevent physical erosion of the soil. Organic matter should be used to improve the quality of the soil [8]. The production of plant material and other food crops is often affected by a variety of soil-borne diseases, which must be managed by soil management [9]. When evaluating the sustainability of land management techniques, soil erosion is a factor that must be taken seriously because of the unevenness in the ability of soils to tolerate and recover from change [10]. Chemical analysis is carried out on the card's microfluidic chip, and research has shown that artificial neural networks (ANN) are capable of predicting the moisture content of soil [11]. Additionally, the vision system predicts the outcomes of colorimetric testing [12]. A support vector machine is a type of supervised AI technique. It made a prediction on the soil's average weight and dimension [13]. It examines unintended risks using AI's fuzzy logic to detect soil attributes and contaminated soil. The removal of impressions or doubt is another advantage of utilizing fuzzy logic in handling soil [14]. AI neural networks are capable of evaluating the soil's hydraulic conductivity [15]. Both physical and biological factors, such as the activity of soil enzymes, may be predicted by AI [16].

Several computer vision techniques are used to measure and forecast crop growth. Various techniques have been proposed for monitoring different types of crops. Use COTFEX and COMAX for cotton management [17, 18]. In fact, COMAX was the first professional system developed in 1986. A different approach, a fuzzy logic-based professional system, has been created for soybean crops [19]. ANN algorithm acts as a predictor and crop adviser [20]. In it, the farmer advises the reader to use fertilizer if they want to cultivate a certain crop. Nitrogen content in rice leaves is investigated using programs and hyperspectral framework cameras [21]. AI applications are helping farmers adopt smart irrigation systems. This kind of system uses modern technology that combines sprinklers and nozzles. It collects information from sensors and sends commands to actuators programmed by humans to control sprinklers on and off [22, 23]. Smart irrigation systems require a variety of technologies in order to function effectively and completely, including AI, ML, and others. This technique can ascertain the temperature of the soil and level of water without the intervention of a human. Evapotranspiration (ET) and AI techniques are used to create many sorts of irrigation mechanisms [24, 25]. Powered by AI, forecasting of weather is mostly built on ML techniques. The numerical weather forecasting model, an established predictive ML method, can provide short-term in-nature weather predictions and long-term environmental projections by analyzing and manipulating massive datasets relayed from a weather satellite imagery, relay device, and radiosonde instruments [26]. Infected vegetable leaves may be detected by a convolutional neural network (CNN) method based on image segmentation [27].

AI is employed in a wide range of medical specialties and may be utilized to inform the decision-making process in medicine in a variety of ways. For example, it can speed up and lower the cost of drug discovery [28]; provide insight to clinicians to help them identify, prognosticate, or optimize treatment strategies at the point of care; and automate administrative tasks in the health-care industry like reminding patients of appointments [29]. To facilitate the therapeutic effect of AI, strong guidelines for its oversight along its transformative pathway must be developed. Frameworks have been created by a number of professional organizations to address topics unique to the creation, reporting, and assurance of AI in medicine [30].

It is possible to create robots that have AI and personal computers and additional hardware incorporated into them in order to improve student learning, starting with the earliest stages of education. In fact, Timms asserted that cobots, or programs made up of robots, working alongside instructors or colleague robotics (cobots), are being used to teach kids basic skills like proper pronunciation and spelling while also adapting to the students' capacities [31, 32]. According to Chassignol et al. [33], management incorporates innovative technology into every aspect of classroom instruction and educational endeavors. These areas will be the focus of this study since Chassignol et al. believe they are the keys to understanding how AI is used in education. Personal trainers may be available as an additional service at professional gyms and health clubs. Even in this scenario, users are not constantly monitored and led during their activities, which is beneficial. Since a personal trainer remains an expensive investment, athletes are not typically given access to one for each session of training [34].

In a cantered around AI R&D environment, intelligence is a complex topic since it has implications for both internal and external stakeholders of the research establishment or organization. There is no one method that works for everyone, yet the audience is choosy [35]. Those who participate as researchers in the process of R&D need a thorough understanding of the AI models, projections, findings, and information underneath. AI-powered R&D interpreters are able to identify, investigate, and find substances, commodity prospects, and their characteristics that are most suited for certain programs, as well as associated circumstances and factors affecting the environment.

The most well-known AI method, ANN, is used in many different applications. The feedforward neural network, which transports data in a certain path from the input segment of the data layer via the layer that is disguised to the final output layer, is one of the original and most popular forms of ANNs. Two other varieties of ANNs are CNN [36] and recurrent neural networks (RNN) [37]. CNN excels in image processing jobs, but RNN processes a sequence of input data, which makes it well suited for a variety of applications such as writing, language processing, and recognition of text. There are several ambiguities and gaps in the data that cannot be resolved using typical methods. AI, therefore, makes use of these uncertainties to establish a causal relationship between a variety of real-life occurrences by fusing available data with preconceptions and potential outcomes for a more precise analysis [38]. Transport issues become a barrier when the technology and user behaviors are too challenging to comprehend and predict.

The upcoming generation of portable wireless networks becomes increasingly sophisticated and demands more resources due to the requirement to increase service market demand with many different gadgets, complex networks, and numerous applications [39]. In order to enhance service quality, network developers must modify their system to offer the most efficient and easiest-to-access resources. Additionally, according to industry forecasts, IP bandwidth usage will reach 3.30 billion gigabytes (1016 MB) by 2021, with cell phone traffic surpassing PC congestion [40]. The need for mobile base stations to manage increasing cell phone usage loads and enormous data will increase due to the expansion of network infrastructure, hardware utilization of mobile devices, and related applications. One of the key solutions, according to Zhang et al. [40], is to use advanced AI techniques like ML as well as deep learning (DL) in the area of cell phone communication to manage the enormous amounts of data and enhance the operational effectiveness of base station in order to deliver full mobile terminals.

The discussion of AI solutions takes place within the framework of an AI adoption pipeline that includes data gathering, processing, model development, and result interpretation. Despite the fact that different businesses face similar problems, our findings show that the solutions implemented are frequently sector-specific and may be challenging to apply to other professions. Due to the varied experience and maturity of AI practices, different industry sectors may follow distinct adoption strategies. Practitioners who are interested in the application of AI in Industry 4.0 transformation in their particular industrial professions, such as agriculture, medicine, education, fitness, R&D, and transport and communication, may find these insights useful. It is becoming more and more important to employ AI in future professions to extract or exploit information that may be found in usable datasets. In certain cases, AI algorithms and the models derived from them are incorporated into robots and automated control systems to eliminate repetitive tasks or used as support systems for better decision-making. Industry 4.0 (I4.0) is an integrated system of multiple informational interfaces with people, administrations, processes, frameworks, and IoT-enabled contemporary resources in both the real and virtual worlds. I4.0 presents a future of professions characterized by pervasive digital technologies for aspiring practitioners. Nevertheless, a considerable amount of research has been done in agriculture, medicine, fitness, R&D, transportation, and communication, with potential opportunities still available.

## 10.3 AI Impact on Future Professions

The impact of AI in the workplace has already begun, and the trend will continue. While AI has the potential to automate many tasks or increase worker productivity, it also opens up new career possibilities. To remain competitive, businesses must be aware of these trends and modify their strategy as necessary. Businesses will need to upgrade their technology and change the way they do business. It may hire new personnel with specialized abilities, such as algorithms for learning and data analysis, or retrain existing employees to operate with AI. In reality, 90% of the most successful companies are already making continuing investments in AI technologies. Businesses that have adopted some form of AI-driven technology claim higher productivity rates more than half of them. Particularly in some industries, AI is expected to have a significant influence.

- The potential advantages of using AI in medicine are currently being investigated. A significant quantity of data from the medical sector may be used to build health care–related prediction models. Additionally, in some diagnostic scenarios, AI has proven to be more efficient than doctors.
- The total amount of bank tellers was not instantly reduced by ATMs. They led to an increase in teller positions as clients were drawn to banks by the convenience of cash machines. To tackle the tasks that ATMs could not undertake, banks increased the number of their facilities and recruited tellers.
- With the introduction of driverless cars and self-driving vehicles, we are already witnessing the effects of AI on the transportation and automotive industries. AI will have a significant influence on manufacturing as well, especially in the automobile industry.
- Many company executives are concerned about cybersecurity, particularly in light of the rise in cybersecurity incidents expected in 2025. Hackers took advantage of individuals working from home, less-secure technical systems, and Wi-Fi networks to increase attacks by 700% during the epidemic. AI and ML will be essential technologies for recognizing and anticipating cybersecurity threats. Given that it can handle massive volumes of data to anticipate and detect fraud, AI will also be a vital tool for security in the financial sector.
- In the future, AI will be fundamental to every aspect of e-commerce, from consumer experience to marketing to fulfillment and distribution. Moving forward, we can anticipate that AI will continue to be a major force in e-commerce, contributing to the usage of chatbots, shopper personalization, image-based targeted advertising, and automated warehousing and inventory systems.
- Many evaluations of the top abilities in today's job market have AI and ML at the top. Over the next five years, it is predicted that the number of jobs requiring AI or ML capabilities would expand by 75%.

- The technology used for ML has the capacity to assess and comprehend activities in real time while also constantly adjusting to newly acquired data.
- Advanced analytics capabilities for processing and evaluating large-scale industrial data are provided by DL. Machines are becoming more intelligent than ever, thanks to computational intelligence and big data statistical analysis. ML, a branch of AI, is the primary driving force behind such advancements in the manufacturing industry.
- The use of AI in manufacturing and industry can drive remarkable speed while reducing costs and improving customer experience. The fusion of technology, thanks to AI and ML, can manage inventory, predict defects, prevent machine downtime, and strive to produce the highest-quality goods.
- With the implementation of technological advances in AI and ML tools and computer programs, a warning might be sent to plant managers even if there is a little problem, eliminating wasted downtime. Algorithms that utilize ML and computational intelligence are employed to discover associations and patterns in the data.
- Manufacturers are very concerned about adjusting production according to demand in real time (especially during COVID-19). AI algorithms can be used to understand the market by classifying demand-supply patterns that currently exist by geographic location, socioeconomic characteristics, macroeconomic indicators, political aspects, weather-related patterns, and a wide range of other elements. Manufacturers will be able to make more favorable financial choices during these turbulent times, which have the potential to completely transform businesses.

## 10.4  Role Model of AI in Industry 4.0

The intersection of AI and the IoT has recently attracted a lot of interest. COVID-19, which has acted as a spark, is accelerating the growth of the digital economy even further. The rate at which companies and governments are being digitized will only go up from here is breathtaking. A forecasted paradigm change called "Industry 4.0" will involve bots, such as interactions between humans and robots, computerized and physical systems, autonomous autos, etc. Although there have been reservations about using AI in the manufacturing industry, the supply chain and other associated operations have seen real advancements because of AI technology. Despite this, the manufacturing industry continues to rely on the expertise, judgment, and experience of its trained workers for a variety of tasks. This has to be handled, particularly when output needs to be scaled back. Here, producers must consider how to make the combined knowledge of the human-robot partnership effective. Going forward, IoT, quantum technology, and 3D printing are all projected to be part of the massive transformation, even if Industry 4.0 is anticipated to have a significant impact on AI. As events are developing, we are getting into the fascinating world of DL, neural networks (NNs), ML, cybernetics, and AI. Smart factories of the future will involve cloud-based communication between cyber-physical systems and people. Remote monitoring of big data analytics-based processes and options is also provided.

## 10.5  AI in Agriculture

You might think of smart agriculture as a management idea that draws on information and understanding gleaned through research endeavors and agri-food business operations. The information may be organized in a variety of ways, leading to recommendations that are sometimes automatically implemented into activities aimed at preserving or boosting agricultural output and food security while contending with a variety of physical and chemical restrictions brought on by changing environmental conditions. AI can be used in smart agriculture to complete tasks that are beyond the capabilities of people. One of the issues for the future is how to handle a vast amount of data and turn it into useful information. Farmers have influence over a number of factors, including the choice of variety, planting thus far, the completion process rate, pesticide treatment rate and intervals, machine use, and consumption. Getting the greatest possible combination of inputs into the process, however, can be extremely difficult due to the interplay of several variables and the constant fluctuation of parameters. The objectives are to guarantee production growth, manage machine costs, and provide predictable market stability. Furthermore, a lot of things, such as environmental conditions, are beyond the control of farmers. Analysis of recent and previous data can provide landowners with a more accurate estimate. External situations that are out of the ordinary make it more difficult to control equipment, make decisions, and assess how these decisions and tools will affect the outcome in terms of profit and environmental implications. Based on the investigation and use of the available data, improvements in several sectors of agriculture are anticipated or promised, as shown in Figure 10.3. Farmers also gain directly from cost savings, crop predictions, and improved ability to make decisions and productivity.

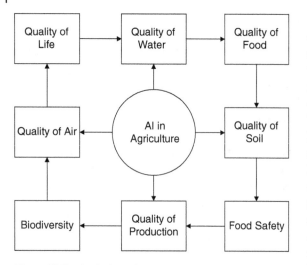

**Figure 10.3** Agriculture improvement areas promised through data extraction.

Greenhouse manufacturing processes are already extremely computerized and regulated, but, like in many other industries, AI systems are increasingly gaining unprecedented power. AI has the power to process enormous volumes of data while making minute ongoing modifications. Systems are emerging to provide conservatory operators with a plethora of production-related benefits. Although the production of greenhouse gases is already highly computerized and regulated, AI systems are rapidly assuming new levels of control, as are many other industries. Because of its capacity to analyze massive quantities of data and execute continual, real-time adjustments, AI systems are beginning to offer greenhouse operators a variety of production-related advantages. AI in protected horticulture can predict output, ensure product quality from seed to harvest, help plan the use of resources, and improve time-to-market and productivity. Therefore, it can assist farmers in generating income and sustaining a level of output. Important factors to take into account include industrial production methods using sizable greenhouses at various locations with isolated volumes, a lack of skilled labor, and a rise in demand for high-quality food close to urban areas. In order to use inputs from the endpoint to modify growth circumstances, it is also required to make clear how growth conditions connect to shelf-life processes.

### 10.5.1 Applying AI in Agriculture

The goal of agricultural practices is to maintain and improve human existence. AI is being used in agriculture through technology, robotics, data analytics, IoT, cameras, low-cost sensors, drones, and the Internet at large. AI algorithms can predict which products are grown in large quantities each year and which diseases affect crops. It has been shown that using technology will lessen the effects on natural ecosystems, promote worker safety, lower the cost of food, and boost food production to fulfill the needs of people everywhere. Poor dietary habits come with a number of concerns that can seriously harm one's health. Seasonal weather variations, fluctuating input prices, crop failure due to pests, soil erosion, and climate change are all factors that affect agriculture. Farmers must manage this ambiguity. Despite the diversity of agricultural practices, this chapter focuses on the critical agrarian components of soil, crops, diseases, and weeds. The use of manufacturing AI to examine agriculture-related issues with soil, crops, alignments, and pest control is crucial (Figure 10.4).

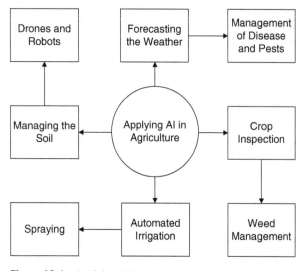

**Figure 10.4** Applying AI in agriculture.

### 10.5.2 Role of AI in Challenges of Agriculture

Even though AI has drastically changed and enhanced agriculture, there are still certain challenges. Lack of information and competence in AI is the first and biggest issue. To effectively apply AI in agriculture, a certain level of proficiency is required. AI technology is a combination of hardware, sensors, software, and various other specialized tools. Farmers must have the necessary training in order to use and operate these equipment. The great majority of farmers, however, lack the time and qualified instructors that are required for training. Another problem that causes issues is the distance between agriculturalists and AI technologists. Agriculturalists often do not study IoT, ML, or AI. On the other hand, engineers seldom conduct research, work in the fields, or engage in agricultural activities. Strange things occasionally happen in farming. It takes time for AI and ML to gather data, do research, analyze it, and find a solution. The situation may have spiraled out of control by then. The cost of technology adds a substantial new issue. There are two different types of costs: equipment costs and maintenance costs. Drones and robotics for agriculture are not cheap. To purchase them, farmers must make a large financial commitment. Because of this, a lot of businesses rent out the aforementioned assets in exchange for a share of the earnings. The use of AI has a lot of problems; thus, more has to be done to improve the farming process. Agriculture effectively employs AI with favorable outcomes. Future advancements could expand the use of AI in agriculture. The market for agricultural technologies and systems, including AI and ML, is anticipated to treble and reach $50 billion by 2030, according to business intelligence company research. To address agricultural issues and guarantee improved yields, AI in agriculture is essential. Farmers depend on the environment and the weather for their harvests. AI and ML act as predictive analysts by analyzing the previously gathered data to determine the ideal time to plant seeds, identify crop possibilities, and choose hybrid seeds to improve production.

The ML model can also suggest altering cropping patterns to boost yields. Microsoft, in addition to International Crop Research Institute for Semi-Arid Tropics (ICRISAR), has developed an AI-sowing. Members of the farming app receive guidance on the best seeds to plant. AI-based advice is expected to increase productivity by 45%. The main problem in agriculture is crop loss due to natural calamities, especially insect infestation. Another effort that has worked with Microsoft to establish a pest risk prediction API that uses AI and ML to estimate the possibility of a pest assault is United Phosphorous Ltd. (UPL). The robots may be outfitted with AI-powered sensing and computations in order to pick fruits and crops more quickly. Using ML and DL, technological devices can make judgments on par with the brain. These decisions are more accurate than those produced by the human brain since they are based on a tremendous quantity of data. One of the primary concerns for farmers is the fluctuation in crop prices. Big data, AI, and ML-based technologies make it feasible to identify pest and catastrophe infestations, calculate agricultural yield, and forecast pricing. With the use of this data, agriculturalists and the government may be advised on future pricing trends, the strength of the demand, and the best sort of crop to produce in order to maximize returns. A tech company named Nature-Fresh, located in the United States, is developing technology to predict how long a crop will take to generate a harvest. This approach might be used to calculate the yield that will eventually be offered for sale. AI can optimize a farm to the last detail, make informed decisions, and efficiently accomplish difficult, time-consuming tasks. For the transformation of traditional agriculture into sustainable programmed agriculture, it is a crucial instrument.

## 10.6 AI in Medicine

The concept of "medical technology" is often employed to describe a wide range of instruments that can assist medical professionals in making earlier diagnoses of patients, preventing complications, optimizing treatment, and/or providing less-invasive options, and cutting the length of hospital stays for both individuals and society at large. With the capacity to house AI-powered instruments in extremely small sizes, the development of sensors, wearables, smartphones, and communication networks has revolutionized medicine. Medical advances were formerly mostly recognized as traditional medical equipment (such as implants, stents, and prosthetics). AI, which is widely recognized as the area of computer science that is capable of handling challenging difficulties with multiple applications in domains with large volumes of data but limited theory, has revolutionized health-care technology. Innovative technological advances have made it feasible for augmented medicine or the use of contemporary technological advances in medicine to improve many aspects of clinical practice. In the last 10 years, a number of algorithms built on AI have received FDA approval; as a consequence, they might be used. Aside from AI-based technology, integrated medicine is made feasible by a number of other digital tools, such as surgical navigational systems for automated technology-assisted surgery and virtual-reality continuum devices for alleviating pain and mental health issues. Although enhanced medicine appears to be a success for individuals, some medical professionals may be reluctant to use it. As a result of the absence of educational opportunities in digital medicine, a number of highly esteemed private health-care institutions are working to prepare the next generation of health-care leaders for the difficult task of enhanced medicine by incorporating digital health literacy and use into updated curricula or by tying the study of the medical field to the research of engineering. This section aims to provide a summary of recent developments in AI in medicine, highlight the key contexts in which AI-powered wellness technologies are currently being employed in hospitals and clinics, and provide viewpoints on the challenges and risks that health-care providers and institutions may run into when carrying out the use of augmented medicine, both in truly patient care and in the educational process of future medical administrators, as shown in Figure 10.5.

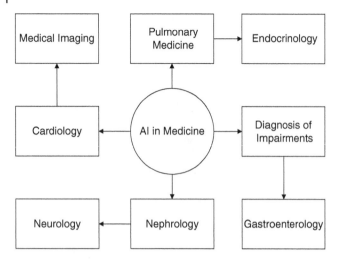

**Figure 10.5** Medicine improvement areas promised through data extraction.

### 10.6.1 Applying AI in Medicine

One of the key challenges of AI's use in medicine over the coming years will be the clinical validation of the core hypotheses meant to alter people's lives in large populations. Continuous surveillance and privacy violations, for instance, might make it more stigmatizing for the chronically ill or less fortunate to be citizens. They might also punish people who cannot follow the new healthy lifestyle recommendations by, among other things, limiting their ability to obtain medical coverage and care. However, these serious and perhaps deadly problems in the creation of health policy have received little to no debate. Some universities have started to create new medical programs that include doctor engineering in an effort to address the need to train future medical professionals for the challenges posed by AI in medicine. The aforementioned "augmented practitioners" would use their clinical expertise and digital proficiency to handle the issues of modern medicine, support the creation of digital approaches for healthcare organizations, manage the digital transition, and train both peers and patients. These experts may serve as a safety net for all treatments, including the deployment of AI in medicine, as well as a spur for creativity and research that would benefit society and health-care institutions.

Furthermore, to basic health-care instruction, ongoing educational programs in the field of digital medicine that are geared toward graduating physicians are required to enable retraining in this developing industry. Despite the fact that AI tools like natural language processing can already help doctors provide thorough medical records, new tools are still needed to solve the issue of the increasing amount of time spent on the indirect treatment of patients. One of the main barriers to doctors embracing modern medical innovations is the concern that medicine could develop into less compassionate. The increased administrative burden put on doctors is mostly to blame for this. The issue related to administrative load will surely be addressed by modern methods like ambient clinical intelligence and natural language processing, allowing professionals to focus more on their patients.

## 10.7 Role of AI in Challenges of Medicine

Despite promising potential, there is yet little actual AI-enabled solution adoption in clinical practice. In addition to privacy issues, AI technology also has additional methodological and technological flaws, as shown in Figures 10.6 and 10.7. The following are the main issues facing AI in medical services:

**Availability of Good Medical Data:** For medical and technical endorsement of AI models, practitioners need excellent-quality datasets. However, gathering patient data and images to test AI algorithms is difficult since medical data is dispersed across numerous electronic health records and different software platforms. Another challenge is that owing to interoperability issues, medical data from one organization could not be interoperable with other platforms. The medical sector must focus on methods for standardizing medical data in order to expand the amount of data accessible for testing AI systems.

**Inappropriate Clinical Performance Indicators:** The success indicators for AI models may not be directly applicable in clinical contexts. The AI chasm is the difference between the actual clinical efficacy seen in practice and the technical accuracy of AI testing. To close this gap, researchers looking at how AI algorithms improve medical care should work with developers and physicians. To achieve this, they can use decision curve analysis to evaluate the correctness of AI models. By comparing the datasets and calculating the likelihood that an AI model would succeed in the real world, this technique enables them to assess the therapeutic utility of a prediction model.

**Flaws in the Research Methodology:** AI in medicine does not yet have a sufficiently established methodology, future research, or peer-reviewed papers. The bulk of research has used historical patient medical information and has been retrospective. However, for doctors to fully understand the true benefit of AI diagnosis in real-world situations,

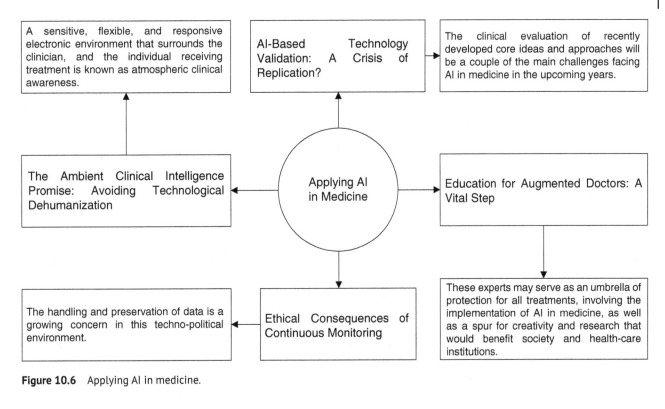

**Figure 10.6** Applying AI in medicine.

**Figure 10.7** Most prevalent AI challenges in the field of medicine.

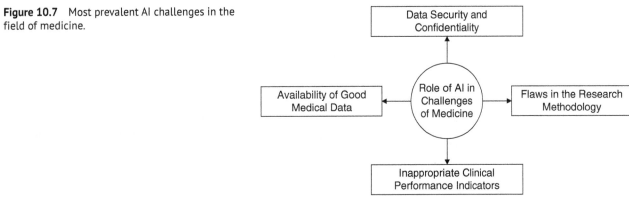

prospective research is necessary to examine existing patients over time. Additionally, for trustworthy prospective research, medical professionals should combine physical examinations with telehealth visits and remote monitoring of modern technologies (trackers and quality sensors) to monitor their patients' health.

**Data Security and Confidentiality:** Large volumes of patient data are needed for the application of AI in health care, which raises questions regarding the security and privacy of such data. A patient's right to determine how their data is used and protected from unauthorized access to that data is crucial.

## 10.8 AI in Education

Every person should place a high value on education since it is essential to living a prosperous life. Around the world, a lot of modifications are always being made to the curriculum and teaching methods in order to better the educational system for the students. Creating novel approaches to teaching and learning for various contexts is one area where AI is set to have an impact on education. AI is now being applied at a variety of colleges and universities throughout the world. The use of AI in education has offered parents, instructors, students, and, obviously, institutions of learning an entirely new way to see education. AI in education implies the application of computer intelligence to support teachers and students and enhance the efficacy of the system of education. It does not refer to the use of robots that are humanoid to educate in place of actual teachers. The next generation of educators will be shaped by a variety of AI technologies that will be integrated into the system. We will talk about the effects and uses of AI for instructional purposes in this subject matter.

A powerful computer immediately springs to mind when AI first arises. A supercomputer, as the name suggests, is a computer with a lot of processing power, adaptive behavior (using sensors, for example), and other features that let it think and act like a person and really improve the interaction between people and supercomputers. The powers of AI have been illustrated in a number of films, including those about smart buildings, where AI-powered devices can regulate the temperature, and the state of the air, and even play entertainment based on people's perceived moods. There has been an increase in the integration of AI within the area of education, moving beyond the conventional concept of AI as a microprocessor to embrace embedded computer systems.

### 10.8.1 Applying AI in Education

The application of AI technology to support teaching, create a smart campus, and realize intelligent educational instruction and administration is the core focus of an increasing amount of research on AI and education. A number of changes are brought about in the professional educational system as a result of the use of image recognition, facial recognition, learning that is adaptive, and other AI technologies, which also increase the productivity of instructors and enhance student learning. Additionally, big data and AI technologies are integrated to thoroughly extract and analyze teaching data, which may also support teaching reform and raise the caliber of instruction. The evolution of robots has increasingly grown sophisticated and humanized in recent years as a result of the evaluation of AI technology, as shown in Figure 10.8. Products that are humanized and intelligent will be more suited for use in classrooms in the future. The findings of the study demonstrate that autistic pupils are extremely engaged and respond well to interactions with tutoring robots.

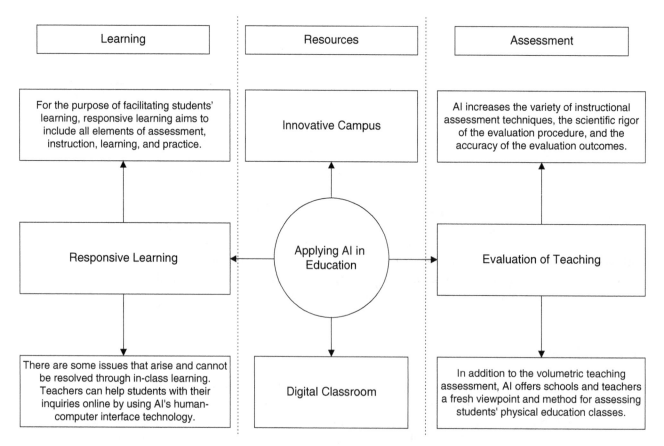

**Figure 10.8** AI employment in education.

### 10.8.2 Role of AI in Challenges of Education

According to the most recent evaluations, the developed world has a greater concentration of AI in education research than the developing world, which is very small. Both nations confront a variety of obstacles that prevent them from employing AI technology to advance their educational institutions. Let us quickly review these restrictions in this case. Equity must be ensured while implementing AI in education. Developing countries are at risk of allowing new technologies to widen gaps in educational attainment as a result of the development of AI. The growing algorithmic divide now threatens to rob many of the educational possibilities offered by AI, much like the digital divide separated those who cannot access the Internet. Most of the AI algorithms originate from developed countries; thus, they do not fully account for the conditions of developing countries and, therefore, cannot be applied directly. For AI to boost learning, the education industry needs to overcome major challenges, such as a lack of infrastructure and basic technologies (Figure 10.9).

In conclusion, the significance of AI technology in education has grown more evident with the thorough growth of economic and technical globalization. Additionally, a lot of nations have made the advancement of AI technology a top priority. Correctness, individualization, and customization of educational services and administration are the primary characteristics of the innovative AI-based learning ecosystem. Schools, instructors, and students are dealing with a number of difficulties and issues brought on by AI as they work to develop an innovative educational ecosystem.

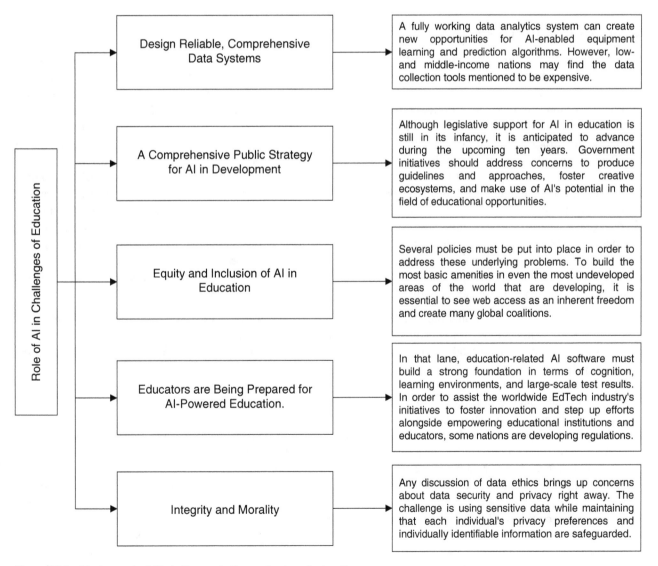

**Figure 10.9** Most prevalent AI challenges in the profession of education.

Instructors, pupils, and other stakeholders in the education ecosystem must collaborate to find solutions to these issues and realize the ideal integration of AI technology and education.

## 10.9 AI in Fitness

"Health is wealth," as we know. This saying has been going on for many centuries. However, only 25–35% of individuals throughout the world genuinely follow healthy habits, and 65% do not undertake any physical exercise save walking. In actuality, those who do not lead healthy lifestyles have health issues from a young age. However, people's concerns about their health and well-being have increased during the past several months. Of course, it is because of the COVID-19 coronavirus! Our schedules get more and more hectic, and we hardly ever have time to regularly practice. Maintaining a healthy diet is much more challenging! Then, several unclear concerns arise, including which diet is the most effective. What strategy is to be decided? These all seem a little ominous. However, do not worry – AI is here to help! In the area of fitness, AI may assist with exercise data analysis, comprehend how consumers train, and design personalized training plans that meet the customer's performance and goals. AI transforms exercise with fitness gear that makes working out at home smarter and healthier. Personal AI trainers are rapidly gaining popularity in addition to smart wearables like the smartwatch, the fitness tracker Fit, and others that may be used to measure your fitness.

### 10.9.1 Applying AI in Fitness

Users can boost their performance and meet their fitness goals in some way. AI applications will be employed in the future to maximize the advantages and potential of fitness AI. Industries or clubs may draw more clients and increase income when AI-based functionalities are included in personalized fitness programs. With features based on AI, you can also make informed selections instantly. Opportunities for AI in the fitness sector are enormous. The leading uses of cutting-edge AI technology are shown in Figure 10.10. People are increasingly relying on technology to solve their health and fitness issues. Numerous new fitness-related applications, devices, and wearables are being introduced to the market and are getting a lot of attention. According to a recent study by Research n Reports, the global market for fitness technology was valued at $28.9 billion in 2022, and by 2030, it is predicted to reach $76.1 billion. You might not be aware of it, but AI is becoming more integrated into our daily lives. The fitness sector is undergoing a significant transition as a result of advancements in fitness product offerings brought about by IoT and AI. The fitness app industry is expected to generate $23.64 billion in yearly revenue by 2030, with 115.2 million active users by 2025, according to research company Reports and Data. The need for applications with AI integration in the fitness, health, and nutrition sectors is growing, and most companies are leaning toward developing fitness apps with AI integration to take advantage of the benefits for their companies. These components, including health and fitness, can be measured using ML as well as AI. Here are a few AI developments for the fitness sector that aim to keep you active and healthy.

- By analyzing your health information, including heartbeats, diabetic signs, pulse rate, calories burned, weight, etc., wearables with AI integration operate more effectively and intelligently. Fitness programs also keep track of your usual exercises, the duration of your daily workouts, and your dietary needs to assist you in performing better activities the following day.
- Several well-known tech firms, like Google, Apple, and Android, want to develop wearables with AI-powered assistants that may virtually aid users in accordance with the gathering of health data. Additionally, AI is being incorporated into workout and fitness equipment so that consumers may utilize it more effectively to keep fit.
- AI-based workout programs might be a huge assistance for novices! Synthetic intelligence is being used by the fitness sectors to build fitness programs that can provide real-time information and individualized instructions. In order to track your motions, several organizations are developing sensors that can be attached to your workout attire. The sensors will send out tailored orders to flow your body in accordance with your actions. They intend to collaborate with businesses to include such sensors in branded clothing. These sensors now function best with yoga, but they will eventually revolutionize other sectors as well. You can benefit much from an AI fitness coach in this way.
- People typically do not have enough time to go to the gym or work out at home in this fast-paced and demanding lifestyle. Because personal trainers are so expensive, not everyone can afford to hire one. But now that fitness apps with AI integration are readily available for smartphones, consumers may exercise whenever and wherever they choose.

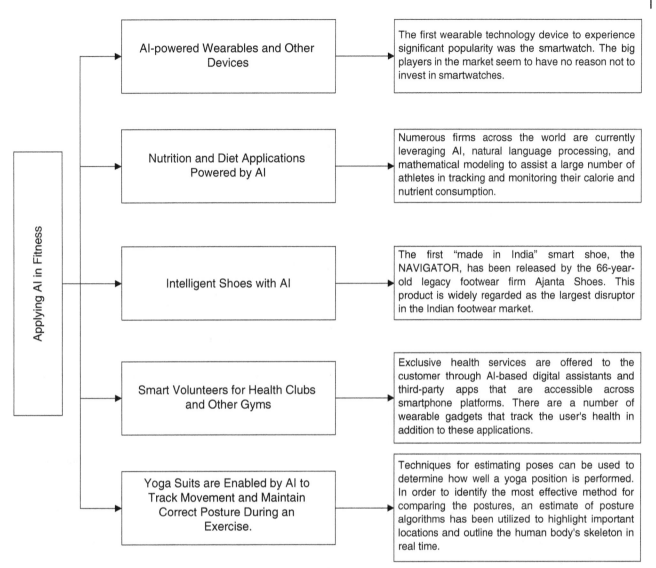

**Figure 10.10** Applying AI in fitness.

Setting your objectives on the app, purchasing a budget-friendly premium plan, and beginning to exercise at home or around the garden while watching instructor videos are all possible options.

### 10.9.2 Role of AI in Challenges of Fitness

The integration of AI and ML in fitness and health applications is already changing the way we think about well-being. Some AI-powered applications keep track of your well-being, while others serve as virtual trainers and may design custom fitness plans. You can already construct healthy meal plans with language learning tools like ChatGPT. The customized nature is now the rule as opposed to the exception in fitness applications due to how well AI is able to improve your experience. Creating a workout and nutrition plan that is unique to you might be difficult because everyone has different demands and interests. It can be challenging to take into account factors like body shape, age, gender, exercise level, dietary limitations, and personal objectives, especially for humans who have limited processing capability. In addition, the bodies of individuals and lifestyles are continually changing – over time, you could have children, lose your muscle mass, or have an injury – so regimens must be continuously modified and improved in order to be as successful as possible, as shown in Figure 10.11.

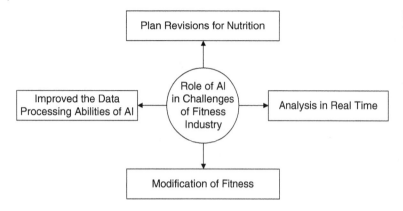

**Figure 10.11** Most prevalent AI challenges in the field of fitness.

## 10.10 AI in R&D

AI-enabled scientific advancements may assist in addressing social problems and launch brand-new industries. These scenarios highlight the value of fundamental research and the need for long-term planning in research strategy. Governments, occasionally aided by foundations dedicated to the public good, have a crucial role to play in providing consistent investment in publicly funded studies with long-term objectives. While the private sector has recently taken the lead in applied AI R&D spending, governments still have a significant role to play. This kind of funding is crucial for fostering and guiding reliable AI innovation as well as guaranteeing favorable outcomes for everybody, especially in regions underserved by being influenced by market investments. A wide spectrum of AI agents and stakeholders can be impacted by challenging technology difficulties that are being addressed through publicly sponsored research. AI research covers a range of topics, including applications like natural language processing, teaching methods like NNs, optimization methods like one-shot understanding, and societal issues like openness and comprehensibility, as well as technological advances to safeguard data integrity. This suggestion also urges funding for cross-disciplinary research on the social, legal, and ethical aspects of AI that are pertinent to public policy because AI has broad application and pervasive repercussions on many sectors of life.

Numerous sectors have been impacted by AI, and it is still accelerating investment, research, technical advancements, and education. AI research is exploding as a result of faster and less-expensive AI system training. The number of AI research articles surged by 350% between 2000 and 2022, based on the AI Benchmark Report. Due to their ability to reduce costs and boost productivity in difficult R&D projects, AI algorithms are crucial to solving research difficulties. Data-driven predictions are a common component of R&D and innovations, and ML may greatly streamline this procedure for researchers. AI technologies improve R&D across a variety of sectors, increasing its appeal to capitalists and philanthropic organizations. The advancement of technology has increased entrepreneurial activity. The size of the worldwide AI market, estimated at $46 billion in 2022, is expected to increase at a rate of 53.2% from 2023 to 2030. At huge cash-flowing digital corporations like Baidu, Amazon, and Google, internal R&D spending accounts for a significant portion of the investment made in AI R&D operations. About 30% of Apple's revenues are consistently used to fund innovative technology. Apple has consistently raised its expenditure in R&D over the past 21 years. It reached $18.7 billion by 2022. Apple's spending for R&D was three times larger than IBM's. Samsung spends over $520 million yearly on research. Additionally, in order to find technologies that they may use, the corporation invests not only in its own institute but also in those of other institutes throughout the globe. Therefore, an AI consulting firm can improve your R&D process if you are preparing an organization's growth strategy.

### 10.10.1 Applying AI in R&D

The accessibility of publicly available, usable, and representative information sets that do not infringe upon privacy and safeguarding of personal and client information, rights to intellectual property, or other crucial rights is a crucial component in ensuring further and better AI R&D due to the significance of data to the lifecycle of AI systems. Governments may assist in reducing the dangers of improper bias in AI systems, even if it may be impossible to create an entirely "bias-free" environment. They can do this by making representative datasets available to the public (and by offering incentives for their creation). For instance, even without the aim of discriminating, AI systems might employ datasets that are insufficiently representative of their intended function. Since ongoing investments in digital technologies, construction projects, and procedures for sharing AI expertise are means to cultivating this digital ecosystem, this suggestion supports the advice on establishing a digital ecosystem for AI. Sharing of AI expertise is made easier, in particular, through investments in publicly available and representative statistics.

Since they aid in various problem-solving efforts and foster creativity, R&D in AI technologies has a unique role in the scientific community. Data analysis, one of the most important problems for researchers, is specifically solved by AI and ML. Every year, over 2 million scientific publications are published. Consider all the additional resources that innovators in the R&D field have mentioned. Finding the pertinent information that researchers need might be challenging due to the abundance of data. Such problems can be swiftly resolved by AI using a single search to identify patterns. Innovation is one of the key catalysts for economic and social progress in the modern world. A product's quality is improved by innovative activity, which also promotes technology being employed in real-world applications to gain competitive advantages, address societal problems, and make money. Businesses are greatly impacted by R&D. Let us examine the following advantages that R&D may provide your business:

- For freshly developed items, you can apply for patents and establish a dominant position in the market.
- R&D initiatives may minimize manufacturing costs and improve product development, enabling you to sell your products for less money and gain a larger consumer base.
- You may make your business adaptable and prepared for the incorporation of novel concepts and procedures by investing in R&D. As a result, your company will be able to respond successfully to market developments.
- Demonstrate your company's unique concept to potential financiers since investors want to invest in businesses that operate in a dynamic manner.
- With a strong foundation in R&D, you will be able to find and hire qualified employees wherever in the globe.

### 10.10.2 Role of AI in Challenges of R&D

ML and AI first emerged in the middle of the previous century. The interest in this technology has peaked recently, mainly due to the development of new basic approaches, the accessibility of enormous amounts of gathered data, and the expansion of hardware capabilities. As a result, there are now the highest hopes for AI, which is seen as a panacea, and efforts are being made to employ it wherever conventional computational approaches may be replaced. Figure 10.12 shows the impact of AI in R&D.

## 10.11 AI in Transport

Whenever the system and users' behavior make it difficult to estimate and anticipate patterns of travel, transportation issues arise. In order to address the issues of rising travel demand, greenhouse gases, safety concerns, and deteriorating the environment, AI is thought to be a strong match for transportation systems. These difficulties result from the continual

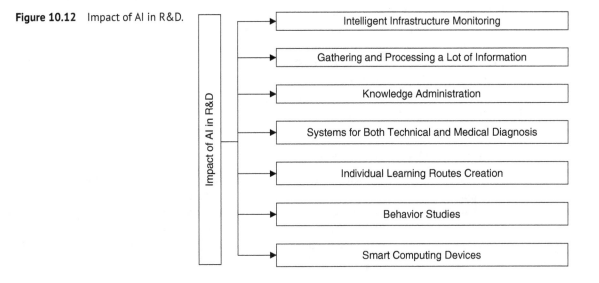

**Figure 10.12** Impact of AI in R&D.

increase of rural and urban traffic brought on by the rising population, particularly in emerging nations. The development and application of AI in transportation have taken several forms.

- The incorporation of AI in planning, management, and decision-making in businesses. This is crucial to solving the problem of a finite quantity of roads and a continually expanding demand. To better estimate traffic volume, traffic conditions, and events, this includes making greater use of precise forecasting and recognition models.
- Also covered are AI applications that attempt to enhance public transit. It is because many view public transit as a sustainable form of transportation.
- The next potential use of AI in transportation is integrated and autonomous automobiles, which seek to improve worker efficiency by decreasing the incidence of fatalities on road networks.

### 10.11.1   Applying AI in Transport

The links between the many aspects of the transportation system are sometimes difficult to completely comprehend; as a result, AI techniques might be suggested as an intelligent solution for such complicated systems that cannot be controlled using conventional techniques. Numerous studies have shown how useful AI is for transportation. Numerous transportation issues result in an optimization issue that demands custom techniques to be easily solved. The term "raster algorithms" refers to extremely sophisticated computing algorithms. One of those techniques is the genetic algorithm. It is based on the idea of biological evolution.

**AI in the Design, Organizing, and Management of Public Transportation Structures:** Planning's goal is to determine the community's requirements and the best strategy for meeting them while taking into account the influence of social, environmental, and economic factors on transportation. The Network design problem includes designing an ideal road system for transportation planning.

**Detection of Incidents:** There have been several attempts to determine the exact moment, place, and degree of an event in order to help traffic management ease congestion. These attempts range from manual reporting through complex algorithms to the use of NNs. Manual reports created by people could take longer to recognize occurrences and might be less economical. However, using data acquired from sensors positioned along the path, computers may compare the flow conditions beforehand and afterward the occurrence.

**Statistical Models:** The demand for cutting-edge techniques to predict traffic data has grown due to the growing adoption of intelligent transportation systems (ITS). These techniques have a significant impact on the effectiveness of ITS subsystems, including complex traveler data systems, advanced traffic management systems, developed public transit systems, and sophisticated commercial vehicle operations. Utilizing previous information gleaned from sensors affixed to highways, intelligent prediction algorithms are created. These data are then used as input by ML and AI algorithms to make short-, medium-, and long-term forecasts.

**AI in Aviation:** AI has been recognized for improving aviation travel management. It may assist with technology (ML and AI), hardware/software, and applications (intelligent maintenance, which is flight route management).

**AI in Collaborative Mobility:** With its potential to offer solutions to reduce the number of single- or low-occupancy personal cars on the road, the collaborative economy in mobility has proven particularly appealing in densely populated metropolitan areas.

**Advanced Urban Mobility:** Autonomous automobiles (AA) rely on DL-based AI software. This method instructs the vehicle on how to sustain safe headways, lane focus, and control, among other things. The use of AAs is expected to significantly alter how transport networks are managed around the world. Their potential to influence travel habits, as well as their impact on traffic safety and congestion, has been forecasted in considerable detail.

### 10.11.2   Role of AI in Challenges of Transport

The establishment of machine intelligence and the accurate comprehension of information that relies on human knowledge make developing an AI-based transportation network highly challenging. Up until now, the use of AI in transport has been restricted to particular ITS applications like data analysis and forecasting future mobility. The process would go more smoothly if AI apps could handle the entire spectrum of tasks. Therefore, it is necessary to fully utilize AI in order to create apps that can function as independent systems. Thus, AI skills must be included in future studies on traffic analysis, data gathering and preservation, making decisions, and optimization modeling. The capacity to predict the immediate as well as long-term flow of traffic is crucial in the transportation industry. Forecasting under unforeseen circumstances and unfavorable weather conditions is difficult. Unfortunately, these occurrences and situations cannot be addressed by the AI systems now in use. To achieve high accuracy, it is crucial to design weather and incident-adaptable algorithms and forecast methods. By implementing DL AI technology, the banking, data mining, and automotive industries are able to keep improving their operations. Additionally, the value of DL will increase quickly, reaching 21.2 USD billion by 2030, as a result of the ability to recognize patterns in data for valuable future predictions. With the use of historical and recent data collected from detector stations both downstream and upstream as well as cameras positioned on the highway, the intention is to anticipate future traffic. The use of AI will enable road authority workers to foresee network interruptions, which will then assist them in executing traffic management plans and measures to offset any negative effects on the motorway network before they happen.

### 10.11.3 AI in Communication

Particularly, AI in the communication system plays a crucial role and offers a potential method to improve performance. In terms of the bigger picture, AI methods greatly aid in the dynamic adaption of mobile communication in a given context. In order to minimize inefficiency and expansiveness, the complex network infrastructure must transition from traditional management and operation approaches to the intelligence approach. The forthcoming generation of cell phone networks is more advanced and demands more resources due to the requirement to increase service demand with diverse devices, complex networks, and varied applications. Network developers must change their system to deliver the greatest and most available resources in order to improve service quality. Additionally, according to industry predictions, the amount of IP traffic used in 2021 would be 4.8 zettabytes, with smartphone traffic expected to surpass PC traffic in the identical year. AI offers to create an adaptable system, improving the performance of the environment and the system. The vast datasets that are now readily accessible through wireless or mobile platforms are a result of the big data age. In other words, the application of AI to mobile phone communication will guarantee that these systems are more effective, perform better, and raise key performance metrics.

### 10.11.4 Applying AI in Communication

There are certain traditional methods of AI, including fuzzy logic and NNs. The NN would then be enhanced to include methods for DL and ML that offer improved performance. The fundamental strategy uses fuzzy logic, which processes any values and produces true and false. Reinforcement learning is a concept used in AI that describes a method for creating computers or other machines that can learn on their own rather than being carefully programmed. NNs are one of the methods. A machine or program that is capable of self-learning to tackle a problem may create this method. The NN technique mimics the brain and human behavior. DL, a more advanced kind of ML, has gained popularity in contemporary concerns. ML and DL are intriguing methods for managing future mobile communication and enhanced network traffic. Mobile communication has employed both unsupervised learning and supervised learning, two forms of AI learning.

**Making Decisions for Mobile Communications:** Researchers have described an AI decision-making process for mobile communication that uses NNs to categorize KPIs and estimate quality of experience (QoE) for various service types, bridging KPIs with QoE in the case of mobile Internet services.

**Mobile Communication Resource Optimization:** In resource optimization, algorithmic evolutionary algorithms have been used to improve the growth of broadcast trees for mobile ad hoc networks. Additional goals of this optimization are restricted end-to-end latency and energy conservation. Another work addressed the peak-to-average power ratio (PAPR) reduction problem, which is relevant to learning through the Internet, by using neural systems and a set-theoretic approach. The effectiveness of OFDM (Frequency Division Multiple Access) channel estimation may be significantly improved by using AI techniques.

**Network Administration in Communication:** An application related to network administration in wireless communication is routing, which is a problem with communications. A few studies in this field have employed AI in the past. For instance, self-configuration and self-optimization for broadcast capabilities and routing are achieved using NNs. In a different research, ML methods were used to address various routing issues from the past. There were three types of routing: dynamic routing, multicasting in-order routing, and shortest-path forwarding. Monitoring different network activities and spotting anomalies, or instances that deviate from typical network behavior, are other network management jobs in wireless communication. AI has also been applied to communication network traffic forecasting. AI technology can provide better reliability, more adaptable systems, and enhanced network performance while reducing the need for conventional networking handling of traffic operations.

**A Network with 5G Speeds:** Fifth-generation (5G) cellular networks may now be proactive and predictive, thanks to advancements in AI, which is essential for achieving the 5G goal. DL and ML are examples of these methods.

### 10.11.5  Role of AI in Challenges of Communication

Future mobile and broadband technologies will rely heavily on AI to manage vast amounts of data, improve analytics on data, and organize a variety of communication devices. The 5G mobile connectivity system will have problems because of the large number of nodes and mobile devices with rapid data exchange and communication. It was anticipated that the network management system for the next wireless communication generation would be more complex. Only a few AI techniques are, however, used for traffic control based on user input on the programs and cognitive layers. Another challenge is figuring out how to include AI in the next wireless communication infrastructure. The mobile communication system must also solve AI issues related to resource optimization and decision-network management. Proactive systems, self-awareness, self-adaptation, predictiveness, effective and economical operation, and optimization are the hallmarks of mobile intelligent communication. Another essential component of mobile communication is how AI may be applied in many wireless networking scenarios, which include power control, radio upper management, assigning resources, mobility administration, and hindrance administration.

## 10.12  AI Market Growth in Future Profession

The rapid adoption of technology advancements and the Internet has significantly assisted the recent growth of the worldwide AI sector. The IT behemoths' astronomical R&D investments are continuously speeding technical advancement across a wide variety of industries. The future growth of the global AI market is anticipated to be significantly fueled by the expanding demand for intelligent technology across a variety of end-use industries, including agriculture, medicine, education, fitness, R&D, transport, and communication, as shown in Figure 10.13. The majority of industries have historically given technical developments a high priority.

Favorable government actions are anticipated to have a good effect on industry expansion. The creation of ML and AI subcommittees within the federal government has increased interest in the AI sector. The Indian government boosted spending on Digital India to $522 million in 2025 in order to advance robots, AI, big data, IoT, and cybersecurity.

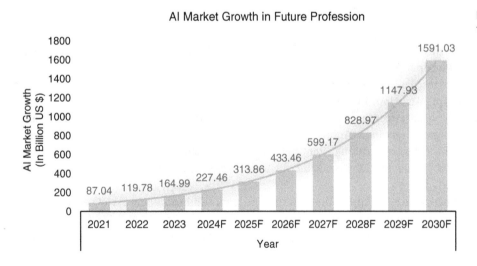

**Figure 10.13**  AI market growth in the future profession.

## 10.13  Conclusion

The objective of this investigation was to determine how AI will affect future professions in fields like agriculture, education, and health, among others. Agriculture, education, medicine, fitness, transportation, and communication are among the fields where AI is having a significant impact today and are all discussed in this chapter. The use of AI in agriculture, education, medicine, and other important fields is discussed in this chapter. In earlier times, work was done manually with many difficulties; however, in the present, the majority of work is completed automatically with the use of AI. Nowadays, AI is a need in the fields of agriculture, education, medicine, fitness, R&D, transportation, and communication; thus, it is important to understand where AI is employed. Finding the area of these vocations where AI is applied is made possible by a methodical examination. Through this methodical investigation, it has become clear how AI is being used in all of these professions, including those of agriculture, education, medicine, fitness, R&D, transportation, and communication. These professions will all be heavily dependent on AI in the future; thus, future studies will look at the social constraints on employing AI through ML. The emergence and usage of computers and associated computer technologies paved the way for research and inventions that eventually led to the creation and use of AI in a variety of fields. AI has been shown to have a significant impact on the professions it permeates, in part because of the advancement of personal computers and subsequent advancements that increased processing and computing power as well as the ability to integrate or incorporate computer technologies in various devices, gadgets, and platforms. In the fields of agriculture, education, medicine, and other relevant fields, AI has been widely embraced and put to use. The investigation concentrated on reviewing how AI has been deployed while evaluating the influence of AI on administrative, instructional, and learning aspects of agriculture, education, medicine, fitness, R&D, transportation, and communication.

## References

1 Zhong, R.Y., Xu, X., Klotz, E., and Newman, S.T. (2017). Intelligent manufacturing in the context of industry 4.0: a review. *Engineering* 3 (5): 616–630.

2 Li, B.H., Hou, B.C., Yu, W.T. et al. (2017). Applications of artificial intelligence in intelligent manufacturing: a review. *Frontiers of Information Technology & Electronic Engineering* 18: 86–96.

3 Daki, H., El Hannani, A., Aqqal, A. et al. (2017). Big data management in smart grid: concepts, requirements and implementation. *Journal of Big Data* 4 (1): 1–19.

4 Joshi, N. (2017). Blockchain meets industry 4.0-what happened next? https://www.allerin.com/blog/5659-2 (accessed 13 January 2024).

5 Eli-Chukwu, N.C. (2019). Applications of artificial intelligence in agriculture: a review. *Engineering, Technology & Applied Science Research* 9 (4): 4377–4383.

6 Oyakhilomen, O. and Zibah, R.G. (2014). Agricultural production and economic growth in Nigeria: implication for rural poverty alleviation. *Quarterly Journal of International Agriculture* 53: 207–223.

7 De Kimpe, C.R. and Morel, J.L. (2000). Urban soil management: a growing concern. *Soil Science* 165 (1): 31–40.

8 Pagliai, M., Vignozzi, N., and Pellegrini, S. (2004). Soil structure and the effect of management practices. *Soil and Tillage Research* 79 (2): 131–143.

9 Abawi, G.S. and Widmer, T.L. (2000). Impact of soil health management practices on soilborne pathogens, nematodes and root diseases of vegetable crops. *Applied Soil Ecology* 15 (1): 37–47.

10 Syers, J.K. (1997). Managing soils for long-term productivity. *Philosophical Transactions of the Royal Society of London. Series B: Biological Sciences* 352 (1356): 1011–1021.

11 Sarkar, A., Maity, P.P., and Mukherjee, A. (2021). Application of AI in soil science. *Technology* 62: 131–136.

12 Sanuade, O.A., Adetokunbo, P., Oladunjoye, M.A., and Olaojo, A.A. (2018). Predicting moisture content of soil from thermal properties using artificial neural network. *Arabian Journal of Geosciences* 11: 1–10.

13 Bhattacharya, P., Maity, P.P., Ray, M., and Mridha, N. (2021). Prediction of mean weight diameter of soil using machine learning approaches. *Agronomy Journal* 113 (2): 1303–1316.

14 López, E.M., García, M., Schuhmacher, M., and Domingo, J.L. (2008). A fuzzy expert system for soil characterization. *Environment International* 34 (7): 950–958.

**15** Ghanbarian-Alavijeh, B., Liaghat, A.M., and Sohrabi, S. (2010). Estimating saturated hydraulic conductivity from soil physical properties using neural networks model. *International Journal of Environmental and Ecological Engineering* 4 (2): 58–63.

**16** Tajik, S., Ayoubi, S., and Nourbakhsh, F. (2012). Prediction of soil enzymes activity by digital terrain analysis: comparing artificial neural network and multiple linear regression models. *Environmental Engineering Science* 29 (8): 798–806.

**17** Lemmon, H. (1986). COMAX: an expert system for cotton crop management. *Science* 233 (4759): 29–33.

**18** Stone, N.D. and Toman, T.W. (1989). A dynamically linked expert-database system for decision support in Texas cotton production. *Computers and Electronics in Agriculture* 4 (2): 139–148.

**19** Prakash, C., Rathor, A.S., and Thakur, G.S.M., (2013). Fuzzy based Agriculture expert system for Soyabean. *International Conference on Computing Sciences WILKES100-ICCS 2013,* Jalandhar, Punjab, India. 113.

**20** Ravichandran, G. and Koteeshwari, R.S. (2016). Agricultural crop predictor and advisor using ANN for smartphones. *2016 International Conference on Emerging Trends in Engineering, Technology and Science (ICETETS)* (ed. S. Sivakumar), 1–6. IEEE.

**21** Arias, F., Zambrano, M., Broce, K. et al. (2021). Hyperspectral imaging for rice cultivation: applications, methods and challenges. *AIMS Agriculture and Food* 6 (1): 273–307.

**22** Mitra, A. and Saini, G. (2019). Automated smart irrigation system (ASIS). *2019 international conference on computing, communication, and intelligent systems (ICCCIS),* (ed. P. N. Astya and M. Singh), 327–330. IEEE.

**23** Jha, K., Doshi, A., and Patel, P. (2018). Intelligent irrigation system using artificial intelligence and machine learning: a comprehensive review. *International Journal of Advanced Research* 6 (10): 1493–1502.

**24** Arif, C., Mizoguchi, M. and Setiawan, B.I. (2013). Estimation of soil moisture in paddy field using artificial neural networks. arXiv preprint *arXiv: 1303.1868.*

**25** Al-Ghobari, H.M. and Mohammad, F.S. (2011). Intelligent irrigation performance: evaluation and quantifying its ability for conserving water in arid region. *Applied Water Science* 1: 73–83.

**26** Al-Yahyai, S., Charabi, Y., and Gastli, A. (2010). Review of the use of numerical weather prediction (NWP) models for wind energy assessment. *Renewable and Sustainable Energy Reviews* 14 (9): 3192–3198.

**27** Mohanty, S.P., Hughes, D.P., and Salathé, M. (2016). Using deep learning for image-based plant disease detection. *Frontiers in Plant Science* 7: 1419.

**28** Chan, H.S., Shan, H., Dahoun, T. et al. (2019). Advancing drug discovery via artificial intelligence. *Trends in Pharmacological Sciences* 40 (8): 592–604.

**29** Malik, P., Pathania, M., and Rathaur, V.K. (2019). Overview of artificial intelligence in medicine. *Journal of Family Medicine and Primary Care* 8 (7): 2328.

**30** Yu, K.H., Beam, A.L., and Kohane, I.S. (2018). Artificial intelligence in healthcare. *Nature Biomedical Engineering* 2 (10): 719–731.

**31** Timms, M.J. (2016). Letting artificial intelligence in education out of the box: educational cobots and smart classrooms. *International Journal of Artificial Intelligence in Education* 26: 701–712.

**32** Fang, Y., Chen, P., Cai, G. et al. (2019). Outage-limit-approaching channel coding for future wireless communications: root-protograph low-density parity-check codes. *IEEE Vehicular Technology Magazine* 14 (2): 85–93.

**33** Chassignol, M., Khoroshavin, A., Klimova, A., and Bilyatdinova, A. (2018). Artificial intelligence trends in education: a narrative overview. *Procedia Computer Science* 136: 16–24.

**34** Lugaresi, C., Tang, J., Nash, H. et al. (2019). Mediapipe: A framework for building perception pipelines. arXiv preprint *arXiv: 1906.08172.*

**35** Jiménez-Luna, J., Grisoni, F., and Schneider, G. (2020). Drug discovery with explainable artificial intelligence. *Nature Machine Intelligence* 2 (10): 573–584.

**36** Krizhevsky, A., Sutskever, I., and Hinton, G.E. (2017). ImageNet classification with deep convolutional neural networks. *Communications of the ACM* 60 (6): 84–90.

**37** Zhang, X.Y., Yin, F., Zhang, Y.M. et al. (2017). Drawing and recognizing Chinese characters with recurrent neural network. *IEEE Transactions on Pattern Analysis and Machine Intelligence* 40 (4): 849–862.

**38** Patterson, D. (1990). *Introduction to Artificial Intelligence and Expert Systems.* Prentice-Hall, Inc.

**39** Kibria, M.G., Nguyen, K., Villardi, G.P. et al. (2018). Big data analytics, machine learning, and artificial intelligence in next-generation wireless networks. *IEEE Access* 6: 32328–32338.

**40** Zhang, C., Patras, P., and Haddadi, H. (2019). Deep learning in mobile and wireless networking: a survey. *IEEE Communications Surveys and Tutorials* 21 (3): 2224–2287.

# 11

# Cybersecurity Issues and Challenges in Quantum Computing

*R. Rahul[1], S. Geetha[1], Soniya Priyatharsini[1], K. Mehata[1], Ts. Sundaresan Perumal[2], N. Ethiraj[1], and S. Sendilvelan[1]*

[1] *Dr. M.G.R. Educational and Research Institute, Chennai, Tamil Nadu, India*
[2] *Universiti Sains Islam Malaysia, Bandar Baru Nilai, Negeri Sembilan, Malaysia*

## 11.1 Introduction

The advent of quantum computing has brought forth a new era of technological advancements and possibilities [1, 2]. With the potential to solve complex problems at an unprecedented speed, quantum computers hold tremendous promise for fields ranging from drug discovery to financial modeling. However, as with any groundbreaking technology, quantum computing also poses significant challenges and concerns, particularly in the realm of cybersecurity. This chapter explores the issues and challenges that arise in the context of cybersecurity as a result of quantum computing. The world of computing has witnessed remarkable advancements over the years, enabling us to solve increasingly complex problems and process vast amounts of data. However, as we approach the limits of classical computing, a new paradigm is emerging that promises to unlock unprecedented computational power. This paradigm is known as quantum computing [1]. Quantum computing represents a revolutionary shift in the way we approach computation by harnessing the principles of quantum mechanics. Classical computers rely on bits, in general, the binary units of data that can exist in either one of two states, zero or one. On the flip side, quantum computers harness quantum bits, termed "qubits," which exhibit the remarkable capability to simultaneously exist in a superposition of both 0 and 1 states. This fundamental distinction forms the basis for the immense computational capabilities of quantum computers.

The underlying concepts rooted in quantum mechanics, like superposition and entanglement, which are fundamental elements, give rise to the extraordinary power of quantum computing. Superposition allows qubits to exist in a combination of states, exponentially increasing the amount of information that can be processed simultaneously. Entanglement, on the other hand, enables qubits to become interconnected along the lines such that the state of one qubit becomes correlated with the state of the other, regardless of the physical separation between them. These properties of superposition and entanglement pave the way for parallelism on an unprecedented scale. While classical computers process information sequentially, quantum computers can perform multiple computations in parallel, leading to exponential speedup for certain algorithms. This remarkable advantage has the potential to turn around the fields such as cryptography, optimization, machine learning, drug discovery, and materials science [3].

Cybersecurity has become a critical aspect of our interconnected world, as individuals, businesses, and governments rely heavily on digital systems and networks to store, transmit, and process sensitive information. From personal banking details to national security secrets, the protection of data and communication channels has become paramount [4]. However, the emergence of quantum computing introduces a level of computational power that has the potential to undermine conventional cryptographic techniques upon which modern cybersecurity relies. By harnessing the principles of quantum mechanics and capitalizing on the distinct characteristics of quantum bits, also known as qubits, quantum computing enables computations to be carried out at an exponentially accelerated pace compared to classical computers [4, 5]. This computational power has profound implications for cryptography, which forms the bedrock of modern cybersecurity.

Most encryption algorithms, such as the widely used RSA and elliptic curve cryptography (ECC), depend on the difficulty of factoring huge numbers or finding solutions for certain mathematical problems. Traditional computers would take an impractical amount of time to crack these codes due to the exponential nature of these problems. However, quantum computers possess the ability to perform these calculations exponentially faster, thus rendering many existing encryption methods vulnerable to attack [6].

One of the fundamental cryptographic algorithms at risk is the RSA algorithm, which is extensively used for secure communication and digital signatures. RSA relies on the assumption that factoring large numbers into their prime factors is estimated to be infeasible. However, Shor's algorithm, developed by Peter Shor [7], demonstrated that a sufficiently powerful quantum computer could solve the factoring problem efficiently, effectively breaking RSA encryption [7]. Likewise, the Diffie-Hellman key exchange and the digital signature algorithm, both extensively utilized algorithms, demonstrate similar characteristics, which could be compromised by quantum computing. These revelations highlight the urgent need for new cryptographic protocols that can withstand the computational power of quantum computers.

Furthermore, quantum computers also pose a threat to symmetric encryption algorithms, which utilize a single key for both encryption and decryption. Grover's algorithm, proposed by Lov Grover [8], demonstrated that quantum computers could perform an exhaustive search of a large solution space in a quadratic time complexity rather than exponential [8]. This means that symmetric encryption keys, which were previously considered secure due to the computational cost of searching the key space, can be easily cracked by a quantum computer. The ramifications of quantum computing on cybersecurity extend beyond encryption algorithms.

Quantum computers have the potential to undermine the integrity of digital signatures, which play a vital role in authenticating the source of information and ensuring its integrity. Digital signatures play a pivotal role in confirming the genuineness and unaltered state of messages, documents, and software. However, the signature schemes that rely on classical cryptographic algorithms may become vulnerable to attacks with the advent of quantum computers [9].

Moreover, the security of blockchain technology, which powers cryptocurrencies like Bitcoin, is also at stake. Blockchain relies on cryptographic hash functions to secure transactions and ensure the immutability of the ledger. However, quantum computers have the potential to break these hash functions, which could lead to the compromise of blockchain systems and the loss of trust in cryptocurrencies. While quantum computing holds tremendous promise for scientific advancements and computational capabilities, it also presents significant challenges in the realm of cybersecurity.

The potential of quantum computers to break widely used encryption algorithms and compromise the security of sensitive information raises concerns about data privacy, integrity, and the trustworthiness of digital communication [6, 10, 11]. As quantum computing continues to advance, it is imperative to address these cybersecurity challenges and develop effective solutions. Researchers, industry experts, and governments are actively engaged in collaborative efforts to address and minimize the potential risks associated with quantum computing. One approach is the advancement of quantum-resistant cryptographic computation that can withstand attacks from both classical and quantum computers. This field of research, known as post-quantum cryptography (PQC), is essential for ensuring the long-term security of confidential information. Figure 11.1 shows the important and major issues and challenges of cybersecurity with reference to quantum computing.

PQC entails the exploration of alternative mathematical problems that are considered to be resilient against attacks from quantum computers. An illustration of this is lattice-based cryptography, which depends on the complexity of solving specific mathematical problems associated with lattices in spaces with higher dimensions. This approach offers a promising avenue for developing secure encryption algorithms that can withstand attacks from quantum adversaries [6, 12]. While progress is being made in the development of post-quantum cryptographic algorithms, the transition from classical to quantum-resistant encryption is not a straightforward process. It requires careful planning, collaboration, and coordination between various stakeholders, including government agencies, standards organizations, and technology providers.

In addition to the challenges associated with cryptographic systems, there are concerns about the security of quantum hardware itself. Quantum computers are highly sensitive machines that require careful protection from physical and software-based attacks. Malicious actors could attempt to tamper with the quantum hardware or exploit vulnerabilities in quantum software, compromising the integrity of computations and potentially compromising sensitive data. To address these concerns, researchers are actively developing methods to ensure the security and integrity of quantum hardware and software. Techniques such as quantum key distribution (QKD) provide a secure way to distribute encryption keys using the principles of quantum mechanics [13]. QKD relies on the fundamental properties of quantum systems to detect any attempts to intercept or eavesdrop on the communication, thereby ensuring secure key exchange.

As the field of quantum computing progresses, it is crucial to establish comprehensive frameworks and standards for quantum-safe cybersecurity. Governments and international organizations play a vital role in coordinating efforts and

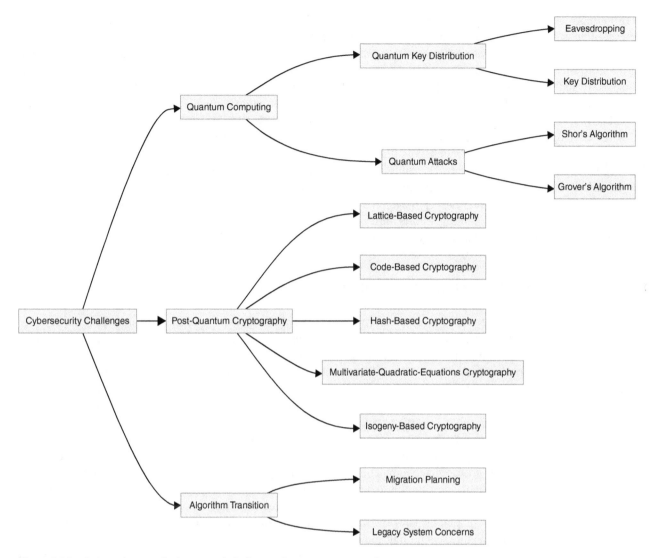

**Figure 11.1** Some cybersecurity issues and challenges in quantum computing.

creating guidelines for the development, implementation, and adoption of quantum-resistant cryptographic algorithms and protocols. Collaboration between academia, industry, and policymakers is key to addressing the challenges posed by quantum computing and ensuring a secure digital future. The rapid advancement of quantum computing poses both promising opportunities and formidable challenges for cybersecurity. While the computational power and speed of quantum computers hold immense potential for solving complex problems and advancing scientific research, they also pose significant risks to the security of data and communications. The ability of quantum computers to break widely used encryption algorithms threatens the confidentiality and integrity of sensitive information. To address these challenges, the field of PQC is actively researching and developing quantum-resistant encryption algorithms that can withstand attacks from quantum adversaries.

However, the transition to quantum-resistant cryptography is not a straightforward process. It requires careful planning, collaboration, and coordination among stakeholders to ensure a seamless and secure transition [9, 14]. The migration of existing systems and infrastructure to quantum-resistant algorithms poses practical and logistical challenges, necessitating significant resources and time. Moreover, the impact of quantum computing on cybersecurity extends beyond encryption algorithms. Additional cryptographic protocols, including key exchange, authentication, and digital signatures, are equally susceptible to assaults originating from quantum computers. Therefore, it is crucial to explore new cryptographic primitives and protocols that can withstand quantum attacks and maintain the integrity and trustworthiness of digital communications.

Furthermore, ensuring the security of quantum hardware and software holds significant significance. Quantum computers are exceedingly delicate machines that demand safeguarding against physical and software-driven assaults. Ongoing endeavors are being made to devise approaches that guarantee the security and authenticity of quantum hardware, including the utilization of QKD. This method uses the principles of quantum mechanics to establish protected key exchange protocols. To address the cybersecurity challenges posed by quantum computing, collaboration and coordination between academia, industry, and policymakers are crucial. Governments and international organizations should play a leading role in establishing comprehensive frameworks and standards for quantum-safe cybersecurity [15]. These frameworks should guide the development, implementation, and adoption of quantum-resistant cryptographic algorithms and protocols. While quantum computing offers immense potential for scientific advancements and computational power, it also presents significant challenges for cybersecurity. The ability of quantum computers to break existing encryption algorithms threatens the security of sensitive information and digital communications. However, through ongoing research and collaboration, the development of post-quantum cryptographic algorithms and protocols is underway to ensure a secure digital future in the era of quantum computing. Efforts must continue to mitigate the risks, address the challenges, and establish robust frameworks to safeguard data privacy, integrity, and trust in a quantum-powered world [6, 16].

## 11.2 Cybersecurity Issues and Challenges in Quantum Computer

Quantum computing issues with respect to cybersecurity refer to the challenges and concerns that arise from the development and deployment of quantum computers in relation to maintaining the security of digital systems and data [17, 18]. Quantum computers have the ability to bring out a solution for certain complex numerical problems at a significantly faster rate compared to traditional computers, through the utilization of quantum phenomena like superposition and entanglement.

The advancements made in quantum computing hold the capacity to uncover susceptibilities within numerous cryptographic algorithms currently utilized to ensure secure communication and data safeguarding. The foundational cryptographic protocols, encompassing public key cryptography and symmetric key algorithms, hinge on the intricacies inherent to distinct mathematical challenges. These intricacies serve to ensure the confidentiality and integrity of data [17]. Figure 11.2 shows the flowchart of cybersecurity issues with reference to quantum computation. The threat posed by quantum computers to cybersecurity arises from their potential to break widely used encryption algorithms, enabling unauthorized access to sensitive information. Consequently, it has become imperative to develop quantum-resistant algorithms and cryptographic techniques in order to effectively address the risks associated with quantum computing. Additionally, ensuring the secure distribution of cryptographic keys and the protection of sensitive data against quantum attacks are significant concerns in the realm of quantum computing and cybersecurity.

Quantum computing challenges with respect to cybersecurity refer to the specific obstacles and issues that arise when considering the impact of quantum computers on the security of digital systems and information [19]. Figure 11.3 shows the details about how the challenges are classified in quantum computing in terms of cybersecurity. Utilizing the principles of quantum mechanics, quantum computers exhibit exponential computational superiority over classical computers, thereby posing a substantial threat to conventional cryptographic algorithms and security protocols.

The primary concern arises from the potential ability of quantum computers to break widely used encryption algorithms, compromising the security of sensitive data against unauthorized access or tampering. Consequently, the development and implementation of quantum-resistant cryptographic techniques become essential to ensure the enduring security of digital communications and sensitive information. Additionally, integrating quantum-resistant algorithms into existing systems, establishing secure key distribution mechanisms, and safeguarding quantum computers against potential attacks are among the other challenges that demand attention in order to uphold robust cybersecurity in the age of quantum computing.

Quantum computing challenges with respect to cybersecurity encompass a range of complex issues and concerns that arise due to the unique capabilities and characteristics of quantum computers. A notable hurdle lies in the potential capacity of quantum computers to breach commonly utilized cryptographic algorithms like RSA and ECC. These algorithms underpin contemporary secure communication and data safeguarding. Their strength rests upon the intricate computational nature of particular mathematical problems, including prime factorization and discrete logarithms. Quantum computers, with the prowess of algorithms like Shor's, can tackle these efficiently. Thus, it's imperative to prioritize the crafting and integration of quantum-resistant or PQC to perpetuate the security of sensitive data. This endeavor involves extensive research and the embrace of novel cryptographic algorithms that can counter quantum attacks. The path to standardize and broadly implement these quantum-resistant algorithms introduces added intricacies, mandating meticulous scrutiny, testing, and validation to ensure their potency and seamless incorporation into existing systems.

**Figure 11.2** Cybersecurity issues in quantum computing.

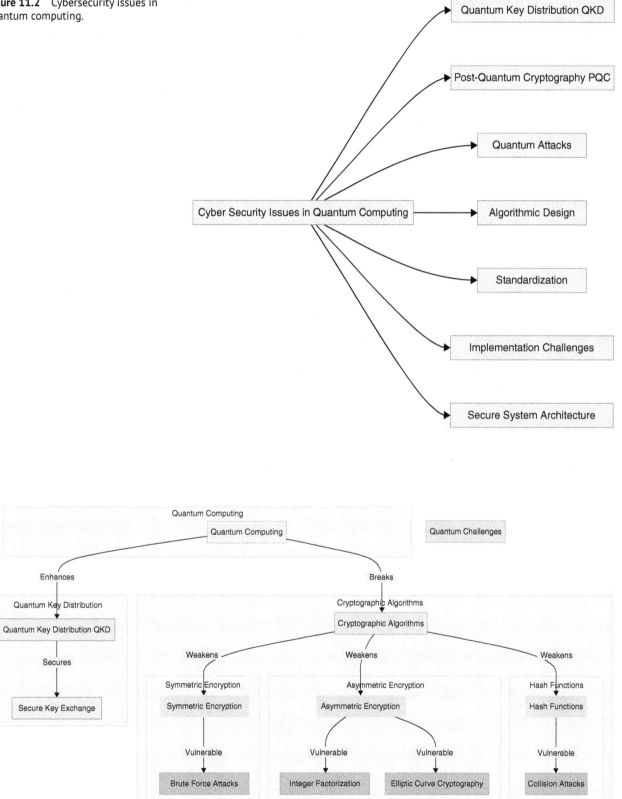

**Figure 11.3** Quantum challenges with respect to cybersecurity.

Another challenge lies in the secure distribution of cryptographic keys. Quantum computers have the potential to undermine the security of key exchange algorithmic protocols, such as the Diffie-Hellman key exchange, which are widely used to establish secure communication channels. QKD offers a potential solution by utilizing quantum principles to securely distribute cryptographic keys. However, deploying and scaling QKD networks while ensuring their practicality, efficiency, and resilience against attacks remain significant challenges. Furthermore, protecting quantum computers themselves from potential attacks is also a critical concern. Quantum computers are highly sensitive to external interference and require precise environmental conditions to function correctly. Ensuring the physical security of quantum computers and safeguarding them against tampering or unauthorized access is paramount [17, 18, 20, 21].Addressing these quantum computing challenges with respect to cybersecurity requires interdisciplinary efforts involving mathematicians, computer scientists, physicists, and engineers. Addressing the cybersecurity implications of quantum computation involves an amalgamation of theoretical research, algorithmic advancements, cryptographic engineering, and hardware security. These collective efforts aim to establish resilient and scalable solutions capable of withstanding the computational power of quantum computers while safeguarding the confidentiality, integrity, and availability of sensitive information in the presence of ever-evolving cyber threats. Various challenges and issues related to quantum computation and its impact on cybersecurity are discussed in detail.

### 11.2.1 Quantum Key Distribution

In the realm of quantum computing, QKD is an approach utilized to securely establish cryptographic keys between two entities. By harnessing the principles of quantum mechanics, QKD guarantees the confidentiality and integrity of these keys. However, QKD is not immune to certain issues. One key challenge is the vulnerability of the transmission channel. QKD relies on the exchange of quantum states, typically photons, between the sender and receiver. Any disruption or interception of these quantum states can compromise the security of the key [22, 23]. Adversaries can employ various techniques, such as eavesdropping or tampering with the transmission, to gain unauthorized access to the key.

Furthermore, the practical implementation of QKD encounters challenges stemming from the delicate nature of quantum states and the introduction of noise within real-world environments. Factors such as photon loss, imperfect detectors, and environmental interference can introduce errors in the key exchange process. Figure 11.4 shows some applicable solutions for the issues and challenges faced by QKD. These errors need to be carefully managed and corrected to ensure the reliability and security of the keys. Efforts are underway to address these challenges through advancements in quantum technologies, error correction techniques, and the development of quantum repeaters to

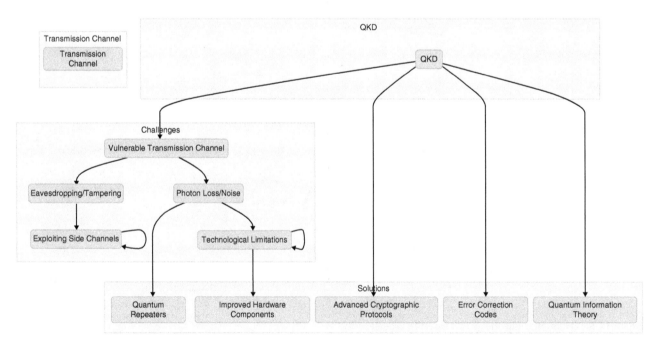

**Figure 11.4** Challenges and issues faced by a QKD and its optimal respective solutions.

extend the range of secure communication [13]. The field of QKD continues to evolve, aiming to provide robust and secure key exchange mechanisms for quantum computing applications.

QKD encounters various technological and experimental challenges that need to be addressed for its effective implementation in quantum computing. One significant issue is the limited transmission distance of QKD systems. Quantum states, such as photons, can suffer from losses as they travel through optical fibers, resulting in a decrease in the signal-to-noise ratio. This limits the maximum distance over which secure key exchange can be achieved. Researchers are working on developing quantum repeaters, devices capable of extending the range of secure communication by amplifying and retransmitting quantum signals.

Another challenge is the susceptibility of QKD to side-channel attacks. Side channels refer to information leakage that can occur unintentionally through physical or technical means. For instance, adversaries can exploit loopholes in the implementation of QKD systems or exploit the vulnerability of auxiliary devices used in the process [24, 25]. Addressing these vulnerabilities and ensuring the overall security of QKD protocols is crucial. Moreover, practical QKD systems must overcome various technological limitations, including the requirement for high-quality single-photon sources, efficient and low-noise detectors, and robust synchronization methods. The implementation of QKD systems faces various technological hurdles, including photon loss, channel noise, and detector inefficiencies. These factors can significantly impact the performance and reliability of QKD systems, making them vulnerable to eavesdropping attacks or rendering them unsuitable for real-world deployment [22, 24, 25].

### 11.2.2 Post-Quantum Cryptography

PQC deals with cryptographic algorithms purposefully designed to endure assaults originating from quantum computers. In contrast, prevalent cryptographic systems such as RSA and ECC are presently pervasive and considered secure; they are susceptible to large-scale quantum computer attacks [26]. The emergence of quantum computers presents a substantial challenge to the security of these conventional systems. Figure 11.5 shows the general challenges faced by PQC.

The main challenge in PQC lies in finding algorithms that can withstand attacks from quantum computers [27]. These algorithms are required to offer an equivalent level of security as existing cryptographic systems while simultaneously maintaining resilience against attacks, even when confronted with immensely capable quantum computers. Scholars are diligently scrutinizing an array of methodologies. These encompass lattice-based cryptography, code-based cryptography, multivariate polynomial cryptography, hash-based cryptography, and other methodologies to address this challenge. Another challenge is the transition from current cryptographic systems to PQC [26, 28]. This transition requires careful planning and coordination, as it involves updating and replacing cryptographic protocols, implementing new algorithms, and ensuring compatibility with existing infrastructure and systems.

PQC is a dedicated field of research centered around the development of cryptographic algorithms capable of withstanding attacks from highly capable quantum computers [29]. Quantum computers have the potential to break current cryptographic systems by exploiting their vulnerability to certain mathematical operations that quantum computers excel at, such as factoring in large numbers and solving discrete logarithm problems [27, 30].

### 11.2.3 Quantum Attacks

Quantum attacks refer to a class of attacks specifically designed to exploit the computational power and unique properties of quantum computers [31]. In contrast to classical computers that operate with bits, quantum computers utilize qubits that can concurrently represent multiple states by virtue of the principles of superposition and entanglement.

Quantum attacks can target various cryptographic algorithms commonly used in classical computing, such as RSA and ECC. Shor's algorithm stands out as the most prominent quantum attack against these algorithms, as it effectively resolves the challenges associated with the problems of integer factorization and the discrete logarithm problem [7]. Since the defense of many encryption schemes relies on the algorithmic complexity of these problems, a sufficiently large and error-corrected quantum computer could break these cryptographic systems.

Furthermore, quantum attacks can also undermine PQC algorithms, which are drafted to resist attacks from quantum computers [26, 32]. These attacks leverage the vulnerabilities in the design or implementation of these algorithms, taking advantage of the novel techniques and computational abilities of quantum computers to compromise their security.

To counter the potential risks posed by quantum attacks, researchers are actively engaged in the rise of PQC. This cryptographic strategy hinges on mathematical challenges that are considered impervious to assaults originating from

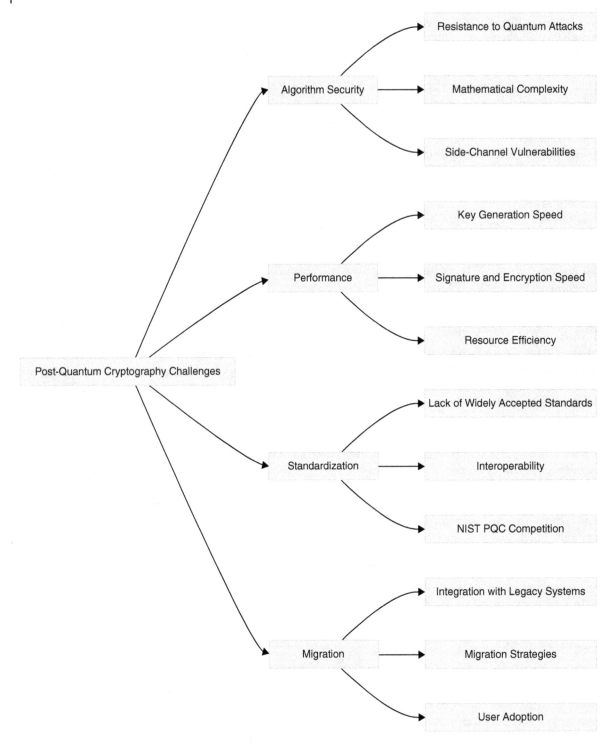

**Figure 11.5** Classification of PQC challenges.

quantum computers. [33]. Figure 11.6 shows the flow of how Shor's algorithm breaks classical encryption and is prone to quantum attack. This entails delving into lattice-based, code-based, multivariate, and alternative cryptographic methodologies impervious to Shor's algorithm and similar established quantum threats. Embracing these post-quantum cryptographic algorithms during the transition empowers organizations to safeguard their sensitive data amidst the quantum computing era [7].

**Figure 11.6** AShor algorithm-based quantum attack.

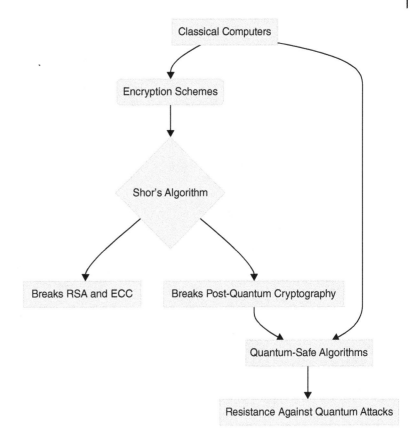

Shor's algorithm emerges as a notable quantum attack that effectively resolves two fundamental numerical problems, namely integer factorization and the enigma of the discrete logarithm. These dilemmas stand as the underlying foundation for many encryption schemes. These problems are considered computationally difficult for classical computers, forming the foundation of widely used cryptographic systems like RSA and ECC. However, Shor's algorithm, when executed on a large and error-corrected quantum computer, can break these cryptographic schemes by quickly finding the prime factors of large numbers or computing discrete logarithms [7, 26].

### 11.2.4 Algorithmic Design

Algorithmic design for quantum computing poses several unique challenges and issues. One primary challenge is the existence of quantum gates, which can perform operations on quantum bits (qubits) in parallel, leading to the potential for exponential speedup [34]. Figure 11.7 shows the general algorithmic design issues and their causes. However, the efficient decomposition of complex problems into quantum gates is nontrivial and requires careful consideration. Another challenge is the presence of quantum errors due to factors like decoherence and imperfect gates. These errors can significantly affect the accuracy of computations, necessitating the implementation of error correction codes and fault-tolerant techniques [35]. Developing efficient, reliable, and optimizable error correction algorithms is essential for scaling up and providing sufficient and necessary security for quantum computers.

Furthermore, quantum algorithms often involve intricate interference patterns, entanglement, and complex probability amplitudes. Designing algorithms that effectively utilize these quantum phenomena and exploit their advantages is a major research area. Additionally, mapping high-level problems onto the limited physical qubit resources of quantum computers is a critical challenge, requiring efficient mapping algorithms.

Finally, there is a need for quantum algorithm designers to understand the limitations and capabilities of quantum hardware to create practical and scalable solutions [34, 36]. Bridging the gap between theoretical algorithms and experimental implementations is crucial for the advancement of quantum computing. Addressing these algorithmic design issues and challenges is vital for harnessing the potential of quantum computing and realizing its impact across various fields, such as cryptography, optimization, and simulation.

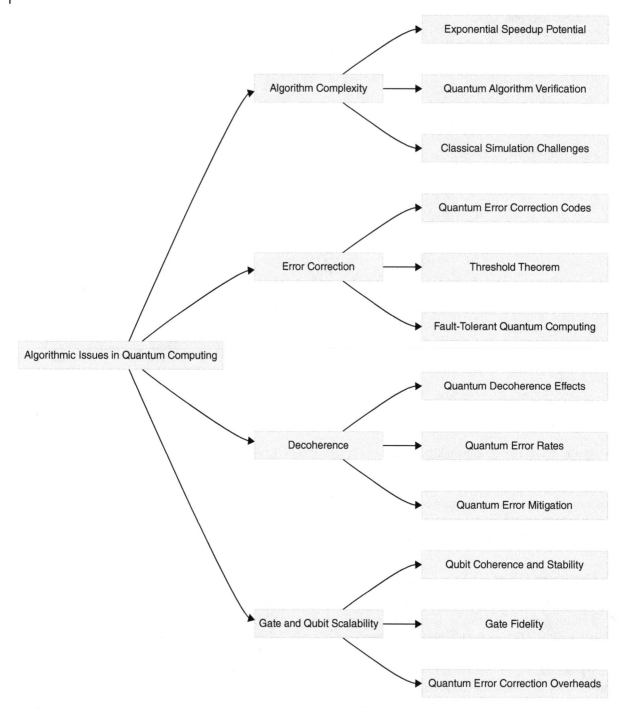

**Figure 11.7** Some general algorithmic issues in quantum computing.

### 11.2.5 Standardization

Standardization in quantum computing refers to the development and establishment of common frameworks, protocols, and methodologies to ensure interoperability and compatibility among different quantum computing systems, software, and applications [37]. Nevertheless, the field of quantum computing remains in its nascent phase, and standardization presents several hurdles. Among these challenges, one prominent issue arises from the diverse array of quantum computing technologies and architectures.

Unlike classical computing, where standardized components and interfaces enable seamless integration, quantum computing platforms vary significantly in their hardware designs, qubit implementations, error correction approaches, and control mechanisms. This heterogeneity makes it difficult to define uniform standards that can accommodate different quantum systems. Without standardization, quantum systems and quantum computers can be prone to vulnerabilities [38].

Another challenge is the rapid pace of technological advancements in the field [6, 38]. Figure 11.8 represents the standardization issues, its impact on quantum computing, and some general solutions to overcome the standardization issue. Quantum computing is a rapidly evolving field, with new hardware designs, error correction techniques, and algorithms being developed regularly. This dynamic nature makes it challenging to establish fixed standards that can keep up with the latest advancements and accommodate future breakthroughs. Quantum computing is heavily reliant on quantum algorithms and software tools. However, there is currently a lack of standardized programming languages, development

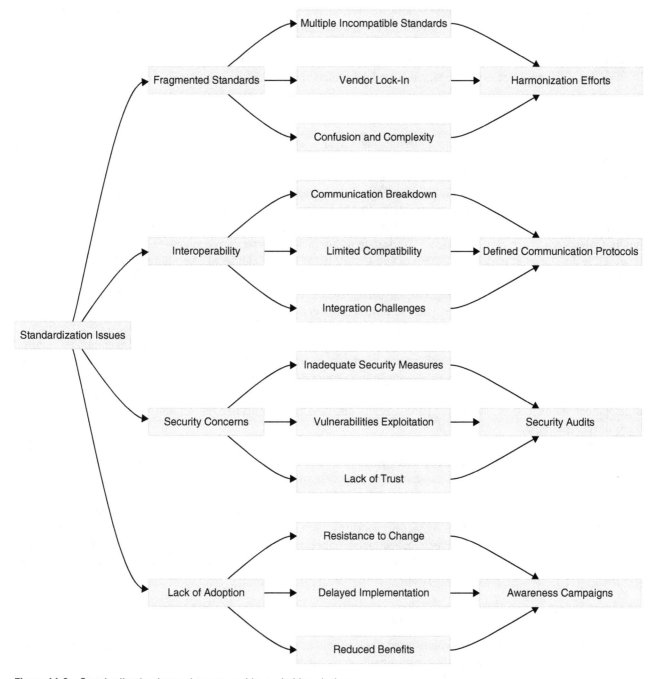

**Figure 11.8** Standardization issues, impacts, and its probable solutions.

frameworks, and libraries for quantum computing. This lack of standardization hinders the portability of quantum programs across different platforms and makes it challenging to foster a thriving quantum software ecosystem.

The absence of standardized metrics for benchmarking and evaluating quantum systems is a significant hurdle. Quantifying and comparing the performance of different quantum computers is a complex task due to factors such as qubit quality, gate fidelity, error rates, and coherence times. Establishing standardized metrics and evaluation methodologies is crucial for objectively assessing the capabilities and progress of quantum computing platforms [39].To address these challenges, the quantum computing community is actively working on standardization efforts. Organizations like the Institute of Electrical and Electronics Engineers and the International Telecommunication Union have initiated projects to develop quantum computing standards.

Additionally, industry consortia, research institutions, and quantum hardware/software vendors are collaborating to define common interfaces, programming languages, and benchmarking protocols. Standardization in quantum computing is essential for enabling interoperability, fostering innovation, and facilitating the development of a robust quantum computing ecosystem [37–39]. However, the diverse nature of quantum systems, the rapid pace of technological advancements, the lack of standardized software tools, and the absence of common evaluation metrics pose significant challenges to standardization efforts. Overcoming these challenges will require collaborative efforts from researchers, industry stakeholders, and standardization bodies to establish a solid foundation for the future of quantum computing [39].

### 11.2.6  Quantum-Aware Infrastructure

Quantum-aware infrastructure refers to the development of an ecosystem that supports the deployment and utilization of quantum computing technologies. As quantum computing evolves, it presents several significant challenges and issues that need to be addressed to create a robust quantum-aware infrastructure.

One of the primary obstacles lies in the necessity for advancing quantum hardware. Quantum computers demand exceptionally specialized and delicate components, such as qubits, which are susceptible to errors stemming from environmental disruptions and decoherence [1, 3]. Building reliable and scalable quantum hardware is a complex engineering task that requires advancements in materials science, fabrication techniques, and error correction methods [40]. Another key challenge is the development of quantum software and algorithms [40, 41]. Traditional computing paradigms and programming languages are not well suited for quantum computing. Quantum algorithms need to be designed to exploit the unique properties of qubits, such as superposition and entanglement. Additionally, quantum software development requires robust tools and frameworks for simulation, optimization, and debugging.

Figure 11.9 shows the classification of challenges faced by quantum computers due to quantum-aware infrastructure issues. Additionally, quantum computing requires substantial computational resources and energy consumption. Quantum systems operate at extremely low temperatures and require sophisticated cooling mechanisms. Security is a critical concern in quantum-aware infrastructure. The advent of quantum computers brings forth the potential to compromise numerous commonly employed encryption algorithms, thereby posing a substantial threat to the security of data [42]. As a result, the development of quantum-resistant encryption methods and protocols is crucial to protect sensitive information in a postquantum computing era.

Furthermore, standardization and interoperability are vital for the widespread adoption of quantum computing. The field is currently fragmented, with different hardware platforms and programming frameworks being developed by various organizations. Establishing common standards and protocols will facilitate collaboration, security, compatibility, and accessibility across different quantum computing systems [41].

Building a quantum-aware infrastructure involves overcoming numerous challenges. These include hardware development, software design, resource scalability, security concerns, and the establishment of standards. Overcoming these obstacles will require interdisciplinary research, collaboration between academia and industry, and sustained investment in quantum technologies [40–42]. By effectively tackling these challenges, we can establish a path toward a future where quantum computing evolves into a practical and transformative technology.

### 11.2.7  Quantum Security Standards and Policy

Quantum security standards and policies assume a pivotal role in confronting the distinct challenges presented by quantum computing, aiming to safeguard the confidentiality, integrity, and authenticity of sensitive information [43]. The proficiency of quantum computers to solve specific mathematical problems exponentially faster can render traditional

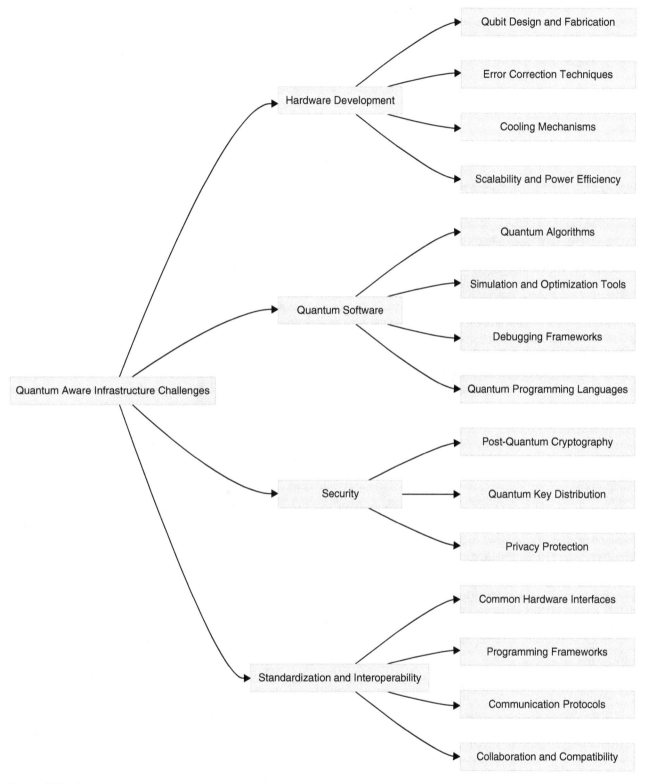

**Figure 11.9** Quantum-aware infrastructure.

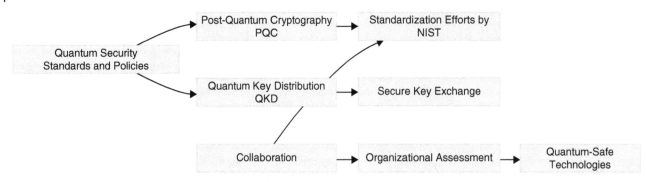

**Figure 11.10** Building quantum-safe technologies.

cryptographic algorithms, which hinge on the complexity of such problems, obsolete. To mitigate this risk, it becomes imperative to develop and implement quantum security standards and policies as a means of protection. Figure 11.10 shows the quantum security standards and policy for building quantum-safe technologies.

One of the basic concepts of quantum security is PQC. PQC involves the design and deployment of cryptographic algorithms that are resistant to attacks from both classical and quantum computers. The National Institute of Standards and Technology (NIST) is leading the standardization effort for PQC, soliciting public input and evaluating candidate algorithms [29]. The goal is to establish a set of quantum-resistant algorithms that can be widely adopted as a new standard to replace the vulnerable classical cryptographic algorithms. Furthermore, QKD emerges as a prospective quantum security technology that harnesses quantum mechanics' principles to ensure the secure distribution of encryption keys. QKD presents a dependable approach to exchanging secure keys, given that any effort to intercept or measure the quantum states engaged in the key distribution procedure would be discernible. This ensures the confidentiality of the exchanged keys [44].

Quantum security policies are crucial for organizations to establish guidelines and best practices to protect their systems and data in the era of quantum computing [45]. These policies encompass various aspects, including key management, data encryption, and secure communication protocols. Organizations need to assess their current security posture, identify vulnerabilities, and develop strategies to transition to quantum-safe technologies. It is essential to note that quantum security standards and policies are still evolving, as quantum computing technology progresses. Collaboration between academia, industry, and government organizations is essential to address the challenges and establish a robust framework for quantum-resistant security [14, 46]. As quantum computers continue to advance, staying informed about the latest developments in quantum security and adopting appropriate standards and policies will be crucial for ensuring the long-term security of sensitive information.

### 11.2.8 Implementation Challenges

As quantum computing continues to advance, it presents both opportunities and challenges for the field of cybersecurity. The potential of quantum computers to break existing cryptographic algorithms places traditional security measures in a vulnerable position [7]. This section explores the implementation challenges associated with integrating quantum computing into cybersecurity systems and discusses the measures required to address these challenges effectively.

- Transitioning to PQC: One of the primary challenges lies in transitioning from classical cryptography to PQC [27, 47]. PQC involves crafting innovative cryptographic algorithms with the resilience to endure assaults from both conventional and quantum computers. However, integrating PQC into existing systems and protocols is a complex task. It requires careful consideration of backward compatibility, performance impact, and the need to maintain a high level of security during the transition process [47].
- Practical Implementation of Quantum Key Distribution: QKD stands as a promising quantum technology that facilitates secure key exchange by leveraging the principles of quantum mechanics. However, implementing QKD on a large scale presents significant challenges [48]. QKD systems are sensitive to environmental noise, such as photon loss and channel disturbances, which can degrade the transmission of quantum states [24]. Ensuring reliable and efficient QKD in real-world scenarios while addressing issues related to distance limitations, compatibility with existing infrastructure, and scalability is a major implementation challenge.

- Integrating Quantum-Safe Encryption: While QKD focuses on ensuring secure key exchange, the need for quantum-safe encryption algorithms arises to safeguard the confidentiality and integrity of the transmitted data itself. Developing and integrating quantum-safe encryption schemes into existing protocols and systems pose significant challenges [49]. It involves ensuring the efficiency, performance, and compatibility of these new algorithms while maintaining a high level of security against quantum attacks [26].

- Infrastructure and Hardware Constraints: The implementation of quantum computing in cybersecurity also faces infrastructure and hardware constraints. Quantum computers require specialized environments, such as reduced temperatures and isolation from external disruptions, to maintain the fragile quantum states necessary for computation [50]. Building and maintaining such infrastructure at scale is a challenge. Additionally, quantum hardware technologies are still evolving, and the availability of reliable, error-corrected quantum systems suitable for practical cybersecurity applications remains limited.

- Standardization and Interoperability: Achieving standardization and interoperability across different systems and platforms is crucial for the effective implementation of quantum computing in cybersecurity. Developing and establishing widely accepted cryptographic standards that are resistant to quantum attacks requires collaboration among researchers, industry experts, and standards organizations [51]. Ensuring interoperability between different software, hardware, and network infrastructure is essential to facilitate seamless integration and compatibility while maintaining high security standards.

## 11.3 General Solutions

Quantum cybersecurity involves addressing the security challenges posed by the development of quantum computers, which have the potential to break traditional cryptographic systems. To mitigate these risks and ensure secure communication in a post-quantum world, several solutions and approaches are being explored:

A) Post-Quantum Cryptography
- Engage in comprehensive mathematical research to unearth problems that remain resistant to quantum attacks.
- Explore various mathematical structures like lattices, codes, multivariate polynomials, hash functions, and isogenies.
- Collaborate extensively to devise cryptographic primitives fortified against quantum adversaries.
- Not only address encryption but also embrace digital signatures, key exchanges, and more under the post-quantum umbrella.

B) Quantum Key Distribution
- Delve into diverse QKD protocols such as BB84, E91, and continuous-variable QKD for their appropriateness across different contexts.
- Tackle practical nuances, including channel noise, losses, and the ever-present threat of eavesdropping.
- Investigate advanced quantum repeater technology and the integration of quantum memory to bolster secure communication ranges.
- Design protocols for managing QKD networks encompassing key renewal, distribution, and maintenance.

C) Quantum-Resistant Cryptography Transition
- Craft meticulous strategies outlining phased migration from classical to post-quantum cryptographic paradigms.
- Develop tools that facilitate the seamless transition of existing cryptographic systems to quantum-safe alternatives.
- Forge collaborations with standardization bodies to ensure seamless interoperability and alignment across a wide vendor spectrum.
- Confront the intricate issues surrounding legacy systems, ensuring their compatibility with emerging post-quantum algorithms.

D) Quantum-Safe VPNs (Virtual Private Networks)
- Engineer robust VPN architectures that intertwine quantum-safe encryption and authentication mechanisms.
- Research the fusion of QKD mechanisms with traditional VPN frameworks to realize cohesive, quantum-resistant communication channels.
- Undertake scalability challenges head-on, addressing large-scale deployment scenarios within cloud environments and expansive networks.

E) Lattice-Based Cryptography
  - Immerse in the study of lattice problems encompassing shortest vector problem, learning with errors, and related conundrums.
  - Gauge the trade-offs involved in different lattice-based cryptographic schemes, evaluating key sizes, computational efficiency, and security margins.
  - Forge partnerships with hardware manufacturers to optimize lattice-based cryptographic operations within specialized hardware accelerators and secure execution domains.

F) Code-Based Cryptography
  - Conduct meticulous analysis of diverse error-correcting codes for their resilience against quantum decoding strategies.
  - Deepen the exploration of the practical feasibility of code-based cryptographic solutions, considering factors such as computation speed and memory consumption.
  - Dive into innovative techniques enhancing the efficiency of code-based encryption, be it through syndrome-based encryption mechanisms or sparse code encryption strategies.

G) Multivariate Polynomial Cryptography
  - Illuminate the intricacies of multivariate polynomial equations, probing their algebraic and computational resilience against quantum threats.
  - Contribute to the security analysis of multivariate polynomial-driven cryptographic components spanning encryption, signatures, and authentication protocols.
  - Innovate efficient algorithms facilitating both the generation and solution of multivariate polynomial equations while preserving overarching security assurances.

H) Hash-Based Cryptography
  - Enrich your understanding of cryptographic hash functions, examining aspects like collision resistance, preimage resistance, and their fortification against quantum attacks.
  - Delve into the nuanced impacts of quantum algorithms on hash-based digital signatures, actively crafting schemes that endure even within the quantum domain.
  - Embark on journeys beyond the realms of traditional Merkle-Damgard construction, venturing into novel constructs such as sponge functions and duplex constructions.

I) Isogeny-Based Cryptography
  - Engage in symbiotic collaborations with algebraic geometers to fathom the bedrock of isogeny-based cryptography along with its underlying security postulates.
  - Develop streamlined algorithms for computing isogenies amidst elliptic curves, addressing real-world implementation constraints to ensure practical viability.
  - Confront potential vulnerabilities in isogeny-based schemes head-on, confronting challenges like small-degree isogenies, while devising strategic mitigation strategies.

J) Secure Quantum Communication Protocols
  - Development of quantum digital signatures that brim with resilience against quantum and classical assaults, carefully balancing variables such as key lengths and collision resistance.
  - Probe the uncharted territories of advanced quantum secret-sharing methodologies, meticulously enabling dynamic secret dissemination while upholding threshold security against prying eyes.
  - Rigorously scrutinize the protocols' endurance against a diverse array of attacks, from quantum side-channel exploits to classical onslaughts, recalibrating designs to meet evolving adversarial strategies.

K) Continuous Monitoring and Research
  - Cultivate an ecosystem of interdisciplinary collaboration, amalgamating the insights of quantum physicists, cryptographers, mathematicians, and computer scientists to forge a holistic defense against emerging quantum threats.
  - Institutionalize research hubs and collaborative platforms fostering perpetual exploration, facilitating the exchange of discoveries, and nurturing the evolution of innovative solutions.
  - Collaborate harmoniously with relevant government entities to ensure the timely sharing of threat insights, aligning research trajectories with national security imperatives.

L) Hybrid Approaches
  - Pioneer the synthesis of hybrid cryptographic systems, harnessing the strengths of classical and quantum domains to address the dual specters of key distribution and data secrecy.

- Traverse the intricate landscape of integrating QKD with classical encryption methodologies, traversing a seam between seamless transition and peak security.
- Strenuously scrutinize the robustness of hybrid strategies against quantum foes, meticulously dissecting attack scenarios that intertwine quantum and classical assault vectors.

M) Quantum-Resistant Algorithms Optimization
  - Immerse in collaborations with hardware manufacturers, conjuring forth optimized incarnations of quantum-resistant algorithms to capitalize on the potential of hardware acceleration.
  - Trail-blaze innovative techniques encompassing parallelization, vectorization, and data structure optimization to amplify the operational efficiency of quantum-resistant algorithms.
  - Compile performance benchmarks and rigorous cross-comparisons delineating the operational excellence of optimized quantum-resistant algorithms across an array of distinct hardware landscapes.

N) Quantum-Safe Standards and Certification
  - Initiate dialogue and engagements with international standardization bodies, injecting your expertise into the crucible of standardized quantum-safe cryptographic algorithms and protocols.
  - Foster the emergence of streamlined certification frameworks that meticulously scrutinize the quantum resilience of cryptographic wares and services, offering end users a verified assurance of security.
  - Forge alliances with regulatory bodies, shaping compliance prerequisites for quantum-safe solutions, particularly within the ambit of critical infrastructures and their security underpinnings.

O) Quantum Threat Modeling and Risk Assessment
  - Emerge as the architect of exhaustive threat models, suffusing these models with a diversity of quantum attacks, intertwining proven quantum algorithms with potential future advancements.
  - Cooperate earnestly with risk assessment luminaries to quantify the prospective magnitude of quantum assaults on critical systems and data of sensitive import.
  - Prioritize remedial actions predicated upon the gravity of risks and the far-reaching consequences hinging upon the success of quantum offensives.

P) Quantum-Resistant Key Management
  - Craft key management protocols that surmount the tribulations of secure key inception, distribution, retention, and revocation in the face of quantum perils.
  - Probe innovative strategies, ranging from quantum-resistant key derivation functions to mechanisms ensuring the quantum-safe storage of these indispensable cryptographic artifacts.
  - Construct best practices manuals and guidelines affording organizations the capacity to uphold the enduring security of cryptographic keys even within the embrace of an increasingly quantum-infused arena.

Q) Quantum-Safe Cryptanalysis
  - Initiate research endeavors centered on unraveling classical cryptographic systems' vulnerabilities in the face of quantum cryptanalysis.
  - Foster partnerships with cryptanalysis virtuosos, refining quantum algorithms calibrated to deftly dismantle classical cryptographic foundations while identifying the spectrum of possible attack vectors.
  - Spearhead the construction of countermeasures meticulously calibrated to mitigate the reverberations of quantum cryptanalysis upon extant cryptographic protocols.

R) Quantum-Safe Hardware and Hardware Security Modules
  - Forge collaborations with hardware luminaries, coalescing your vision into quantum-resistant hardware modules spanning quantum-safe processors and memory constructs.
  - Embark upon deep investigations into tamper-resistant architectures, erecting bastions that shield quantum-resistant hardware from physical attacks and the predations of invasive probing.

## 11.4  Conclusion

The survey on cybersecurity issues and challenges in quantum computing sheds light on the potential threats and vulnerabilities that arise with the advent of this revolutionary technology. Quantum computing has the potential to disrupt traditional cryptographic systems, making them susceptible to attacks that were previously considered impractical or impossible. The survey underscores the pressing requirement for resilient and quantum-resistant cryptographic algorithms in order to protect sensitive information in the post-quantum era. One of the survey's key discoveries emphasizes the susceptibility of commonly employed encryption methods, such as RSA and ECC, to quantum attacks. As quantum computers continue to evolve, these algorithms will become increasingly obsolete, necessitating the development and adoption of new encryption schemes that can withstand quantum attacks.

Standardization efforts by organizations such as NIST are underway to identify and promote post-quantum cryptographic algorithms. Additionally, the survey highlights the potential impact of quantum computing on other areas of cybersecurity, including authentication, integrity verification, and secure communication protocols. Quantum technologies have the capability to enhance these areas through the development of quantum-resistant authentication methods, QKD protocols, and quantum-resistant hash functions. However, the survey also points out the challenges in implementing quantum-resistant solutions. These challenges include the complexity of transitioning existing systems to PQC, the need for significant computational resources to run quantum-resistant algorithms, and the uncertain timeline for the arrival of large-scale quantum computers. Collaboration between academia, industry, and government entities is crucial to address the emerging cybersecurity challenges posed by quantum computing. By proactively addressing these challenges, we can ensure a secure and resilient digital infrastructure in the period of quantum computing.

# References

**1** Padmavathi, V., Sujatha, C.N., Sitharamulu, V. et al. (2023). *Introduction to Quantum Computing*. Wiley Online Library.

**2** Bhogaraju, V., Jain, R., and Vijayaraghavan, A. (2021). Quantum computing: a comprehensive survey. *IEEE Access* 9: 9685–9715.

**3** Harrow, A.W. and Montanaro, A. (2021). Quantum computational supremacy. *Nature Reviews Physics* 3 (3): 204–214.

**4** Catal, C., Ozcan, A., Donmez, E., and Kasif, A. (2023). Analysis of cyber security knowledge gaps based on cyber security body of knowledge. *Education and Information Technologies* 28: 1809–1831.

**5** Gupta, R., Wilson, D., and Patel, S. (2020). Quantum computing and security: areview. *IEEE Potentials* 39 (4): 32–37.

**6** Le, D.N., Kumar, R., Mishra, B.K. et al. (2019). *Cyber Security in Parallel and Distributed Computing: Concepts, Techniques, Applications and Case Studies*. Wiley Online Library.

**7** Shor, P. W. (1994), Algorithms for quantum computation: Discrete logarithms and factoring. *Proceedings of the 35th Annual Symposium on Foundations of Computer Science*, 124–134.

**8** Grover, L. K. (1996), A fast quantum mechanical algorithm for database search. *Proceedings of the Twenty-eighth Annual ACM Symposium on the Theory of Computing*, 212–219.

**9** Pachos, J.K. (2018). Quantum cybersecurity. *Reports on Progress in Physics* 81 (1): 1–22.

**10** Sharma, G., Sharma, D.K., and Kumar, A. (2023). Role of cybersecurity and blockchain in the battlefield of things. *International Journal of Advanced Computer Science and Applications* 6 (3): 124–153.

**11** Fahmi, N., Hastasakti, D.E., Zaspiagi, D., and Saputra, R.K. (2022). A comparison of blockchain application and security issues from bitcoin to cybersecurity. *Blockchain Frontier Technology* 3 (1): 81–88.

**12** Easa, R.J., Yahya, A.S., and Ahmad, E.K. (2023). Protection from a quantum computer cyber-attack: survey. *Technium* 5: 1–12.

**13** Thomas, E., Johnson, R., and Lee, C. (2020). Quantum key distribution: acomprehensive review. *IEEE Communication Surveys and Tutorials* 22 (2): 1074–1108.

**14** Zhao, G., Zhang, S., and Li, Y. (2016). A lattice-based quantum-resistant digital signature scheme. *9th International Symposium on Computational Intelligence and Design*, 242–246.

**15** Smith, J., Johnson, S., and Brown, M. (2020). Quantum computing and cyber security: challenges and opportunities. *IEEE Transactions on Dependable and Secure Computing* 19 (2): 470–480.

**16** Ford, P. (2023). Thequantum cybersecurity threat may arrive sooner than you think. *IEEE – Computer* 56 (2): 72–75.

**17** Brown, A., Lee, C., and Patel, S. (2020). Quantum computing and its implications for cybersecurity. *IEEE Security and Privacy* 18 (2): 26–33.

**18** Faruk, M. J. H., Tahora, S., Tasnim, M. et al. (2022), A Review of Quantum Cybersecurity: Threats, Risks and Opportunities. *1st International Conference on AI in Cybersecurity (ICAIC)*, 1–6.

**19** Chen, J., Luo, X., and Zhuang, W. (2020). Quantum computing and cybersecurity: challenges and opportunities. *IEEE Network* 34 (3): 28–34.

**20** Lee, C., Garcia, M., and Wang, W. (2021). Quantum computing threats to cybersecurity: a review. *IEEE Access* 9: 39891–39903.

**21** Lee, C., Johnson, R., and Brown, A. (2018). Quantum computing and cybersecurity: a comprehensive survey. *IEEE Communication Surveys and Tutorials* 20 (4): 3207–3234.

**22** Bennett, C.H. and Brassard, G. (1984). Quantum Cryptography: Public Key Distribution and Coin Tossing. *Proceedings of the IEEE International Conference on Computers, Systems, and Signal Processing*, 175–179.

**23** Bennett, C.H. and Brassard, G. (2017). Quantum key distribution and the security of cryptographic systems. *IEEE* 105 (4): 660–671.

**24** Scarani, V., Bechmann-Pasquinucci, H., Cerf, N.J. et al. (2009). The security of practical quantum key distribution. *Reviews of Modern Physics* 81 (3): 1301–1350.

**25** Brown, A., Garcia, M., and Wang, W. (2016). Quantum security: challenges and solutions for quantum key distribution systems. *IEEE Communication Surveys and Tutorials* 18 (2): 1412–1433.

**26** Biasse, J.-F., Bonnetain, X., Kirshanova, E. et al. (2023). Quantum algorithms for attacking hardness assumptions in classical and post-quantum cryptography. *IET Information Security* 17 (2): 171–209.

**27** Dulek, Y. and Waters, B. (2020). Asurvey of post-quantum cryptography. *Journal of Cryptology* 33 (1): 3–86.

**28** Sharma, S., Ramkumar, K.R., Kaur, A. et al. (2023). Post-quantum cryptography: asolution to the challenges of classical encryption algorithms, modern electronics devices and communication systems. *Lecture Notes in Electrical Engineering* 948: 23–38.

**29** National Institute of Standards and Technology (NIST) (2021). NISTIR 8309: Report on Post-Quantum Cryptography.

**30** Smart, N.P. (2018). Post-quantum cryptography. *Journal of Cryptographic Engineering* 8 (3): 147–187.

**31** Ding, J. et al. (2020). Quantum attacks on post-quantum cryptosystems. *Nature* 587 (7833): 547–553.

**32** Kuang, R. and Barbeau, M. (2022). Quantum permutation pad for universal quantum-safe cryptography. *Quantum Information Processing* 21: 147–187.

**33** Johnson, R., Thomas, E., and Wilson, D. (2021). Survey on quantum computing threats and vulnerabilities. *IEEE Communication Surveys and Tutorials* 23 (4): 2505–2535.

**34** Yan, Z., Zhang, L., and Chen, Y. (2012). Security challenges in quantum cryptography and information security. *4th International Conference on Intelligent Networking and Collaborative Systems*, 282–287.

**35** Mosca, M., Stebila, D., and Wallden, P. (2021). Quantum-safe cryptography. *Nature Reviews Physics* 3 (7): 453–471.

**36** Schöffel, M., Lauer, F., Rheinländer, C.C., and When, N. (2022). Secure IoT in the era of quantum computers – where are the bottlenecks? *Sensors* 22 (8): 24–37.

**37** Mosca, M. and Stebila, D. (2019). The case for post-quantum standardization. *Nature Electronics* 2 (1): 26–30.

**38** Kim, S. and Hong, S. (2023). Design and security analysis of cryptosystems. *Applied Sciences* 13: 14–27.

**39** Kampanakis, P. and Lepoint, T. (2023). Vision Paper: Do We Need to Change Some Things?, Lecture Notes in Computer Science, 13895.

**40** Purohit, A., Kaur, M., Seskir, Z.C. et al. (2023). *Building a Quantum-Ready Ecosystem*. Wiley Online Library.

**41** García de la Barrera, A., García-Rodríguez de Guzmán, I., Polo, M., and Piattini, M. (2023). *Quantum Software Testing: State of the Art*. Wiley Online Library.

**42** Ren, P., Xiaozhuo, G., and Wang, Z. (2023). *Efficient Module Learning with Errors-Based Post-Quantum Password-Authenticated Key Exchange*. Wiley Online Library.

**43** Serrano, M.A., Sanchez, L.E., Santos-Olmo, A. et al. (2023). *Towards a Quantum World in Cybersecurity Land*. Wiley Online Library.

**44** Verma, G. and Kumar, A. (2022). *Novel Quantum Key Distribution and Attribute Based Encryption for Cloud Data Security*. Wiley Online Library.

**45** Joshi, A. (2023). *The Impact of Quantum Computing on Cybersecurity*. Wiley Online Library.

**46** Zeydan, E., Turk, Y., Aksoy, B., and Ozturk, S.B. (2022). Recent advances in post-quantum cryptography for networks: a survey. In: *Conference on Mobile*, 17–25. IEEE.

**47** Wang, X., Chen, G., and Zhang, Z. (2020). Quantum-resistant authentication and digital signature schemes: a survey. *Journal of Computer Science and Technology* 35 (4): 679–700.

**48** Pappa, A., Rass, S., Papadakis, M., and Kikiras, P. (2017). Challenges in the practical implementation of quantum key distribution. *International Journal of Quantum Information* 15 (03): 174–187.

**49** Gheorghiu, V., Wallden, P., and Crépeau, C. (2019). Quantum cryptography: overview and challenges. *IEEE Communication Surveys and Tutorials* 21 (3): 2622–2653.

**50** National Institute of Standards and Technology (NIST). (2021). Post-quantum cryptography. https://csrc.nist.gov/Projects/post-quantum-cryptography (accessed 3 June 2023).

**51** Chen, L., Gao, F., Ma, J. et al. (2016). Quantum key distribution network: security and challenges. *IEEE Communication Surveys and Tutorials* 18 (2): 1203–1223.

# 12

# Security, Privacy, Trust, and Other Issues in Industries 4.0

*Ambeshwar Kumar[1], Manikandan Ramachandran[2], M. Manjula[3], Monika Agarwal[3], Pooja[3], and Utku Köse[4]*

[1] *Computer Science and Engineering, GITAM University, Visakhapatnam, Andhra Pradesh, India*
[2] *School of Computing, SASTRA Deemed University, Thanjavur, Tamil Nadu, India*
[3] *Computer Science and Engineering, Dayananda Sagar University, Bangalore, Karnataka, India*
[4] *Computer Engineering, Suleyman Demirel University, Kaskelen, Kazakhstan*

## 12.1   Introduction

It is not unexpected that the concept of Industry 4.0 originated in Germany given that the country has one of the most competitive manufacturing sectors in the world and even leads the globe in the production of equipment. The German government's strategic programming of Industry 4.0 has historically given the industrial sector strong support. The correct management of the mechanization, electrification, and IT phenomena led to the first three industrial revolutions. The Fourth Industrial Revolution, known as "Industry 4.0," has been greatly aided by the recent Internet of Things (IoT) movement, which has had a significant impact on the manufacturing environment. Future firms will undoubtedly establish global networks and integrate their equipment, storage systems, and manufacturing facilities into internet-enabled cyber-physical system (CPS). These CPSs, which are used in manufacturing, include intelligent machines, storage systems, and production facilities that can communicate, take independent actions, and control each other. Industry 4.0 is a relatively new concept that relates to information and communication technologies (ICT) [1]. It changes especially the incorporation of information technology into the manufacturing procedures. Industry 4.0 is the fourth technological revolution in a series that began with the first industrial revolution, which saw the development of the mechanization of the industry at the end of the eighteenth century as a result of the steam engine and increased usage of water and steam power. The advent of electricity and the assembly line at the beginning of the twentieth century helped to foster the second industrial revolution known as "mass production." The value proposition of Industry 4.0 is directly related to the end-to-end digitalization of all physical assets and the integration of all value chain participants into digital ecosystems.

### 12.1.1   Integration of Modern Technologies

The goal is to enable autonomous decision-making processes, equally real-time connected value creation networks, and real-time asset and process monitoring through early stakeholder involvement, vertical and horizontal integration, and real-time asset and process monitoring. Industry 4.0 is a concept, policy, and vision that is active, with definitions, reference architectures, and standardization all in transition. The end-to-end digital supply chain, suppliers, and the origins of the materials and components required for different types of smart manufacturing, as well as the end consumer, who serves as the final destination of all manufacturing and production regardless of the number of intermediary steps and players, are all crucial to understanding Industry 4.0 [2]. It is feasible to completely integrate the full value chain, from design to realization, while optimizing with a continuous flow of data, by fusing the physical and digital worlds. A truly digital enterprise may take advantage of the unbounded power of data by getting insightful knowledge that helps them make quick, confident decisions as well as best-in-class products through effective manufacturing (Figure 12.1).

*Topics in Artificial Intelligence Applied to Industry 4.0*, First Edition. Edited by Mahmoud Ragab AL-Refaey, Amit Kumar Tyagi, Abdullah Saad AL-Malaise AL-Ghamdi, and Swetta Kukreja.
© 2024 John Wiley & Sons Ltd. Published 2024 by John Wiley & Sons Ltd.

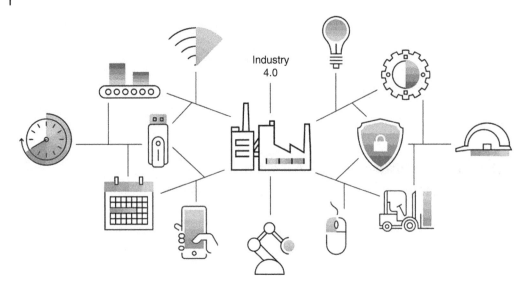

**Figure 12.1** Industry 4.0 transformation.

### 12.1.2 Globalization and Emerging Issues

Corporate globalization strategies are a key element of Global Population Network (GPNs). In this context, the globalization strategies as actions within GPNs that involve outsourcing, insourcing, offshoring, and re-shoring. The terms "outsourcing" and "insourcing" relate to, between the business and outside stakeholders, a change of ownership and management of activities [3]. In order to confront the future in a positive and cooperative manner, it is essential to concentrate on the implications of the current transformation in the economic, social, geopolitical, and environmental contexts. We are living through the so-called Fourth Industrial Revolution, which is being fueled by new, cutting-edge technology. This revolution sees the convergence of the physical and digital worlds with inventions that are unprecedented in the history of civilization. But what precisely do we mean when we discuss new digital technologies that will contribute to Industry 4.0? We must assess some of the technologies that are seriously disrupting institutional and corporate business operations in order to provide an answer to this topic.

Technologies have the power to fundamentally alter how an organization operates, introducing elements of innovation that go well beyond basic digitalization and dematerialization. The forms of issue solutions that are usual in a simple or complex organizational structure might benefit significantly from technological advancements. For the sake of concision, we will not analyze them all but will instead focus on the ones that, according to their impact, we believe to be the most disruptive: lockchain, IoT, robotic process automation, and artificial intelligence.

The introduction of digital technologies has forced businesses to adopt a new method of communicating with both internal and external stakeholders. A business uses a variety of technical resources to meet the demands of its human resources in order to generate goods or services. In the past, individuals who wrote or read the data that would later be aggregated for statistics were primarily responsible for managing the information flow produced by these operations. We may now achieve a fully creative flow, thanks to the IoT, where devices communicate with one another directly without the need for human intermediaries. Direct communication between the machine and the workpiece yields the results (Figure 12.2).

The following are the common obstacles that industries encounter when undergoing a digital transformation, per a recent study:

- Insufficient digital proficiency among factory personnel;
- IT problems like difficulties with cybersecurity, improperly configured networks, and system failures;
- Business model alignment and change management issues related to digital transformation present leadership challenges;
- Finances, particularly the first investment required for digital transformation.

Problem and suggested solution for the following term:

**Problem:** The issue is a shortage of qualified staff to handle sophisticated 4.0 components. Working with digital tools through interfaces might be difficult for employees.

**Figure 12.2** Globalization 4.0.

**Solution:** Before hiring someone, evaluate their digital aptitude. Examine your employees' need for retraining. Employ software development teams that can produce business solutions with clear and user-friendly layouts while taking into account industrial workers.

**Problem:** The requirement for more online connectivity and interorganizational data sharing will make industrial organizations more susceptible to cyberattacks. Additionally, limitations for Industry 4.0 will include IT network concerns, including incorrect command, configuration, and software failure.

**Solution:** Your IT infrastructure needs to be able to handle the extra connectivity required for your digital transformation. Employing IT and data security solution providers can help businesses analyze enterprise system vulnerabilities and machine-level operational failure threats.

**Problem:** Industry 4.0 adoption is expensive. Large corporations have an advantage because they can invest in the development and research of unique digital solutions.

**Solution:** It is not always necessary to spend a fortune to transform a company. Ministry of Micro Small and Medium Enterprise (MSMEs) can take baby steps on their road toward digital transformation. Here is a guide on how to successfully implement the digital transformation.

### 12.1.3 Cybersecurity 4.0

For all businesses committed to the Industry 4.0 paradigm, cybersecurity provides a complex challenge. Cybersecurity is crucial in Industry 4.0 environments for preventing the decline in business competitiveness [4]. In fact, a variety of cyberattacks that have the potential to completely disrupt the business model can currently harm vital industrial equipment.

Once the industrial assets most frequently implicated in cybersecurity challenges in Industry 4.0 contexts have been determined (i.e., what needs to be safeguarded), the next steps involve the following:

- the description of the system's inherent weaknesses that endanger their security;
- the system-affecting cyber-threats;
- the dangers brought on by cyberattacks;
- the defenses against cybersecurity problems.

These components are all connected to the idea of cybersecurity.

The technological advancements that form Industry 4.0's foundation also give rise to plenty of security problems. Although still connected to the traditional computer and network security perspective, cyber-threats in an industrial context create a number of specific features.

Cybersecurity Principles: Systems should be protected from the effects of security incidents by the deployment and management of cybersecurity. Activities could be planned out in accordance with the stages mentioned earlier. It is a continual process that needs constant work.

- Increasing employee awareness
- Managing assets and analyzing risks; incident detection
- Prevention: the idea of "defense-in-depth"
- Alert-chain and incident management
- Threats and vulnerability monitoring
- Business continuity and disaster recovery plans.

## 12.2  Security Fog Computing

### 12.2.1  Fog Computing in Industrial Internet of Things

It provides a novel idea called "fog computing" for managing data locally at the network edge. The problems of latency and bandwidth efficiency have been solved by fog computing, which is a supplement to the current cloud architecture [5]. It requires a strong check on Quality of Service (QoS) requirements because of its distributed architecture in order to be useful. Distributed networking and all-encompassing pervasive computing form the foundation of fog. It consists of small data centers or a collection of computers called "cloudlets" (fog clouds) that offer services to nearby devices [6, 7]. When compared to the cloud, initial installation costs, latency, and energy usage are significantly lower, but operational costs can vary.

In fog computing, processing data on a single server (the fog cloud) facilitates dependable, in-the-moment communication. It returns personal data protection and safety to our facilities. Fog computing also offers a cost-effective method for lowering data transmission and storage costs based on service locations. Fog computing, therefore, has the potential to offer cost-effective solutions for major IIoT projects. Fog computing allows for the option to select any hardware from information technology (IT) solutions as opposed to being limited to only one pricey cloud connection. It supports all legacy devices that are currently in use as well as non-IIoT devices that were never meant to be used with IIoT applications. This is both more affordable and versatile. It enables real-time processing, and secure management of fog is possible from a distance. It is dynamically scalable and updatable. It offers increased security, enhanced performance, and cost savings. Fog utilizes the advantages of the cloud and offers advantages that could help with future IIoT applications.

Fog is a relatively recent paradigm that presents difficulties for scalable and effective network architecture. It is anticipated that it will evolve gradually over the upcoming years in order to realize Industry 4.0. Future automation requires resolving unresolved concerns, including energy conservation, real-time communication, effective spectrum utilization, cache memory on edge devices, and resource allocation optimization. Guaranteed QoS requirements for IoT devices might not be met in the absence of such factors. For the industrial revolution to advance in the future, researchers will need to offer solutions to these problems.

The use of intelligent sensors and actuators to improve manufacturing and industrial processes is known as "the industrial Internet of Things" (IIoT). IIoT, sometimes referred to as "Industry 4.0" or "the Industrial Internet," makes use of real-time analytics and smart devices to make the most of the data that "dumb machines" have been producing in industrial settings for years. The underlying tenet of IIoT is that intelligent machines are superior to people at both data collection and real-time analysis, as well as in conveying critical information that can be utilized to make decisions about the conduct of business more quickly and accurately.

Connected sensors and actuators enable companies to pick up on inefficiencies and problems sooner and save time and money while supporting business intelligence efforts. The IIoT has enormous potential for improving supply chain efficiency, supply chain traceability, sustainable and green industrial practices, and quality control in particular. IIoT is essential to operations like predictive maintenance (PdM), improved field service, energy management, and asset tracking in an industrial setting.

#### 12.2.1.1  Working of Industrial Internet of Things
IIoT is a network of intelligent devices connected to form systems that monitor, collect, exchange, and analyze data. Each industrial IoT ecosystem consists of

- interconnected gadgets that can perceive, transmit, and store data about them;
- public and/or private data communications infrastructure;

- programs that use analytics to create business intelligence from raw data;
- storage for the data that is generated by the IoT devices and people.

These edge devices and intelligent assets transmit information directly to the data communications infrastructure, where it is converted into actionable information on how a certain piece of machinery is operating. This information can be used for PdM, as well as to optimize business processes.

### 12.2.2 Plant Safety and Security Including Augmented Reality and Virtual Reality

ICT safety and security can be viewed in terms of identity protection, online fraud, and cybersecurity [3]. Safety and security are discussed in terms of workplace safety, employee ergonomics, and task performance, as well as safety training, in sectors like maintenance or construction.

This impact can be achieved by exploring the immersive environment that virtual reality (VR) and augmented reality (AR) offer. Due to its flexibility, realism of the experience, and ability to safely recreate hazardous procedures and scenarios, VR, an electronic simulation of settings, provides a suitable virtual environment (VE) for education and training, according to prior studies [8]. A technique known as "AR" mixes actual and virtual content with real-time interaction and 3D registration [9]. AR provides the chance to supplement safety procedures in the real world, enhancing productivity, quality, and worker safety, and lowering physical and mental workload [10].

### 12.2.3 Augmented Reality Security and Privacy Issues

#### 12.2.3.1 Augmented Reality Concerns

One of the main privacy concerns with AR is its perceived risk. Because AR technologies can observe what a person is doing, their privacy is in danger. In comparison to other types of technology, such as social media networks, AR gathers a lot of data on the user, including who they are and what they are doing. This raises concerns and questions:

- If hackers gain access to a device, the potential loss of privacy is huge.
- How do AR companies use and secure the information they have gathered from users?
- Where do companies store AR data – locally on the device or in the cloud? If the information is sent to a cloud, is it encrypted?

Although third-party manufacturers and applications produce and transport the material, AR browsers help with the augmentation process. Given that AR is a relatively new field and means for creating and transmitting verified content are still developing, this poses the issue of dependability. Complex hackers may replace a user's AR with their own, deceiving users or disseminating false information.

Even if the source is legitimate, many cyber dangers could render the content unreliable. These consist of data modification, sniffing, and spoofing. Due to the possible unreliability of material, AR systems can be a useful tool for social engineering assaults that aim to trick people. For instance, hackers may manipulate users' perceptions of reality by using false signs or displays to persuade them to do activities that are advantageous to the hackers.

**Malware:** By using advertising, AR hackers can insert harmful content into the program. Unaware consumers may click on adverts that take them to hostage websites or AR servers that are compromised with malware and contain inaccurate visuals, weakening the security of AR.

**Network Credentials Theft:** Criminals may steal network credentials from Android-powered wearables. Hacking may be a security risk for merchants who employ AR and VR purchasing apps. Many clients have already stored their card information and mobile payment options.

**Denial of Service:** This type of security breach in AR is also possible. An illustration would be users who depend on AR for their jobs being abruptly disconnected from the information stream they are now receiving. This would be especially troubling for professionals who rely on technology to complete duties in urgent situations where a lack of information could have grave repercussions. A driver abruptly loses sight of the road because his AR windscreen turns into a black screen or a surgeon suddenly loses access to crucial real-time information on their AR glasses.

**Man-in-the-Middle Attacks:** Network intruders have the ability to eavesdrop on conversations between an AR browser and its service provider, channel owners, and outside servers.

### 12.2.3.2 Virtual Reality Concerns

Privacy is a huge issue with VR, just like it is with AR. The very personal nature of the data obtained, such as biometric information like iris or retina scans, fingerprints and handprints, face geometry, and voiceprints, is a major concern for VR privacy.

### 12.2.3.3 Examples

**Finger Tracking:** A user might utilize hand motions in the virtual world much like they would in the actual one, such as typing a code on a keypad with their fingers. However, doing so results in the system recording and sending finger-tracking data that shows a PIN being typed. An attacker will be able to create a user's PIN if they can obtain that information.

**Eye-Tracking:** Some VR and AR headsets might also have eye-tracking technology. Malicious actors might find further value in this material. Knowing exactly what a person is looking at might provide an attacker access to important information, which they could then use to reenact human behaviors [11].

### 12.2.4 Safety Application

- Pressure-relief valves
- Rupture disks
- Corrosion/erosion monitoring
- Steam traps
- WirelessHART and analytics

### 12.2.4.1 WirelessHART and Analytics

The devices discussed so far should be regularly monitored from a central location to maximize the benefits of safety and reliability. Although they can be linked together using a conventional point-to-point cable, using standards-based wireless technologies, like WirelessHART, is a far more economical option. As a result, extra junction boxes, cable trays, control system terminations, and I/O cards are not required. For data dependability comparable to that of traditional wiring, WirelessHART uses a self-organizing mesh network. It is shielded by multilevel, active security.

### 12.2.4.2 Maximize Plant Safety

Instrumentation is frequently considered in the context of specialized systems like the Safety Instrument System (SIS) and Basic Process Control System (BPCS) by plant designers. However, any business that wants to maximize plant safety needs to implement additional levels of security, such as physical protection and plant emergency response, on top of these existing systems. Using new IIoT technology, this is now simpler and more affordable than before [12].

## 12.3 IoT Challenges

Privacy and security are the most important challenges of IoT. Lack of robust and efficient security protocols, improper device updates, famous active device monitoring, and user unawareness are among the major challenges that IoT is facing. IoT bought users huge benefits but with some challenges like cybersecurity and privacy risks. These two challenges are the primary concerns of researchers and security specialists, which also pose a considerable predicament for many public organizations as well as business organizations. In IoT technologies, the vulnerability exists because of prevalent high-profile cybersecurity attacks. This vulnerability is produced because of the network's interconnectivity in the IoT that brings along accessibility from anonymous and untrusted Internet, requiring novel security solutions.

### 12.3.1 Privacy and Security Concerns of Industrial Internet of Things

The IoT differs from traditional computers and computing devices; it is more vulnerable to security threats in a variety of ways.

- Typically, IoT deployments consist of a collection of comparable or nearly identical appliances with similar properties. This commonality magnifies the impact of any security flaw that may affect a large number of them.

- Similarly, several institutions have developed risk assessment guidelines. This phase implies that the number of networked IoT devices is likely to be unparalleled. It is also obvious that many of these gadgets may automatically create connections and talk with other devices in an erratic manner. These necessitate an evaluation of the available technologies, strategies, and tactics linked to IoT security.
- Most IoT devices are designed for large-scale deployment. Sensors are a good illustration of this.

Even though the concern about security in the information and technology sector is not new, the widespread use of IoT has introduced novel issues that must be addressed. Consumers must trust that IoT devices and services are safe against flaws, especially as this technology becomes more passive and integrated into our daily lives. With poorly secured IoT devices and services, this is one of the most important routes for cyber assaults as well as the disclosure of user data by leaving data streams unprotected sufficiently [13].

One of the most common IoT attacks is the man in the middle, in which a third party may hijack a communication channel to spoof the identities of the nodes engaged in network exchange. Because the adversary does not need to know the identity of the presumed victim, the man-in-the-middle attack effectively causes the bank server to recognize the transaction as a genuine event. The effectiveness of the IoT is reliant on how effectively it can respect people's privacy preferences. Concerns about privacy and other hazards associated with IoT may be relevant in delaying the full implementation of IoT. It is critical to understand that privacy rights and user privacy respect are critical in ensuring users' confidence and self-assurance in the IoT, linked devices, and associated services offered. A lot of work is being done to guarantee that IoT is redefining privacy problems such as increased surveillance and tracking.

Industry 4.0 has been recognized as a transformational move that combines data, connection, and autonomy to generate the Fourth Industrial Revolution. However, these CPSs face many kinds of critical challenges. Significant cyber-physical attacks have been identified over the years, and there are likely many more that have not been revealed or even detected.

The Maroochy Water Services assault in Australia in 2000 was one example, in which the sewage system faced a sequence of failures, the pumps did not operate when they were intended to, and alarms were deactivated. This was exacerbated by a breakdown in the connection between the central computer and the individual pumping stations. Stuxnet, meanwhile, had a swift and major impact on Iran's nuclear industry [13]. Recent strikes include an attack on a German steel factory in 2014, as well as interruptions to the Ukrainian electricity network.

### 12.3.2   Security-Related Issues in Industry 4.0

- **Liability over Products**
  A vast number of stakeholders are engaged in the process of creating a smart product that can be connected to the Internet via the IoT. These stakeholders include the manufacturing brand, the companies that supply one of the many parts that are being assembled to manufacture that specific smart device, the security teams working on the security systems installed in that smart device, the software teams that program that device, and many more in the supply chain [14]. This large number of stakeholders ends up allocating responsibility difficulties in the event of a security incident using that smart device, which might be a big hurdle in the deployment of Industry 4.0.
- **Technical Constraints of the Devices**
  Although Industry 4.0 is intended to be the digitization of current industrial processes, rather than a completely new system, the same activities that were done manually will now be done digitally. This indicates that digital platforms will be built on top of current platforms, which is the only option since otherwise we would have to create a whole new sector from scratch. So, while this is a simple approach to adopting Industry 4.0, it is not without obstacles.

  Most of these devices, which are currently in use or are still being manufactured using outdated processes, have technological limits that pose significant security issues.

  The second most significant technical constraint is that all of these presently operational gadgets lack the infrastructure to enable any form of protective function. All of these gadgets were designed with one single operation in mind, with little regard for the protective aspect of these devices [14]. Furthermore, many new gadgets being manufactured today lack this essential defensive architecture.
- **Lack of Policies and Funds to Focus on Security**
  The industry at the leading edge of the Fourth Industrial Revolution is nevertheless hesitant to emphasize the need for security. This is evident from the absence or lack of policies and organizational regulations that ensure that privacy and security are given top priority while developing a product. Regardless of how much we emphasize it, the market has yet

to recognize privacy and security as a fundamental necessity. To this day, most operators consider it an add-on or a luxury option, which leads us to our next topic.

For the exact same reason that companies and operators refuse to prioritize security over other aspects, they have cut down on funding and are particularly unwilling to invest or spend more on R&D connected to current security concerns. This creates a lot of issues since cybersecurity technology will no longer disseminate and improve as it should.

An excellent instance of an industry not spending enough money on security solutions is when an industry decides to move its information systems to the cloud rather than storing them locally. Now, such migration occurs when an industry realizes the number of resources that are being spent on keeping information locally, and one of the main motivators for this move is generally the financial savings that can be realized by migrating to cloud services. So how much money would a firm be ready to spend on cloud security if the only purpose for moving its information systems to the cloud was to save money?

- **Lack of IT/OT Security Expertise and Awareness**

  This issue is due to human beings who are either engaged with the changing manufacturing processes or in the security of the digital component of it. There is a major lack of information security knowledge in both circumstances. People involved in the manufacturing process usually have no idea what security measures must be taken, and those dealing with the security aspect of the process do not fully understand the manufacturing process in order to secure it so that it is not vulnerable to outside attacks [14].

  To fully comprehend the process, a person must be qualified in several areas, including IT and OT security, embedded systems, and network security. Instead of considering security as a feature, make it a priority. To compound matters, most corporations experiencing the Fourth Industrial Revolution have chosen the simpler option of deploying security services on top of their processes rather than educating their present manufacturing staff.

  There are several causes for this, including employees' inability to understand and learn present security systems and the excessively expensive cost of training. These days, the quantity of cybersecurity training is decreasing and getting increasingly pricey in terms of cost.

### 12.3.3 Side-Channel Attacks

Side-channel attacks are among the increasingly critical hardware-based IoT attacks. IoT devices are especially sensitive to these attacks because of restricted resources such as battery power, storage, and computing power, which opens the door for side-channel attacks and makes the detection of these malicious programs difficult.

#### 12.3.3.1 Types of Side-Channel Attacks

Along with thermal imaging attacks, there are still a variety of other side-channel attacks that might be used, including timing attacks, power analysis attacks, electromagnetic analysis attacks, fault analysis attacks, optical side-channel assaults, traffic analysis attacks, and acoustic attacks [15].

- **Fault Analysis Attacks**

  The attacker inserted a crypto node with a defect and then assessed the difference between proper and faulty text in order to obtain the cryptographic key value in fault analysis attacks. The invader needed particular expertise in the design of hardware components to perform this assault. Attackers employ a variety of ways to inject the fault, including voltage glitching, clock pin manipulation, electromagnetic (EM) disturbances, and laser glitching.

- **Power Analysis Attacks**

  Power analysis involves attackers diligently measuring the power spent throughout the different cryptographic hardware parts of IoT devices and then analyzing electric current changes to recover crucial data contained in devices. Power analysis attacks are further categorized into three subcategories: basic power analysis attacks (BPA), differential power analysis attacks (DPA), and correlation power analysis attacks (CPA).

- **Timing Attacks**

  Timing attacks are carried out by utilizing timing variations such as overclocking, which is widely used to inject malicious nodes or other IoT devices' failures in order to leak critical information. These attacks can track how long it takes a program to complete various activities and then use that information to acquire highly confidential information such as bank account numbers, PIN codes, passwords, and cryptographic keys. The goal of side-channel timing attacks is to obtain the key to encryption methods [15].

- **Electromagnetic Analysis Attacks**

  Attackers record and interpret electromagnetic radiation in order to acquire confidential data from IoT device hardware gadgets such as display panels. In other circumstances, attackers put a micro antenna nearer to the integrated circuit (IC) to record electromagnetic signals. These electromagnetic attacks have been employed in military activities.

### 12.3.4 Spear Phishing

In the past few years, phishing attacks have increased security experts' vigilance. In 2018, the FBI received over 100 complaints. Health care, air travel, education, and many other areas have been targeted as the most primitive. As a result, they ended up incurring a loss of around $100 million USD. This was an email phishing structure designed to get employees' personal information and credentials [16]. These credentials were used to gain access to their payroll system, and attackers modified some criteria to those transactions, resulting in employees receiving no warning about account changes.

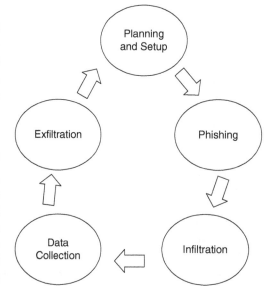

**Figure 12.3** Life cycle of phishing attacks.

Phishing attacks may be avoided by adopting alert behaviors, raising awareness, continuing to search when browsing the internet, and accessing the link if the source is trustworthy. Browser extensions, for example, inform users when they submit their credentials on a phishing site, passing their credentials to the malicious intent user. Other techniques enable the network to grant access to only whitelisted websites while blocking everything else (Figure 12.3).

Spear phishing is the most specific sort of phishing attack. Blasting millions of individuals with email campaigns is considered spam in social engineering. Spear phishing accounts for 91% of all phishing attempts. It is a targeted method of attack. Attackers customize their accounts with your company, position, name, work phone number, and other internet credentials. Users appear to have fun and know one another well. They want to catch you by clicking on a malicious email attachment or URL that supplies the attacker with all of your personal information [16]. A spear-phishing email would be one from the HR department asking you to verify your details for insurance benefits. Success in spear phishing is dependent on the following aspects of human psychology:

- Authority
- Liking
- Commitment
- Reciprocity
- Social Proof
- Scarcity

### 12.3.5 Cyberterrorism

One of the most significant issues for industries approaching the Industry 4.0 paradigm is cyberattacks. The size and range of cyberattacks have increased tremendously in the era of Industry 4.0 when working machines are connected to the network and to one another by means of smart devices. The term "Industry 4.0" refers to the use of intelligent, interconnected CPSs with the goal of automating all phases of industrial operations (from design and manufacturing to supply chain and service maintenance) [17]. In other words, Industry 4.0 connects manufacturing to ICT, combines product and process data with machine data, and allows machines to share information with one another.

Industry 4.0 modifications have the ability to add value to businesses by increasing efficiency by 15–20%. In particular, Industry 4.0 benefits for performance are achieved by (a) optimizing asset utilization and reducing machine downtime through remote monitoring and preventative maintenance, (b) increasing labor efficiency through manual labor automated processes, and (c) reducing the amount of inventory and improving service and product quality through real-time analysis of information produced by machine sensors [17].

The existence of linked CPSs in industrial scenarios, on the other hand, creates an important threat to security because the majority of systems of this sort were not developed with cybersecurity in mind.

## 12.4 Security Threats and Solutions of Industrial Internet of Things

The collected data may now be stored remotely and accessed from anywhere for analysis, thanks to cloud computing. The sort of cloud service utilized will determine whether extra vulnerabilities are introduced by cloud services. The following types of attack are used to exploit vulnerabilities related to the various IIoT and cloud computing:

1) **Malware:** In this attack, malicious software is used to infect a victim's computer, causing it to run slowly or shut down entirely, and allowing attackers to steal important data. The code is straightforward, which allows it to spread quickly and even infect whole file systems or networks. The more time this code spends on a user's computer, the more damage it does. Malware could perhaps affect the devices that are located there, causing them to malfunction. Spyware, viruses, and worms are the three most prevalent types of malware. The information entered by the victim is collected by spyware, which is used to spy on a victim. A computer virus is a program that, when run, starts reproducing itself and substitutes malicious code for the original code. Computer worms are similar to computer viruses, with the exception that worms can propagate more quickly and can survive without a host. A hacker in cloud computing uses malware and implements a malicious service or virtual machine. Once the malware has been created, it is added to the cloud computing infrastructure. Once the virus is there, the attacker must mislead the cloud into believing it is legitimate and send users to that instance. Any users who log into that system once the VM has been compromised are also impacted [18]. It is a cybersecurity strategy that focuses on safeguarding IoT-connected devices and networks with various processes, tools, technologies, and methods.
The IT business faces a severe problem with data privacy; the data is at risk every time a smart gadget gathers it and sends it to an important place.
**Solutions**
   - Integrate data encryption, privacy, and categorization technologies that restrict access to only authorized personnel.
   - To solve problems with data protection, use cryptography. Enable access controls such as multifactor authentication, role-based controls.
2) **Authentication Attacks:** A user must submit valid credentials to access cloud services. However, the techniques and authentication procedures are quite weak and frequently targeted. Many cloud services still use single-factor authentication and require only a login and password. Attackers take advantage of the vulnerability when attempting to interrupt services or steal information from a cloud-based organization [19].
Insecure authorization and authentication are a threat. The security of the device must be authorized and authenticated.
   - The device serves as a frequent access point for hackers and other cybercriminals, leaving users open to attacks via the internet (Figure 12.4).

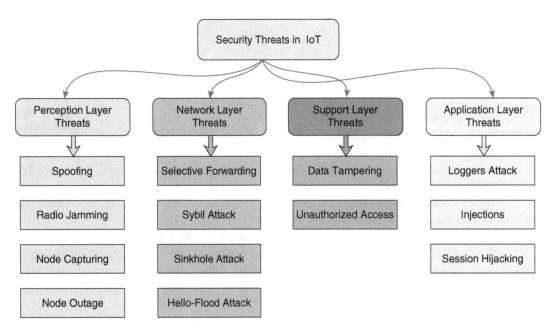

**Figure 12.4** IoT security threats.

### 12.4.1 Spoofing

There are various scenarios that can apply to attacks against biometric sensors; attacks include presenting a registered deceased person or dismembered body part, forcing a registered individual to verify or identify themselves, utilizing a genetic clone, and introducing bogus biometrics spoofing or sample use. The following lists several of these situations, many of which could benefit from potential fixes. Forceful attacks and the use of genetic clones are the exceptions. This risk exists even with the security mechanisms now in use for forceful attacks. Things such as cameras, "panic" buttons, and alarms are utilized to reduce risk. Biometric samples from individuals, including identical twins, still typically vary in the case of genetic clones. Although clones of DNA would be identical, DNA has not yet been produced for use in a scenario involving verification or identification. It may be useful to clarify how the false accept ratio, a popular evaluation indicator for biometric devices, is related to spoofing before going into more depth about the phenomenon. A submitted sample that is wrongly matched to a template enrolled by another user is known as "a false accept." This solely applies to a zero-effort attempt or an unauthorized individual attempting to access a system using their own biometric [20]. The likelihood of a specific user with criminal intent matching another template is extremely unlikely if the false accept ratio is maintained low. Several pieces in the popular press that criticize biometric technology have been released as a result of these reported incidents. While others have gone as far as to claim that these studies have entirely discredited the business and that biometric security systems are useless, these claims are ludicrous. Even though someone could take my office key and duplicate it to enter without authorization, this does not invalidate the use of keys. Studies illustrating the sensitivity to spoofing draw attention to a number of crucial considerations when thinking about the use of biometric sensors. The false accept ratio does not reveal a system's susceptibility to spoof attacks. People have frequently complained to IoT about receiving "undeliverable" message notifications from addresses to which they did not send a message. Most likely, a SPAMMER is to blame.

Here is a typical instance:

a) A website that lists you as a contact for Indiana gives a SPAM sender access to your email address.
b) The "spoofing" process involves a SPAM sender preparing a message to be sent to hundreds of recipients and inserting your address in the "From:" field.
c) The message is sent to thousands of addresses while pretending to be from you.
d) A couple of the addresses are wrong.

If the receiving mail system believes the message (yours) was sent incorrectly, it responds to the address with its response5. To report the email, you contact IoT customer service or IoT security (Figure 12.5).

### 12.4.2 Data Tempering

Tampering is the act of modifying or erasing a resource without permission. A web application is a programming that can be viewed online using a web browser. Web application data tampering is the process by which a hacker or other malicious person enters a website and modifies, deletes, or accesses unauthorized files. Using a script attack, a hacker or malicious user might also interfere inadvertently by disguising their actions as user input from a page or a web link [21].

**Example**: Data tampering is typically the result of competitor businesses attacking one another to obtain crucial information about the other, such as sales files or new projects on prototype products that are taken for the benefit of the other businesses. Due to the destruction of the firm's critical file archives, which are required to maintain the company operating its day-to-day operations, this typically results in one company closing down (Figure 12.6).

### 12.4.3 Malicious Code Injection

Malicious codes come in a wider variety and are more prevalent, thanks to the Internet's rapid development. Malicious code, usually referred to as "malware," is a catch-all word for software such as viruses, worms, Trojan horses, and other malicious programmers that carry out the malicious intent of the attacker. Malicious code's writing difficulty and skill have been regularly increased in order to prevent the destruction of antivirus software and increase the survival rate of core code. Attackers frequently use the so-called remote code injection technology to remotely inject their malicious malware's main function code into other processes so that it can run. This helps to keep the code stable and prevent it from being fully found. Malicious malware frequently employs remote code injection, a crucial hiding mechanism that

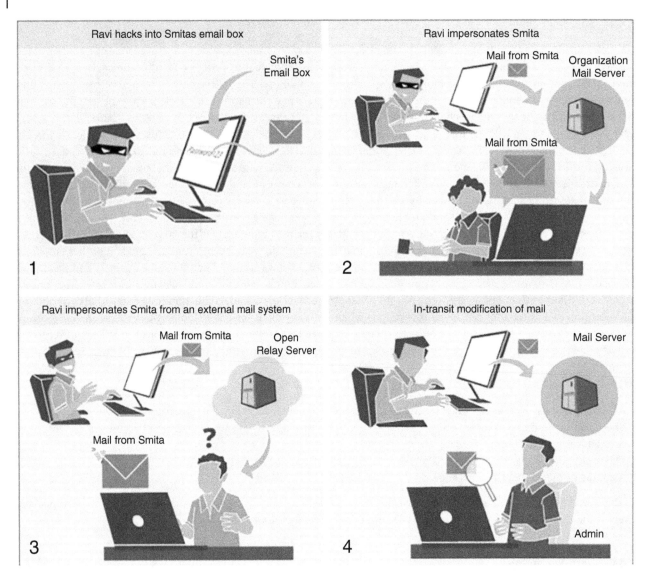

**Figure 12.5** Spoofing.

is highly deceptive and unlikely to be verified and requires no changing of any crucial system data. A malicious code injection involves changing lines of original code and replacing it with code that may damage the user's system. The malicious code is injected into the wearable device's application, allowing an attacker to exploit the system's vulnerability. An attacker can use this method to gain complete access to the wearable device's system threatening both user's privacy and confidentiality by potentially using the access to steal the user's personal data. Attacks involving the injection of binary code into an application to change how it runs and executes erroneous compiled code are known as "binary code injection" attacks. When the boundaries of memory locations are not examined and the program has access to areas outside of these boundaries, attacks like these may happen. Based on this, unscrupulous individuals can inject extra data, overwriting the data already present in nearby memory. From there, they can hijack a program or cause it to crash. Format string flaws are an additional attack method. The unanticipated behavior of functions with variable arguments is the root cause of this flaw. A function handling many arguments typically needs to read them from the stack. If we give "printf" a format string that expects two and if we only supply one argument while there are integers on the stack, the second parameter must be an additional integer. Attackers could read from or write to any address in memory if they had control over the format string. Malicious code injection represents the second class of security flaws. A compiler or interpreter runs the code as soon as the learner clicks a submit button in auto-grade systems, as opposed to

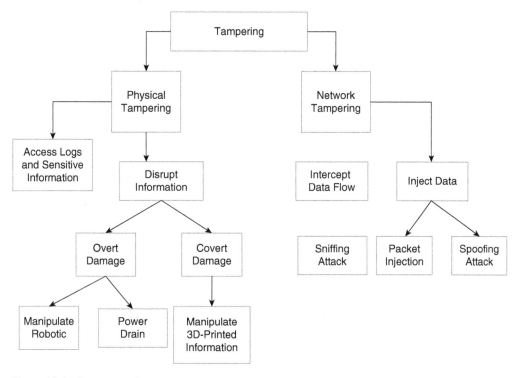

**Figure 12.6** Data tampering.

**Figure 12.7** Malicious code injection.

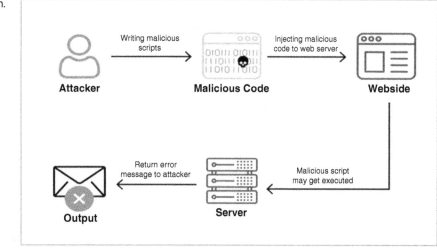

traditional assignment submission when code is submitted as an email to the person handling the grading. By attempting to upload executable code or a portion of such code to the server, where it can be run to retrieve some sensitive data or interfere with server functioning, these vulnerabilities can be exploited. Cross-side scripting (XSS) and SQL injection attacks are just two examples of these assaults [21, 22].

A different kind of attack involves leveraging the platform for improper purposes, such as using and utilizing the platform's resources by using an infinite loop (on recursion error) to run code access to external services. The main defense against these kinds of attacks is to use a container like a Docker to enclose the web application; that is, running the auto-grade by combining internet and physical space, Society 5.0 envisions a human-centered community that can strike a balance between societal obligations and economic proliferation [23] (Figure 12.7).

## 12.5  Conclusion

The aim of this chapter has been to contribute to understanding the concepts of security, threats, and privacy issues in Industry 4.0. The major points of the overall conclusion are as follows:

- Technologies have the power to fundamentally alter how an organization operates, introducing elements of innovation that go well beyond basic digitalization and dematerialization.
- IIoT is a network of intelligent devices connected to form systems that monitor, collect, exchange, and analyze data. Each industrial IoT ecosystem consists of connected devices that can sense, communicate, and store information about them.
- Industry 4.0 has been recognized as a transformational move that combines data, connection, and autonomy to generate the Fourth Industrial Revolution. However, these CPSs face many kinds of critical challenges. Significant cyber-physical attacks have been identified over the years, and there are likely many more that have not been revealed or even detected.
- The main defense against these kinds of attacks is to use a container like a Docker to enclose the web application that is running the auto-grade.

## References

1  Pereira, T., Barreto, L., and Amaral, A. (2017). Network and information security challenges within Industry 4.0 paradigm. *Procedia Manufacturing* 13: 1253–1260.

2  Gorkhali, A. (2022). Industry 4.0 and enabling technologies: integration framework and challenges. *Journal of Industrial Integration and Management* 7 (03): 311–348.

3  Sivertsson, M. and Utz, J. M. (2021). The influence of Industry 4.0 on globalisation strategies of multinational enterprises: A qualitative study of MNEs and their business decisions regarding offshoring and reshoring strategies. Dissertation.

4  Lezzi, M., Lazoi, M., and Corallo, A. (2018). Cybersecurity for Industry 4.0 in the current literature: a reference framework. *Computers in Industry* 103: 97–110.

5  Bonomi, F., Milito, R., Zhu, J., and Addepalli, S. (2012). Fog computing and its role in the internet of things. *Proceedings of the first edition of the MCC workshop on Mobile cloud computing*, 13–16.

6  Verma, M., Bhardwaj, N., and Yadav, A.K. (2016). Real time efficient scheduling algorithm for load balancing in fog computing environment. *International Journal of Information Technology and Science* 8 (4): 1–10.

7  Puliafito, C., Mingozzi, E., Longo, F. et al. (2019). Fog computing for the internet of things: a survey. *ACM Transactions on Internet Technology (TOIT)* 19 (2): 1–41.

8  Steuer, J., Biocca, F., and Levy, M.R. (1995). Defining virtual reality: dimensions determining telepresence. *Communication in The Age of Virtual Reality* 33: 37–39.

9  Billinghurst, M., Clark, A., and Lee, G. (2015). A survey of augmented reality. *Foundations and Trends® in Human–Computer Interaction* 8 (2–3): 73–272.

10  Burova, A., Mäkelä, J., Hakulinen, J., et al. (2020). Utilizing VR and gaze tracking to develop AR solutions for industrial maintenance. *Proceedings of the 2020 CHI Conference on Human Factors in Computing Systems*, 1–13.

11  Kaspersky Lab Resource Centre (2023). https://www.kaspersky.com/resource-center/threats/security-and-privacy-risks-of-ar-and-vr (accessed 13- January 2023).

12  International Society of Automation (2023). https://www.isa.org/intech-home/2018/march-april/features/iiot-in-safety-applications (accessed 13- January 2023).

13  Maple, C. (2017). Security and privacy in the internet of things. *Journal of Cyber Policy* 2 (2): 155–184. https://doi.org/10.1080/23738871.2017.1366536.

14  Mentsiev, A.U. et al. (2020). *Journal of Physics: Conference Series* 1515: 032074. https://doi.org/10.1088/1742-6596/1515/3/032074.

15  Aqeel, M., Ali, F., Iqbal, M.W. et al. (2022). A review of security and privacy concerns in the internet of things (IoT). *Journal of Sensors* 2022, Article ID: 5724168, 20 p. https://doi.org/10.1155/2022/5724168.

16  Ashina, S., Muhammad, A., and Rizwan, A.B. (2021). *Human Behavior and Emerging Technologies* 3 (2): https://doi.org/10.1002/hbe2.301.

**17** Vats, P. (2016). A comprehensive review of Cyber Terrorism in the current scenario. *2016 Second International Innovative Applications of Computational Intelligence on Power, Energy and Controls with their Impact on Humanity (CIPECH)*, Ghaziabad, India, 277–281, http://dx.doi.org/10.1109/CIPECH.2016.7918782.

**18** Shah, Y. and Sengupta, S. (2020). A survey on Classification of Cyber-attacks on IoT and IIoT devices. *2020 11th IEEE Annual Ubiquitous Computing, Electronics & Mobile Communication Conference (UEMCON)*, 0406–0413. IEEE.

**19** Kozlov, D., Veijalainen, J., and Ali, Y. (2012). Security and privacy threats in IoT architectures. *BODYNETS*, 256–262.

**20** Wu, Z., Zhang, Y., Yang, Y. et al. (2020). Spoofing and anti-spoofing technologies of global navigation satellite system: a survey. *IEEE Access* 8: 165444–165496.

**21** Shrivastava, R. K., Mishra, S., Archana, V. E., and Hota, C. (2019). Preventing data tampering in IoT networks. *2019 IEEE International Conference on Advanced Networks and Telecommunications Systems (ANTS)*, 1–6. IEEE.

**22** Tyagi, A.K., Dananjayan, S., Agarwal, D., and Thariq Ahmed, H.F. (2023). Blockchain – internet of things applications: opportunities and challenges for Industry 4.0 and Society 5.0. *Sensors* 23 (2): 947. https://doi.org/10.3390/s23020947.

**23** Nair, M. M., Tyagi, A. K. and Sreenath, N. (2021). The Future with Industry 4.0 at the Core of Society 5.0: Open Issues, Future Opportunities and Challenges. *2021 International Conference on Computer Communication and Informatics (ICCCI)*, 1–7. http://dx.doi.org/10.1109/ICCCI50826.2021.9402498.

# 13

# Designing a Quantum Computer to Gear up Artificial Intelligence for Industry 4.0

*K. Pradheep Kumar[1], Neha Sharma[2], and Juergen Seitz[3]*

[1] *Department of CSIS, BITS Pilani, Pilani, Rajasthan, India*
[2] *Tata Consultancy Services, Pune, Maharashtra, India*
[3] *Duale Hochschule Baden-Württemberg, Wirtschaftsinformatik, Heidenheim, Germany*

## 13.1  Introduction

With the explosion of huge volumes of data in the digital world, it has become essential to have technologies and hardware to process the data in a reliable and timely manner. However, the existing hardware and technologies pose numerous challenges to process the voluminous data. To add to this digital evolution, most of the applications to be used in Industry 4.0 require sophisticated hardware and processing algorithms to address several challenges that exist in routine activities. Despite providing several trade-off optimizations in the hardware and also providing compiler-based optimizations, either the reliability of the inference or timeliness of the inference for several critical applications poses a threat. Besides, with the increase in the number of computations, the processing hardware tends to get heated up or ultimately stop. This is evident from the fact that with decrease in the size of the components, the processing power also gets limited.

The aforementioned reasons are the motivation to design a quantum computer that relies on nanoscience technology. Currently, there is a need to have an advancement of nanoscience technology, which relies on atomic particles called "excitons." The excitons are basically a combination of photons and electrons that transfer most of the current by way of light carriers. The phenomenon is used basically for the transmission of large currents and power and is known as "superconductivity." To achieve the same, the material needs to be cooled below an absolute zero, and the temperature should be maintained to ensure smooth operation.

The fundamental element of the quantum computer is a qubit or "quantum bit." The "quantum bit" is essentially a "Josephson junction." Basically, a Josephson junction has two superconducting electrodes separated by a barrier. It has a memristor, an inductor, and a capacitor in a shunt configuration. The qubits are fabricated, which consist of a polarized light source and cryo-cooling arrangements. The qubits are classified based on the temperature as very cool, cool, normal, hot, and very hot. They are generated and stored in a qubit container. The quantum computer comprises of quantum data plane, which requires 75% cold qubits and 25% warm qubits for functioning. This combination is made randomly and cyclically from the qubit container. The quantum data plane is cooled by cryo-cooling arrangement. The cryo-cooling unit and the quantum data plane are controlled by a quantum control plane. It could be observed for an average of 547 qubits and 40 complex datasets out of 59 datasets, the processing time reduces by 34%, compared to simple datasets.

The chapter is organized as follows: Section 13.2 reports the literature; Section 13.3 explains the proposed model for the quantum computer, describing the various components and an algorithm for the working. Section 13.4 discusses the simulation results mapping the qubits and datasets. Section 13.5 concludes the chapter, and Section 13.6 discusses the future direction of research.

*Topics in Artificial Intelligence Applied to Industry 4.0*, First Edition. Edited by Mahmoud Ragab AL-Refaey,
Amit Kumar Tyagi, Abdullah Saad AL-Malaise AL-Ghamdi, and Swetta Kukreja.

## 13.2 Literature Survey

This section presents the study of literature related to various approaches and architecture toward the making of quantum computers. Trapped ion (TI) qubit features have been elaborately explained by Upadhyay et al. [1]. These qubits are identical and have a shuttle exchange of ions. They are also scalable. Multi-trap mechanism has been used for scaling up. Initially a greedy mapping has been used. Qubit mapping mechanisms and algorithms have been explained with adaptive scheduling strategies. An optimal mapping mechanism has also been developed with these strategies. Multi-tenant architecture strategies have been discussed by Mi et al. [2]. This has been discussed by illustrating a quantum internet. The quantum internet is a combination of multiple quantum clouds. Initially, a single tenant uniform batch (STUB) is created and then the same has been replicated. To introduce additional complexity, heterogeneity has been incorporated to classify the clouds based on performance and functionality. Heterogeneous-based scheduling could be enabled based on a set of optimality criteria.

Solving complex differential equations using a quantum computer has been explained by Shao [3]. These are modeled using datasets as matrices. The eigenvalues of the matrices are analyzed using specific boundary conditions. Sparse matrices were simplified using quantum linear solver techniques. Quantum circuits and qubit reliability using test points have been discussed by Acharya et al. [4]. These test points are created by monitoring an error factor due to noise in the quantum circuits. Based on the computed error factors, a reliability model has been designed to ensure the quantum circuits created are error free.

Quantum neural networks interconnecting multiple qubits have been explained by Samoylov et al. [5]. The neural networks designed have multiple layers with different thresholds. Artificial intelligence, machine and deep learning algorithms have been implemented across the layers of the quantum neural networks. The layers form a quantum bed. An optimal threshold computation for the quantum layers has been computed iteratively. Solving wave equations and mathematically complex problems using a practical quantum computing approach has been discussed by Suau et al. [6]. The quantum wave equation has been modeled using datasets. These datasets are then stored as matrices. Using quantum wave equation solver approach, the matrices are classified as sparse matrices. These sparse matrices are solved using a quantum linear solver approach. The mapping of the hardware and software for quantum computers has been designed by Li et al. [7]. The heuristic approach has been described using a block-wise architecture. The qubits are embedded in these blocks, and this has been extended to gate-level synthesis. Several controlled not (CNOT) trees have been illustrated, mapping the qubits to quantum gate blocks. Hamido et al. [8] help in modifying the states of qubits using Bell state contribution. It also shows how the states of qubits are linked to quantum gates for the processing of datasets. The superposition of the states is modified according to the requirement to obtain an optimal state.

Dilip [9] explains quantum block chains and the hardware required for the same. It explains the optimizations required to facilitate the multi-tenant architecture of the states. The smart contracts associated with the same aim to give a decentralization flavor to the quantum internet. Node selection and Hyperledger entry check using the qubits are also taken care of by the states optimized. New qubits and the mapping mechanism using the multi-programming feature using community detection-assisted partitioning. A task scheduler has been designed to facilitate the optimal mapping for the states. Training iteratively for binary neural networks has been illustrated extensively using Dou and Lou in [10] and Fawaz et al. [11]. Quantum machine learning algorithms with optimal thresholds have also been indicated. Appropriate activation function choice according to the training dataset and identifying an optimal bound has been explained. The landscape state for the optimal training has been explained with multiple datasets.

Sargaran and Mohammad Zadeh have explained a scalable architecture in Sahar and Naser [12]. A nearest neighbor architecture of the nodes and the states has been discussed. Quantum information and the qubits associated have been explained with the processing. The architecture has been made flexible. This has been implemented using a scaffold programming language. Multi-qubit operations have been explained with examples. Murali et al. [13] explained full stack architectural comparison by explaining matrix computation using qubit mapping. The gate mapping implementation and optimal mapping to handle qubits have been explained based on a benchmark study.

The qubit movements and the architectures have been explained by Almeida et al. [14] using permutations. The optimal permutation model has been obtained using an integer linear program model. Potok et al. [15] explained high-performance, neuromorphic computing. This is achieved using convolution neural networks. Multiple query optimizations using an adiabatic quantum computer are explained by Trummer and Koch [16]. A logical mapping between the blocks and qubits was created. A formal analysis of the same has been explained. Homulle et al. [17] have explained the cryo-CMOS hardware technology. The error correction mechanism has been illustrated, and the possible solutions have been explained. The different miniaturization perspectives have been explained.

Ahsan and Kim [18] explain the quantum architecture model. The required baseline and device parameters have been obtained by a benchmark application. The qubit delay has been reduced. Ahsan et al. [19] explain how universal quantum gates could be used for qubit operations. The quantum hardware and the architecture models have been explained in detail. The design flow and tool components have been explained. A fault-tolerant circuit generator has been explained to analyze the data flow and error rate. Van and Horsman [20] had explained a blueprint of a quantum computer. The architecture and stack machines have been explained in this model. The micro-architecture model has been illustrated with different optimizations. Quantum programming for the qubits processing has also been explained. David [21] explains quantum computer science using different quantum algorithms. The quantum computation steps have been explained. Spector [22] explains the evolutionary quantum algorithms by using experiments like using a beam splitter and interferometer. A photon-triggered bomb facilitates beam blasting on a mirror. Thaker et al. [23] explain the parallelism in quantum memory hierarchies. The different optimizations of cache memory and main memory in quantum computers have been discussed showing the various memory enhancements in Ben [24]. A quantum computing survey has been made by Kasivajhula [25]. The qubit representation using quantum algorithms and hardware has also been explained.

After having a fair understanding of various methodologies by different researchers toward quantum computers, the authors present a blueprint of the version of quantum computers that can gear up artificial intelligence for Industry 4.0.

## 13.3 Proposed Work

### 13.3.1 Motivation for Quantum Computers in Industry 4.0 Revolution

Industry 4.0 requires creation of numerous data science applications that could process huge volumes of data in a timely manner. Most of the applications require instantaneous solutions to problems that become impossible with the existing hardware and artificial intelligence algorithms. The quantum computer model has five major components, as shown in Figure 13.1.

a) Cryo-cooling unit
b) Qubit container
c) Qubit functional unit
d) Quantum data plane
e) Quantum control plane

#### 13.3.1.1 Cryo-cooling Unit

This unit is responsible for maintaining the temperature of all components of the quantum computer below absolute zero. It has an alarm indication system to raise an alert to the qubit container to provide cool qubits when the qubits in the qubit functional unit get heated up. The cryo-cooling unit ensures that optimal cooling is provided to all qubits and that the quantum gates and qubits are functional.

#### 13.3.1.2 Qubit Container

The qubit container has different qubits, which are of the following categories:

Very cold (VC)
Cold (C)
Normal (N)
Hot (H)
Very hot (VH)

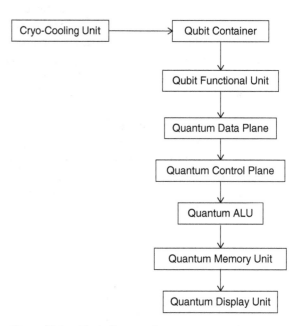

**Figure 13.1** Block diagram of a quantum computer.

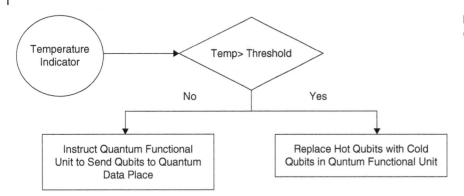

**Figure 13.2** Cryo-cooling unit for quantum computer.

The cryo-cooling unit decides the optimal cooling temperature to be maintained based on the qubit category. For example, the cooling rate for very hot qubits would not be the same as very cold qubits. The qubit container has various strategies to ensure that qubits are operational and the cooling is optimal for all qubits. The cryo-cooling unit has been illustrated in Figure 13.2.

### 13.3.1.3 Qubit Functional Unit

This unit is the main functional unit that decides on the operational qubits and the quantum computer's functionality. Ideally, for the quantum computer to be operational, the proportion of the qubits in this unit should be 75% cold qubits and 25% warm qubits. When the qubits temperature increases, it raises an alert to the cryo-cooling unit indicating the number of qubits that are hot and need to be replaced. The cryo-cooling unit, in turn, raises an alert to the qubit container to replace the hot qubits with cold qubits. There is a computational logic by way of "algorithmic cooling" to replace the qubits, which would be discussed. The qubit functional unit provides appropriate qubits to the quantum data plane based on the type of datasets and the processing requirements. Hence, the optimal combination of qubits needs to be supplied to the quantum functional unit and the quantum data plane. But there are exceptional conditions, which have to be closely monitored by the qubit functional unit and quantum data plane. The qubit functional unit and the proportion of qubits have been illustrated in Figure 13.3.

### 13.3.1.4 Quantum Data Plane

The quantum data plane does a mapping of the datasets to the qubits chosen in the qubit functional unit. This unit is responsible for organizing the datasets required for the processing to be done by the quantum control plane. This unit also has a metadata table of the datasets available for processing. The quantum data plane also needs to analyze the datasets and the operations to be used. It needs to assign the appropriate datasets to the qubits. When complex operations are involved, an appropriate sensing mechanism needs to be provided to ensure that optimal processing is ensured. The quantum data plane is shown in Figure 13.4.

### 13.3.1.5 Quantum Control Plane

The quantum control plane is responsible for the processing of the quantum datasets. It chooses the required quantum logic gates required to process the datasets in the quantum data plane. When a query is raised by the user, the query is decoded by the quantum control plane into gate-level operations, and then a mapping is done between the qubits and the appropriate quantum gates. It then does the processing using the quantum arithmetic and logic unit (ALU). The quantum control plane is shown in Figure 13.5.

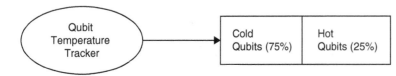

**Figure 13.3** Qubit functional unit.

#### 13.3.1.6 Quantum ALU

The quantum ALU is responsible for carrying out the partial operations decided by the quantum control plane. It then combines the partial results to provide the combined result. The quantum ALU is shown in Figure 13.6.

#### 13.3.1.7 Quantum Memory Unit

The quantum memory stores all results of the quantum ALU using the memresistor technology. The quantum operations could be complex, and these instructions should be performed by the quantum ALU. The quantum memory is partitioned into main memory and cache memory. The immediate computations carried out are stored in the cache memory. The finalized results are then written into the main memory. The quantum memory unit is shown in Figure 13.7.

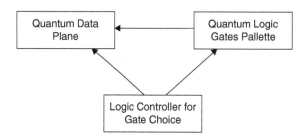

**Figure 13.4** Quantum data plane.

#### 13.3.1.8 Quantum Display Unit

The quantum display unit displays the results or inferences obtained from the quantum memory unit. It uses an optical matrix to display results from the quantum memory unit. An optical matrix embedded in quantum dots is used to display the required information on processing the qubits. The quantum display unit is shown in Figure 13.8.

### 13.3.2 Working of the Quantum Computer

The quantum computer needs to be maintained at a temperature below absolute zero. The cryo-cooling unit manages this temperature and ensures that the qubits are maintained at this temperature.

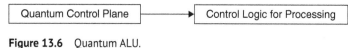

**Figure 13.5** Quantum control plane.

**Figure 13.6** Quantum ALU.

When the cryo-cooling unit is not functional, the quantum computer would not start. Hence, proper backup needs to be provided for power for the same. The qubits are extracted from the qubit container and placed in the qubit functional unit. The proportion of cold qubits to hot qubits is decided by the quantum control unit. These qubits are then sent to the quantum data plane. The datasets to be processed for a particular task are fetched using the metadata table available in the quantum data plane. After this, the qubit mapping is being done between the qubits and datasets. The quantum control plane based on this decides the quantum logic gates required for the processing operation of qubits and then instructs the quantum ALU to process the same. The results of the quantum ALU are then sent to the quantum memory unit. An algorithm for the functional operation of the quantum computer has been presented in the next section.

#### 13.3.2.1 Algorithm

- Identify a use case application.
- Decompose the same into tasks.
- Identify the required datasets for each task.
- Map each task to a qubit.
- For each qubit,
  - Process the relevant datasets.
- Combine the results of all qubits.
- End the procedure.

The various qubit proportion combinations are discussed in the next section.

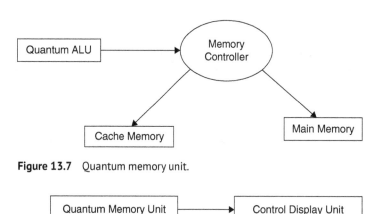

**Figure 13.7** Quantum memory unit.

**Figure 13.8** Quantum display unit.

#### 13.3.2.2 Qubit Proportion Combinations

For the normal function of the quantum computers, 1,000 qubits would be required. These are grouped into five categories as very cold (VC), cold (C), normal (N), hot (H), and very hot (VH). Since 75% of cold qubits are required for normal working, the VC and C categories are split accordingly, and the remaining are split as follows: 5% for normal; H and VH would be 10% each. The qubit container would store the qubits in the proportions as indicated in Table 13.1. Table 13.1 shows the split-up of each category of qubits and also indicates the exact number of qubits.

The qubit functional unit would operate with 75% cold qubits and 25% warm qubits. The possible combinations of proportions are given in Table 13.2 using 75:25% proportion

The exact number of qubits required is computed and rounded off to integers and indicated in Table 13.3. On average, 366 cold qubits and 33 hot qubits would be used for the normal functioning of the quantum computer. Total number of qubits would be 399. So the number of datasets to be processed would be $2n$, where $n$ is the total number of qubits. The quantum control plane decides on the qubit proportion based on the following parameters:

- Temperature indication of qubits in the quantum functional unit
- Optimal threshold to be maintained.

**Table 13.1** Qubit category-wise proportion.

| Qubit category | Proportion (%) | Number of qubits |
| --- | --- | --- |
| VC | 37.50 | 375 |
| C | 37.50 | 375 |
| N | 5 | 50 |
| H | 10 | 100 |
| VH | 10 | 100 |

**Table 13.2** Cold versus hot qubit proportion combinations.

| Qubit combinations | Cold qubits (75%) | Warm qubits (25%) |
| --- | --- | --- |
| 1 | 0.75*VC | 0.25*VH |
| 2 | 0.75*(VC+C) | 0.25(VH+H) |
| 3 | 0.75*(VC+C+N) | 0.25*(VH+H+N) |
| 4 | 0.75*(VC+N) | 0.25*(VH+N) |
| 5 | 0.75*(C+N) | 0.25*(H+N) |
| 6 | 0.75*C | 0.25*H |
| 7 | 0.75*C | 0.25*N |
| 8 | 0.75*VC | 0.25*N |

**Table 13.3** Actual number of cold versus hot qubit combinations.

| Qubit combinations | Cold qubits (75%) | Warm qubits (25%) | Total qubits |
| --- | --- | --- | --- |
| 1 | 281 | 25 | 306 |
| 2 | 563 | 50 | 613 |
| 3 | 600 | 63 | 663 |
| 4 | 319 | 38 | 357 |
| 5 | 319 | 38 | 357 |
| 6 | 281 | 25 | 306 |
| 7 | 281 | 13 | 294 |
| 8 | 281 | 13 | 294 |
| **Average** | **366** | **33** | **399** |

For the efficient operation of the quantum computer, we need to design the same with 1,000 qubits. Hence, each category VC, C, N, H, and VH would have 200 qubits each, and the proportion of qubit combinations would be as shown in Table 13.3.

The computation of datasets and qubits for the processing of the datasets is explained in the next section.

### 13.3.3 Datasets and Qubit Processing Requirement

$2^n$ qubits could be used to store the datasets. Hence, the number of datasets that could be processed is given by Eq. (13.1):

$$m = 2^n,$$ (13.1)

where $m$ is the number of datasets, and $n$ is the total number of qubits.

#### 13.3.3.1 Quantum Datasets

Quantum datasets are defined from conventional datasets. Each data point is modeled as a qubit, which has a tag ID, phase angle, and required data value. Each data point could be visualized to be a qubit. A quantum dataset is the combination of a group of qubits in proper sequence. Each quantum dataset could be modeled as a vector. The combination of these vectors together forms a lattice.

#### 13.3.3.2 Lattice Cube Formation

Each vector has a magnitude. A ranking is provided based on the magnitude. The vector set is then partitioned as horizontal and vertical vectors. This is done by computing the sum of all the magnitudes of vectors and then dividing the same by 2. The lower partition is chosen as a horizontal vector set, and the upper partition is chosen as the vertical vector set. Multiple lattice cubes could be generated as inputs to the quantum computer. Here is an example illustrating the qubits, quantum dataset, and lattice:

Consider a set of qubits with phase angles 0, 45, 90, 135, 180, 225, 270, 315, and 360.

The magnitude of each quantum dataset modeled as a vector is computed using Eq. (13.2):

$$mag(qd) = \sqrt{\left( (qbph_1)^2 + (qbph_2)^2 + \ldots\ldots + (qbph_n)^2 \right)}$$ (13.2)

where $mag(qd)$ is the magnitude of the quantum dataset vector, and $qbphn$ is the qubit phase angle of each qubit. Table 13.4 illustrates the different combinations of quantum datasets for the qubits.

**Table 13.4** Quantum datasets for the various qubits.

| Quantum dataset | Qubit 1 phase | Qubit 2 phase | Qubit 3 phase | Qubit 4 phase | Qubit 5 phase | Qubit 6 phase | Qubit 7 phase | Qubit 8 phase | Qubit 9 phase | Magnitude |
|---|---|---|---|---|---|---|---|---|---|---|
| 5 | 180 | 225 | 270 | 315 | 360 | 0 | 45 | 90 | 135 | 736.68 |
| 4 | 135 | 180 | 225 | 270 | 315 | 360 | 0 | 45 | 90 | 715.77 |
| 3 | 90 | 135 | 180 | 225 | 270 | 315 | 360 | 0 | 45 | 697.14 |
| 2 | 45 | 90 | 135 | 180 | 225 | 270 | 315 | 360 | 0 | 680.97 |
| 1 | 0 | 45 | 90 | 135 | 180 | 225 | 270 | 315 | 360 | 667.46 |
| 9 | 360 | 0 | 45 | 90 | 135 | 180 | 225 | 270 | 315 | 656.75 |
| 8 | 315 | 360 | 0 | 45 | 90 | 135 | 180 | 225 | 270 | 649.00 |
| 7 | 270 | 315 | 360 | 0 | 45 | 90 | 135 | 180 | 225 | 644.30 |
| 6 | 225 | 270 | 315 | 360 | 0 | 45 | 90 | 135 | 180 | 642.73 |
| **Total** | | | | | | | | | | **6,090.80** |

The sum of the magnitudes of the quantum dataset vectors is computed to be 6090.80. The quantum dataset vectors are sorted and ranked in descending order of magnitudes shown in Table 13.5.

Table 13.6 shows the quantum datasets ranked based on the magnitude. The ranking has been done in descending order of magnitude.

The horizontal vector range would be from 642.73 to 667, and vertical vector range would be from 667.46 to 736.68. The qubit ranges for each category of qubit like VC, C, N, H, and VH have been modeled. Each qubit category is designated an appropriate range based on the data given earlier in Table 13.6 and shown in Table 13.7.

The average optimal value of the qubit category is also computed from the average of mid-values, which is 674.68 for the optimal working of the quantum computer. Hence, the lattice cube would contain four horizontal and four vertical vectors. The mid-vector partitioning of the cube would have a magnitude of 667.46 as indicated in Table 13.7.

**Table 13.5** Quantum datasets sorted based on magnitude.

| Quantum dataset | Qubit 1 phase | Qubit 2 phase | Qubit 3 phase | Qubit 4 phase | Qubit 5 phase | Qubit 6 phase | Qubit 7 phase | Qubit 8 phase | Qubit 9 phase | Magnitude |
|---|---|---|---|---|---|---|---|---|---|---|
| 1 | 180 | 225 | 270 | 315 | 360 | 0 | 45 | 90 | 135 | 736.68 |
| 2 | 135 | 180 | 225 | 270 | 315 | 360 | 0 | 45 | 90 | 715.77 |
| 3 | 90 | 135 | 180 | 225 | 270 | 315 | 360 | 0 | 45 | 697.14 |
| 4 | 45 | 90 | 135 | 180 | 225 | 270 | 315 | 360 | 0 | 680.97 |
| 5 | 0 | 45 | 90 | 135 | 180 | 225 | 270 | 315 | 360 | 667.46 |
| 6 | 360 | 0 | 45 | 90 | 135 | 180 | 225 | 270 | 315 | 656.75 |
| 7 | 315 | 360 | 0 | 45 | 90 | 135 | 180 | 225 | 270 | 649.00 |
| 8 | 270 | 315 | 360 | 0 | 45 | 90 | 135 | 180 | 225 | 644.30 |
| 9 | 225 | 270 | 315 | 360 | 0 | 45 | 90 | 135 | 180 | 642.73 |
| **Total** | | | | | | | | | | **6,090.80** |

**Table 13.6** Quantum datasets ranked based on the magnitude.

| Quantum dataset | Qubit 1 phase | Qubit 2 phase | Qubit 3 phase | Qubit 4 phase | Qubit 5 phase | Qubit 6 phase | Qubit 7 phase | Qubit 8 phase | Qubit 9 phase | Magnitude | Rank |
|---|---|---|---|---|---|---|---|---|---|---|---|
| 5 | 180 | 225 | 270 | 315 | 360 | 0 | 45 | 90 | 135 | 736.68 | 1 |
| 4 | 135 | 180 | 225 | 270 | 315 | 360 | 0 | 45 | 90 | 715.77 | 2 |
| 3 | 90 | 135 | 180 | 225 | 270 | 315 | 360 | 0 | 45 | 697.14 | 3 |
| 2 | 45 | 90 | 135 | 180 | 225 | 270 | 315 | 360 | 0 | 680.97 | 4 |
| 1 | 0 | 45 | 90 | 135 | 180 | 225 | 270 | 315 | 360 | 667.46 | 5 |
| 9 | 360 | 0 | 45 | 90 | 135 | 180 | 225 | 270 | 315 | 656.75 | 6 |
| 8 | 315 | 360 | 0 | 45 | 90 | 135 | 180 | 225 | 270 | 649.00 | 7 |
| 7 | 270 | 315 | 360 | 0 | 45 | 90 | 135 | 180 | 225 | 644.30 | 8 |
| 6 | 225 | 270 | 315 | 360 | 0 | 45 | 90 | 135 | 180 | 642.73 | 9 |
| **Total** | | | | | | | | | | **6,090.80** | |

**Table 13.7** Qubit categorization and range with a mid-value bound.

| Qubit category | Range lower bound | Range upper bound | Mid-value |
|---|---|---|---|
| VC | 642.73 | 644.3 | 643.52 |
| C | 649 | 656 | 652.50 |
| N | 656.75 | 667.46 | 662.11 |
| H | 680.97 | 697.14 | 689.06 |
| VH | 715.77 | 736.68 | 726.23 |
| **Average** | **669.04** | **680.32** | **674.68** |

### 13.3.3.3 Threshold of Quantum Cube

The threshold of the quantum cube is defined as the minimum number of quantum dataset vectors required by the machine learning algorithm to arrive at an optimal solution. The optimal solution ensures that the data model is completely trained. The threshold is computed using Eq. (13.3):

$$T = \left( \frac{\sum_{i=1}^{i=n}(Qdno \times Mag)}{\sum_{i-1}^{i=n} Qdno} \right) \tag{13.3}$$

where $T$ is the threshold of the quantum cube, $Qdno$ is the quantum dataset number, and $Mag$ is its corresponding magnitude.

### 13.3.3.4 Randomized Quantum Dataset Generator

A random quantum dataset could be generated using a phase shift gate and Hadamard gate. The phase shift gates generate different qubit phase angle combinations. A randomization of these qubit phase angles is obtained using a Hadamard gate. The Hadamard gate selects a set of optimized states from the entire sample space of states. This could be helpful in creating a large combination of randomized quantum cubes.

### 13.3.3.5 Significance of Threshold Computation

The threshold of the quantum cube indicates the minimum number of quantum dataset vectors required to ensure optimal learning.

- When the value of T > 1000, it indicates that datasets have missing attributes, and adequate data cleaning is required.
- When the value of T is <1000, it indicates that the dataset is optimal, and learning could be completed rapidly.

*Example of Threshold Computation* The threshold computation for the quantum cube has been obtained using Eqs. (13.4) and (13.5), respectively, from Table 13.6.

$$\sum_{i-1}^{i=9} Qdno \times Mag = 30136.56 \tag{13.4}$$

$$\sum_{i=1}^{i=9} Qdno = 45 \tag{13.5}$$

Hence, threshold T = 30 136.56/45 = 669.70, which is nearly equal to the average mid-value obtained in Table 13.7. When a number of such quantum cubes are available, the overall optimality could be rated using Eq. (13.6):

$$overopt = \frac{(\alpha \times Thr(opt) + ((1-\beta) \times Thr(nopt)))}{(\alpha + (1-\beta))} \tag{13.6}$$

where *overopt* is the overall optimality, $\alpha$ is the proportion of optimal quantum cubes, *Thr(opt)* is the threshold of the optimal quantum cubes, $\beta$ is the proportion of nonoptimal quantum cubes, and *Thr(nopt)* is the threshold of the nonoptimal quantum cubes. Ideally, for quick inferences, $\alpha$ should be 75%, $\beta$ should be 25%, and the overopt should be 0.7 for effective learning to take place. The following are the bounds of the significance optimization:

- When *overopt* is less than 500, the learning process would be slow, and more optimizations are required.
- If *overopt* is greater than 800, the learning process would be very rapid.
- Hence, ideally we prefer *overopt* to be greater than or equal to 500 and less than 800.

If in the example given earlier, $\alpha = 0.75$ and $\beta = 0.25$ *Thr(opt)* = 669.7 and *Thr(nopt)* = 1000 – 669.7 = 330.3. Hence, *overopt* would be $(((0.75*669.7)+((1-0.25)*330.3))/(0.75+(1-0.25))) = 500$.

Hence, this is an optimal learning rate, and the datasets are complete. Several quantum lattice cubes using the aforementioned combinations could be generated, which would act as inputs to the quantum computer. Ideally, a quantum lattice cube comprises nine vectors.

There are several optimizations, which could be made to enhance the performance of the quantum computer. Some of the optimizations have been discussed in the next section.

### 13.3.4 Optimizations for Quantum Computer

The following optimizations could be made to facilitate the normal functioning of the quantum computer:

- Cryo-cooling unit backup power
- Qubit balancing based on dataset volume
- Quantum memory enhancement

#### 13.3.4.1 Cryo-Cooling Unit Backup Power

The cryo-cooling unit is provided with appropriate power backup to ensure continuous working to maintain the required temperature. In case of power failure or required power being low, an appropriate backup power source ensures that the quantum computer is continuously powered up.

#### 13.3.4.2 Qubit Balancing

The qubits are balanced when there is a huge dataset-processing activity. Accordingly the qubits are balanced between the qubit container and the qubit functional unit. The number of states could also be enhanced based on the complexity of the datasets to ensure optimal processing power of the tasks. The qubit complexity is enhanced using the following expression mentioned in Eqs. (13.7) and (13.8):

$$QBenh = \left(\frac{Cplxdata}{Tdata}\right) * QBtot \tag{13.7}$$

$$QBopt = QBnor + QBenh \tag{13.8}$$

where *QBenh* is the number of qubits enhanced, *Cplxdata* is the number of complex datasets, *Tdata* is the number of normal datasets, *QBtot* is the total number of datasets, *QBopt* is the number of optimal qubits, and *QBnor* is the number of normal qubits.

There are two types of qubits classified based on datasets:

- Complex qubits
- Simple qubits

Complex Qubits: These qubits have special functions with additional features in the functional modification of the superposition and entanglement of states. These qubits provide special functions to process complex datasets. Overhead in processing is minimized, and hence, datasets need to be screened to avoid qubits getting heated beyond the threshold.

Simple Qubits: These qubits have the regular superposition of states to handle the normal processing of simple datasets. These operations do not incur additional processing overhead.

***Quantum Logic Gates*** Some of the quantum logic gates are Pauli gate, Hadamard, phase, CNOT, SWAP, and Toffoli gate.

a) Pauli Gate: This gate helps in obtaining an exclusive state out of the group of states offered by a qubit. This gate circuitry is useful while working with complex datasets.

b) Hadamard Gate: The Hadamard gate has a special function which has a function similar to a rolling die. When the datasets are modeled as a hybrid of simple and complex datasets, the gate logic needs to process the data in a hybrid manner. Hadamard gate is used as an aggregation function that is a combination of simple and complex magnitudes of the quantum datasets under consideration.

c) SWAP: SWAP gate helps to interchange values and do the aggregation when the quantum cube gets complex. It also helps when different combinations of datasets are required from multiple quantum cubes.

d) CNOT: The CNOT gate provides a controlled inversion operation. This provides Hermitian state matrices to invert states and their functions.

e) Phase Shift Gate: The phase shift operation provides a sequential phase angle addition. This is very useful for the random quantum dataset generation.

f) Toffoli Gate: This is a double CNOT gate that reverts back the original state of the Hermitian matrix. The datasets generated are mapped to multiple states of the qubit, and they are later aggregated.

#### 13.3.4.3 Quantum Memory Optimization

The quantum memory is again balanced between the main memory and cache memory based on the storage volume of datasets. When there is an increase in the complexity of datasets, a trade-off factor is provided between the main and cache memory. The trade-off factor is given by Eq. (13.9):

$$T = \left( \frac{Cplxdata}{Tdata} \right) \times Tmem \tag{13.9}$$

where $T$ is the trade-off factor, *Cplxdata* is the number of complex datasets, *Tdata* is the total number of datasets, and *Tmem* is the total memory required.

#### 13.3.4.4 Significance of Overall Optimality

When the overall optimality is within the prescribed limits, it ensures sparse operations on the quantum data plane and the quantum ALU. The operations and performance are well balanced, and reliability of the system would be high. On the contrary, when the overall optimality is reduced, additional optimizations using scheduling policies would be required to ensure that the quantum cubes are balanced on the quantum data plane, thus ensuring optimal communication and computation costs by the qubits.

The scheduling policy should ensure there is minimal task and data migration on the quantum data plane. This would reduce the impact to a large extent for the quantum ALU. The optimization also ensures that the quantum memory is optimally used and balanced between the main memory and cache memory. The cache and memory optimizations would also be handled based on the profiling carried out. The scheduling policies could incorporate slack resource strategies to facilitate exceptional qubits where the resource consumption would be large. This could be facilitated by making an assessment of the exact quantum of resources required by the qubits, and the remaining resources could be cumulated as slack resources.

When exceptional qubits arrive with inadequate resources, the qubits could be prioritized based on the criticality and proportionate to the criticality and required resources; slack resources could be utilized to ensure optimal resource allocation to qubits. Qubits based on the criticality could be categorized as critical, noncritical, and exceptional. Qubits that are critical and exceptional may have higher priority and need immediate access to slack resources to ensure all qubits are processed by the quantum logic gates. Exceptional qubits, in particular, need to have more priority as they are unexpected and have increased priority compared to critical qubits. Noncritical qubits could be granted lesser access to the cumulated slack as this does not affect the feasibility and optimality of the schedule.

Further, all qubits need to have an ideal utilization of resources. Qubits with under and overconsumption of resources should be assessed, and optimal allocation needs to be guaranteed to ensure all quantum gates have optimal work allocation. Qubits that tend to arrive on the fly need to be initially assessed, and then iteratively resources should be allocated after checking for optimality. To achieve this, qubit utilization needs to be assessed in a timely manner prior to the commencement of scheduling, which also would optimize the computation and communication costs.

In some cases, qubits could be grouped after assessing the dependencies based on functionality and performance, and then qubit clusters could be formed. These clusters could be ranked and then processed to minimize the communication costs, which depend partly on the topological interconnect among qubits and also the number of migrations based on the requirement.

The computation and communication costs always affect the resource optimization and need to be closely monitored for effective resource utilization and optimality. When optimality is affected, suboptimal but feasible schedules could be obtained to ensure the effectiveness and reliability of the quantum computer. The functionality and performance of the qubits are also assessed to assign the appropriate qubits to the respective dataset, which has a higher impact on the performance of the quantum computer.

Quantum logic gates operate on these quantum lattice cubes. There are several quantum logic gates.

### 13.3.4.5 Complexity in Quantum Cubes
The quantum cubes could be made to handle hybrid datasets that are combinations of simple and complex datasets by using the SWAP quantum logic gates. Complexity in the quantum cubes could be generated by using a combination of Hadamard and SWAP quantum gate circuit. This circuit helps to handle hybrid datasets. Hence, the randomness introduced in the quantum cubes enables.

The complexity of datasets based on the processing has been explained in the next section.

### 13.3.5 Dataset Complexity

The datasets could be simple and complex. For normal machine learning applications where data has only a few attributes, the modeled data points and datasets would be easier to process. They could be easily grouped. For some applications where there are multiple attributes, and the volume of data is large, the modeling of data points and datasets becomes complicated. Hence, the hardware should have additional number-crunching capabilities, and vector and matrix operations support need to be available.

Such complex datasets are modeled as vectors and matrices. Datasets and data points modeled would be obtained using a convolution operation for the respective vectors and matrices. The convolution operation could be given by Eq. (13.10):

$$Convec = Xvec \otimes Yvec \tag{13.10}$$

where *Convec* is the convoluted vector, *Xvec* and *Yvec* are the respective vectors whose product needs to be obtained.

The simulation results based on the complexity of the datasets discussed have been explained in the next section.

### 13.3.6 Green Quantum Computing and Its Importance for Industry 4.0

Quantum computing provides a renewable source of energy, which is eco-friendly and does not generate large quantity of heat. Quantum computers operate at cryogenic temperatures; hence, they do not cause destruction to the environment. They are a likely and potential candidate for server farms for huge data processing. In Industry 4.0, they are very useful in designing modern data-driven applications where speed of processing and reliability of inferences are mandatory requirements.

## 13.4 Simulation Results

The quantum computer simulation was carried out using IBM Qiskit development kit varying the number and complexity of the qubits along with the number and complexity of the datasets. The output screenshot illustrating the control and data path signals is shown in Figure 13.9.

The processing has been done based on two metrics:

- Processing time
- Memory consumption.

**Figure 13.9** Output of IBM Qiskit illustrating the control and data path signals.

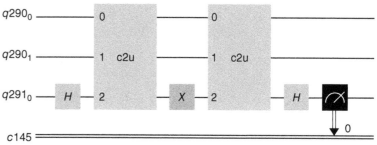

### 13.4.1 Processing Time

It is defined as the time to analyze a query decoding the same, identifying the required datasets, mapping the datasets to qubits, and processing them using quantum logic gates. The simulation results are given here.

The number of qubits and number of datasets have varied from 50 to 1,000 and from 15 to 98, respectively. The proportion of complex datasets has been varied from 10% and incremented by 5% gradually up to 98%. The processing times obtained have been tabulated in Table 13.8.

**Table 13.8** Comparison of processing times (simple versus complex datasets).

| | | | | | | | Processing time | | |
|---|---|---|---|---|---|---|---|---|---|
| Total number of qubits | Proportion of complex tasks (%) | Number of complex qubits | Rounded complex qubits | Total number of datasets | Number of complex datasets | Rounded number of datasets | Simple datasets | Complex datasets | Reduction in processing time (%) |
| 50 | 10 | 5 | 5 | 15 | 1.5 | 2 | 18.67 | 12.54 | 32.83 |
| 150 | 15 | 22.5 | 23 | 20 | 3 | 3 | 21.45 | 14.56 | 32.12 |
| 200 | 20 | 40 | 40 | 25 | 5 | 5 | 25.67 | 16.78 | 34.63 |
| 250 | 25 | 62.5 | 63 | 30 | 7.5 | 8 | 29.65 | 19.56 | 34.03 |
| 300 | 30 | 90 | 90 | 35 | 10.5 | 11 | 31.45 | 20.53 | 34.72 |
| 350 | 35 | 122.5 | 123 | 40 | 14 | 14 | 35.65 | 23.34 | 34.53 |
| 400 | 40 | 160 | 160 | 45 | 18 | 18 | 38.45 | 25.12 | 34.67 |
| 450 | 45 | 202.5 | 203 | 50 | 22.5 | 23 | 41.56 | 27.32 | 34.26 |
| 500 | 50 | 250 | 250 | 55 | 27.5 | 28 | 48.65 | 31.76 | 34.72 |
| 550 | 55 | 302.5 | 303 | 60 | 33 | 33 | 52.56 | 34.24 | 34.86 |
| 600 | 60 | 360 | 360 | 65 | 39 | 39 | 56.74 | 37.34 | 34.19 |
| 650 | 65 | 422.5 | 423 | 70 | 45.5 | 46 | 61.56 | 40.54 | 34.15 |
| 700 | 70 | 490 | 490 | 75 | 52.5 | 53 | 63.56 | 41.45 | 34.79 |
| 750 | 75 | 562.5 | 563 | 80 | 60 | 60 | 65.67 | 43.24 | 34.16 |
| 800 | 80 | 640 | 640 | 85 | 68 | 68 | 68.54 | 44.56 | 34.99 |
| 850 | 85 | 722.5 | 723 | 90 | 76.5 | 77 | 71.56 | 46.54 | 34.96 |
| 900 | 90 | 810 | 810 | 95 | 85.5 | 86 | 73.56 | 48.42 | 34.18 |
| 950 | 95 | 902.5 | 903 | 97 | 92.15 | 92 | 76.54 | 50 | 34.67 |
| 1000 | 98 | 980 | 980 | 98 | 96.04 | 96 | 78.56 | 51.46 | 34.50 |
| **Average** **547.37** | **54.89** | **376.18** | **376.42** | **59.47** | **39.88** | **40.11** | **50.53** | **33.12** | **34.31** |

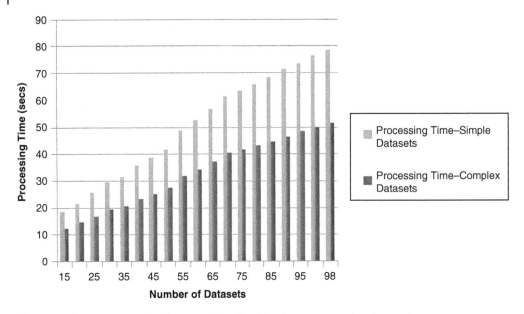

**Figure 13.10**  Plot comparing the processing time (simple versus complex datasets).

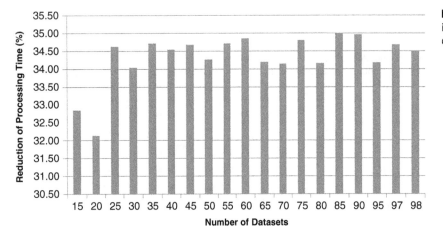

**Figure 13.11**  Plot showing the reduction in processing time for complex datasets compared to simple datasets.

It could be observed that for an average of 547 qubits and 40 complex datasets out of 59 datasets, the processing time reduces by 34%, compared to simple datasets. The average processing time for complex datasets is 33 seconds, compared to simple datasets, which have a processing time of 33 seconds. The same has been shown in Table 13.8 and Figures 13.10 and 13.11.

### 13.4.2  Memory Consumption

It is defined as the memory consumed to store the results obtained after the quantum logic gate processing of the qubits. The number of qubits and number of datasets have varied from 50 to 1,000 and from 15 to 98, respectively. The proportion of complex datasets has been varied from 10% and incremented by 5% gradually up to 98%. The memory consumption has been tabulated in Table 13.9.

It could be observed that for an average of 547 qubits and 40 complex datasets out of 59 datasets, the processing time reduces by 34%, compared to simple datasets. The average memory consumption for complex datasets is 33 seconds, compared to simple datasets, which have a memory consumption of 33 seconds. The same has been shown in Table 13.9 and Figures 13.12 and 13.13.

**Table 13.9** Comparison of memory consumption (simple vs complex datasets).

| | | | | | | | Memory consumption | | |
|---|---|---|---|---|---|---|---|---|---|
| Total Number of qubits | Proportion of complex tasks (%) | Number of complex Qubits | Rounded complex qubits | Total number of datasets | Number of complex datasets | Rounded number of datasets | Simple datasets | Complex datasets | Reduction in memory consumption (%) |
| 50 | 10 | 5 | 5 | 15 | 1.5 | 2 | 19.54 | 13.54 | 30.71 |
| 150 | 15 | 22.5 | 23 | 20 | 3 | 3 | 23.56 | 16.48 | 30.05 |
| 200 | 20 | 40 | 40 | 25 | 5 | 5 | 28.65 | 19.85 | 30.72 |
| 250 | 25 | 62.5 | 63 | 30 | 7.5 | 8 | 34.45 | 23.89 | 30.65 |
| 300 | 30 | 90 | 90 | 35 | 10.5 | 11 | 39.56 | 27.54 | 30.38 |
| 350 | 35 | 122.5 | 123 | 40 | 14 | 14 | 41.56 | 29 | 30.22 |
| 400 | 40 | 160 | 160 | 45 | 18 | 18 | 44.67 | 31.2 | 30.15 |
| 450 | 45 | 202.5 | 203 | 50 | 22.5 | 23 | 48.56 | 33.54 | 30.93 |
| 500 | 50 | 250 | 250 | 55 | 27.5 | 28 | 51.56 | 35.67 | 30.82 |
| 550 | 55 | 302.5 | 303 | 60 | 33 | 33 | 56.54 | 39.24 | 30.60 |
| 600 | 60 | 360 | 360 | 65 | 39 | 39 | 61.56 | 43 | 30.15 |
| 650 | 65 | 422.5 | 423 | 70 | 45.5 | 46 | 63.67 | 44.56 | 30.01 |
| 700 | 70 | 490 | 490 | 75 | 52.5 | 53 | 68.76 | 48.1 | 30.05 |
| 750 | 75 | 562.5 | 563 | 80 | 60 | 60 | 71.45 | 49.34 | 30.94 |
| 800 | 80 | 640 | 640 | 85 | 68 | 68 | 75.67 | 52.45 | 30.69 |
| 850 | 85 | 722.5 | 723 | 90 | 76.5 | 77 | 78.56 | 54.56 | 30.55 |
| 900 | 90 | 810 | 810 | 95 | 85.5 | 86 | 81.45 | 56.34 | 30.83 |
| 950 | 95 | 902.5 | 903 | 97 | 92.15 | 92 | 86.53 | 60.23 | 30.39 |
| 1000 | 98 | 980 | 980 | 98 | 96.04 | 96 | 89.45 | 62.45 | 30.18 |
| **Average** | **547.37** | **54.89** | **376.18** | **376.42** | **59.47** | **39.88** | **40.11** | **56.09** | **39.00** | **30.48** |

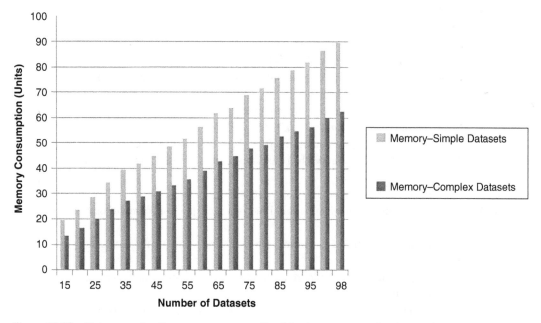

**Figure 13.12** Plot comparing the memory consumption (simple versus complex datasets).

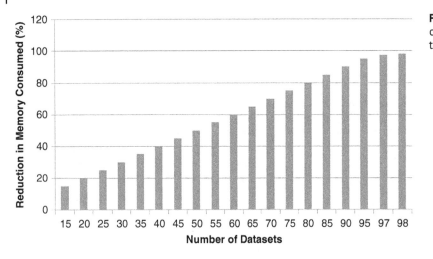

**Figure 13.13** Reduction in memory consumption for complex datasets compared to simple datasets.

## 13.5 Conclusion and Future Work

In this chapter, a complete design and blueprint of a quantum computer have been shown, explaining the underlying architecture and the various design optimizations to be made to handle complex datasets and improve the performance. It has been observed that there is a significant reduction in processing time and memory consumption while processing complex datasets.

The quantum computer could be designed with the design and optimizations mentioned, and the same could be extended to the design of an atomic clock by mainly utilizing the time crystal feature for several machine and deep learning applications.

## References

**1** Upadhyay, S., Saki, A.A., Topaloglu, R.O., and Ghosh, S. (2022). A shuttle-efficient qubit mapper for trapped ion quantum computers. *GLSVLSI'22*, 305–208

**2** Mli, A., Deng, S., and Szefer, J. (2022). Securing reset operations in NISQ quantum computers. *ACM Journal on Emerging Technologies in Computing Systems* 2279–2293.

**3** Shao, C. (2022). Computing eigen values of diagonalizable matrices on a quantum computer. *ACM* 3 (4): 1–21.

**4** Acharya, N., Urbanek, M., De Jong, W., and Saeed, S.M. (2021). Test points for online monitoring of quantum circuits. *ACM Journal on Emerging Technologies in Computing Systems* 18 (1): 1–14.

**5** Samoylov, A., Gushanskiy, S., and Potapov, V. (2021). Research on quantum neural networks and development of a single-qubit model of neuron. *ACM* 381–385.

**6** Suau, A., Staffelbach, G., and Calendra, H. (2021). Practical quantum computing: solving the wave equation using a quantum approach. *ACM Transactions on Quantum Computing* 2 (1): 1–2.

**7** Li, G., Anbang, W., Shi, Y. et al. (2021). On the co-design of quantum software and hardware. *NANOCOM'21*, 1–7.

**8** Hamido, O.C., Cirillo, G.A., and Giusto, E. (2020). Quantum synth: a quantum computing-based synthesise. *ACM* 2 (1): 265–268.

**9** Krishnaswamy, D. (2020). *Quantum Blockchain Networks*, vol. 1(2), 327–332. ACM.

**10** Dou, X. and Lou, L. (2020). A new qubits mapping mechanism for multiprogramming quantum computing. *PACT'20*, 349–350.

**11** Fawaz, A., Klein, P., Severini, S., and Mountney, P. (2019). Training and Meta-training binary neural networks with Quantum Computing. *KDD'19*, 1674–1681

**12** Sargaran, S. and Mohammadzadeh, N. (2019). SAQIP: a scalable architecture for quantum information processors. *ACM Transactions on Architecture and Code Optimisation* 16 (2): 1–12.

**13** Murali, P., Linke, N.M. (2019). Full-stack, real system quantum computer studies: architectural comparisons and design insights. *ISCA'19*, 527–540.

**14** de Alexandre A.A.A., Dueck, G.W., and da Silva, A.C.R. (2019). Finding optimal qubit permutations for IBM's quantum computer architectures. *SBCCI'19*, 1–6.

**15** Potok, T.E., Schuman, C., Young, S. et al. (2018). A study of complex deep learning networks on high-performance, neuromorphic and quantum computers. *ACM Journal on Emerging Technologies in Computing Systems* 14: 1–19.

**16** Trummer, I. and Koch, C. (2016). Multiple query optimisation on the D-wave 2x adiabatic quantum computer. *Proceedings of the VLDW Endowment* 9 (9): 648–659.

**17** Homulle, H., Visser, S., Patra, B. et al. (2016). *CryoCMOS Hardware Technology a Classical Infrastructure for a Scalable Quantum Computer*, 283–287. ACM.

**18** Ahsan, M. and Kim, J. (2015). Optimisation of quantum computer architecture using a resource-performance simulator. *EDAA*, 1108–1113.

**19** Ahsan, M., Van Meter, R., and Kim, J. (2015). Designing a million qubit quantum computer using a resource performance simulator. *ACM Journal of Emerging Technologies in Computing Systems* 12 (4): 1–39.

**20** Van Meter, R. and Horsman, D. (2013). A blue print for building a quantum computer. *Communications of the ACM* 56 (10): 84–93.

**21** Mermin's, D. (2010). Quantum computer science: An introduction. *ACM SIGACT*, 39–44.

**22** Spector, L. (2010). 'Evolution of quantum algorithms', genetic and evolutionary computation, *GECCO'2010*, 2739–2768

**23** Thaker, D.D., Metodi, T.S., Cross, A.W. et al. (2006). memory hierarchies: efficient designs to match available parallelism in quantum computing. *International Symposium of Computer Architecture ISCA'06*, 1–12.

**24** Travaglione, B. (2003). Designing and implementing small quantum circuits and algorithms. *DAC'2003*, 894–899.

**25** Kasivajhula, S. (2006). Quantum computing: A survey. *ACM SE'06*, 249–253.

# 14

## Opportunities in Neural Networks for Industry 4.0

*Rodrigo de Paula Monteiro, José P.G. de Oliveira, Sérgio C. Oliveira, and Carmelo J.A.B. Filho*

*Polytechnic School of Pernambuco, University of Pernambuco, Recife, Pernambuco, Brazil*

## 14.1 Introduction: Why Is Machine Learning Interesting to Industry 4.0?

The Fourth Industrial Revolution, that is, Industry 4.0, promoted significant changes in industrial organizations worldwide. Those changes were supported by technologies such as the Internet of Things (IoT), robotics, cloud computing, and artificial intelligence (AI), which led to improvements in the whole productive chain [1].

Nowadays, industrial plants use complex systems consisting of sensors, networks, computers, and cloud platforms to monitor industrial processes [2]. Sensors, for example, are used to collect various information about those processes, for example, temperature, current, and vibration. This information is commonly sent to storage structures like local servers and cloud computing storage platforms, where they accumulate and can be posteriorly used for decision-making. It is an increasing trend among industries since the more information there is, the more accurate decisions can be.

On the other hand, those improvements in information generation and storage gave rise to many challenges, such as the interoperability among the different devices and systems involved, cybersecurity, and large amounts of information to be analyzed [3].

In this context, machine learning (ML) algorithms arise as an attractive supporting tool for analyzing large amounts of information. Those are successful algorithms in many tasks due to their automation, scalability, and good predictive power. Regarding human experts, for example, ML algorithms can provide faster and more accurate responses, improving the performance of processes such as failure prediction of industrial equipment and management of resources, among others. Due to those reasons, ML-based solutions are increasingly present in Industry 4.0 [1].

## 14.2 Machine Learning

ML is a field of AI that focuses on developing algorithms and techniques that allow computers to learn from data and improve their performance at a specific task over time without the need to be explicitly programmed for each step [4]. ML is based on algorithms that learn to identify and recognize patterns in historical data to make predictions or make decisions based on new data. There are three main types of ML: supervised, unsupervised, and reinforcement learning.

In supervised learning [5], the algorithm is fed with labeled data and trained to make predictions or classify new data based on these labels. Supervised ML is widely used in several areas to solve classification, regression, and prediction problems. Some main supervised ML applications are identifying objects in images, natural language processing, time-series forecasting, recommender systems, fraud detection, and medical diagnosis.

In unsupervised learning [6], the algorithm is fed with unlabeled data and trained to find patterns and structures in the data. Unsupervised ML is primarily used for clustering and dimensionality reduction problems. Some applications are dimensionality reduction, social network analysis, consumer segmentation, identification of frequent patterns, and data preprocessing.

*Topics in Artificial Intelligence Applied to Industry 4.0*, First Edition. Edited by Mahmoud Ragab AL-Refaey, Amit Kumar Tyagi, Abdullah Saad AL-Malaise AL-Ghamdi, and Swetta Kukreja.

In reinforcement learning [7], the algorithm learns to make decisions based on rewards or punishments received for its actions in a dynamic environment. Some main applications of reinforcement ML are games, robotics, process control, online advertising, recommender systems, and health-care assistance.

Different machine-learning approaches can be used depending on the type of functional elements deployed to build the model. The most common families are based on prototypes, decision trees, support vectors, and artificial neural networks (ANN).

ML based on prototypes generally classifies new data points based on their similarity to existing prototype examples. The prototype examples are a subset of the training data representing the present classes. The algorithms identify a set of prototypes from the training data. Each prototype is then associated with a class, and when new data is presented to the algorithm, the similarity between the new data and each prototype is calculated. The new data is then assigned the label of the prototype to which it is most similar. The limitations regard the unsuitability for high-dimensional data and the dependency on the initially selected prototypes' quality. The K-Nearest Neighbors is the most used algorithm of this family, but currently, it is not helpful for challenging problems [8].

Decision trees [9] are commonly used for classification and regression tasks and are essentially a tree-like model of decisions represented by nodes and branches. Each node corresponds to a test on one of the input variables, while each leaf node represents a predicted output or decision. The goal of a decision tree is to create a model that can predict the class or category of a new observation based on the values of its input features. The tree is built by recursively splitting the training data into subsets based on the values of the input variables according to metrics to maximize information gain. Two widely used approaches are random forest and gradient boosting machines (GBMs). Random forests are an ensemble of decision trees, where multiple trees are trained on different subsets of the training data and different subsets of the input features. GBMs are another type of ensemble of decision trees. However, unlike random forests, they use a gradient-based optimization algorithm to add decision trees to the model iteratively.

The key idea behind a support vector machine (SVM) is to transform the input data into a higher-dimensional feature space using a kernel function and find a hyperplane that separates the data into classes with the broadest possible margin [10]. The margin is the distance between the hyperplane and the nearest data points from each class. The advantages of SVMs are that they can handle both linearly separable and nonlinearly separable data using different kernel functions, such as linear, polynomial, and radial basis function (RBF) kernels. SVMs are also robust to overfitting, as they try to find the hyperplane that maximizes the margin and thus tend to generalize well to new data. However, SVMs can be computationally intensive, especially for large datasets, and the choice of kernel function and the model's hyperparameters can significantly impact the algorithm's performance. Additionally, the interpretability of the model can be limited as the SVM focuses on finding the optimal hyperplane and may not provide insights into the underlying relationships between the input features and the output.

ANNs [11] were inspired by the structure and function of biological neural networks in the brain. The critical element is the artificial neurons, which take a weighted sum of their input signals and apply an activation function to produce their output signal. The weights and biases of the neurons are learned during the training process, where the algorithm adjusts the parameters to minimize a loss function between the predicted output and the actual output. The ANN models can be classified according to the architecture and connectivity of the neurons. In the feedforward neural networks (FNN), the neurons are arranged in layers, where each layer receives input signals from the previous layer and produces output signals to the next layer. The first layer is called "the input layer," the last layer is called "the output layer," and the intermediate layers are called "hidden layers." The most used FNN is the multi-layer perceptron (MLP).

On the other hand, recurrent neural networks (RNNs) have recurrent connections between the neurons to propagate information across time steps and maintain a memory of previous inputs. RNNs have many applications for problems where one needs to predict time series. The most known architectures are long short-term memory (LSTM) and gated recurrent unit (GRU) [12].

The ANN architectures can be shallow or deep. In shallow architectures, it is common to have inputs defined by relevant features and only one or a few hidden layers between the input and output layers. Shallow ANNs have a simple architecture that is easy to train and computationally efficient, but they may have limited representation power, generalization, and flexibility. On the other hand, in deep learning approaches, there are many layers between the input and output layers. Deep neural networks can automatically learn hierarchical representations of data, allowing them to extract complex features and patterns from the input data. Because of this, deep architectures are generally applied for raw data as input signals, such as temporal signals, images, and videos. Convolutional neural networks (CNNs) [13] are a famous architecture of deep learning and have gained much attention recently. Beyond the conventional artificial neurons, CNNs use convolutional layers to process local features from the input data and pooling layers to reduce the spatial dimensionality of the features. Some famous CNN models are VGGNet, Resnet, YOLO, Faster R-CNN, and U-Net, among many others.

## 14.3 Challenges in Industry 4.0 That Can Benefit from Using Machine Learning

Many tasks in Industry 4.0 can benefit from using ML and especially neural networks. Some representative examples are (i) fault detection and diagnosis, (ii) prediction of the remaining useful lifetime (RUL) of industrial equipment, (iii) predictive maintenance, (iv) optimizing energy consumption, (iv) cybersecurity, and (vi) virtual sensors. Those applications are discussed in the following subsections.

### 14.3.1 Fault Detection and Diagnosis

Fault detection and diagnosis play a critical role in ensuring the correct operation of industrial processes [14]. Early identification of malfunctions allows for precise intervention, reducing downtime and preventing further damage. ML techniques, especially those based on deep learning, have shown tremendous success in detecting and diagnosing faults in complex systems. However, the successful integration of ML requires careful attention to data quality, algorithm selection, and interpretability of results [15]. In smart factories equipped with sensors and IoT devices, large volumes of real-time data are generated. ML algorithms can be trained on this data to learn the normal behavior of machines and systems. When deviations from the norm occur, ML models can detect these anomalies and alert operators or trigger automated actions. This proactive approach not only saves time and resources but also improves overall product quality and customer satisfaction.

Furthermore, ML techniques can go beyond simple fault detection and provide valuable insights into fault diagnosis. By analyzing patterns and correlations within the data, ML models can identify the root causes of faults and provide actionable recommendations for maintenance or repair [16]. This enables more efficient troubleshooting and reduces the reliance on manual expertise, ultimately leading to cost savings and improved operational efficiency.

However, it is important to acknowledge that the successful implementation of ML in Industry 4.0 requires careful consideration of data quality, model training, and deployment. High-quality data is essential for training ML models, and organizations need to invest in robust data collection and preprocessing mechanisms. Additionally, the selection and optimization of ML algorithms, as well as the interpretability [17, 18] and explainability [5, 19] of their results, are critical factors to ensure trust and acceptance of ML-driven solutions.

### 14.3.2 Predicting Remaining Useful Lifetime

Predicting the RUL of industrial assets has become a critical challenge in the context of Industry 4.0. As machines and equipment play a critical role in manufacturing processes, the ability to estimate their remaining operational lifespan is of paramount importance for optimizing maintenance strategies, reducing costs, and ensuring uninterrupted production. The concept of RUL refers to the remaining time until a machine or component fails or becomes inefficient, rendering it unsuitable for its intended purpose. Traditional maintenance approaches, such as preventive or scheduled maintenance, are often based on predefined maintenance intervals, irrespective of the actual condition of the equipment. This approach can result in unnecessary maintenance activities, leading to increased costs and reduced operational efficiency.

The benefits of accurate RUL prediction are significant. By identifying potential failures in advance, maintenance activities can be scheduled proactively, reducing unplanned downtime and production losses. Furthermore, optimizing maintenance interventions based on RUL predictions can lead to cost savings by avoiding unnecessary replacements or repairs.

ML models are trained with historical data collected by sensors to predict when a given equipment is likely to fail. ANNs are examples of algorithms used for that purpose [20]. By leveraging historical data collected from machines, ML models can learn patterns, correlations, and degradation trends. These models can then be used to forecast the RUL, providing valuable insights for maintenance decision-making. One of the primary advantages of using ML for RUL prediction is its ability to handle complex and nonlinear relationships between multiple variables. ML algorithms can identify hidden patterns and extract valuable features from large datasets, enabling accurate predictions. Furthermore, ML models can adapt and improve over time as new data becomes available, enhancing their predictive capabilities.

Several ML algorithms have been successfully applied to RUL prediction tasks, such as improved variational mode decomposition [21], support vector regression [22], and random forest [23]. These models aim to establish a relationship between input features, such as sensor readings, operating conditions, and historical performance data, to predict the remaining lifespan. Another promising approach for RUL prediction is the use of RNNs and LSTM networks [24, 25]. RNNs are designed to handle sequential data, making them suitable for analyzing time-series data collected from sensors. LSTM networks, a type of RNN, can capture long-term dependencies and dynamic patterns in the data, enabling accurate RUL predictions.

Furthermore, it is noteworthy to highlight that in the context of Industry 4.0, the accessibility of real-time sensor data and the uninterrupted monitoring of machine performance greatly facilitate the deployment of RUL prediction models. Through the seamless integration of ML algorithms into the data acquisition and processing systems, the capability to generate real-time predictions is attained, thereby enabling proactive maintenance interventions.

### 14.3.3 Predictive Maintenance

Traditional maintenance practices in industrial plants often rely on scheduled maintenance routines and even reactive measures when a breakdown occurs. Although those are recurrent practices, they present several drawbacks.

The preventive maintenance exemplifies this context. This strategy relies on performing regularly scheduled maintenance routines on industrial equipment to prevent unexpected failures. The preventive practices improve the safety of industrial processes and help reduce unplanned interruptions in the manufacturing line. However, this kind of practice usually is expensive, demands extensive planning, and does not meet the real needs of the industrial plant [26].

Another recurrent practice is the corrective maintenance. Unlike the preventive strategy, the corrective approach focuses on identifying, isolating, and repairing failures, restoring the industrial equipment to its expected operational condition. This practice can reduce maintenance costs, but only noncritical equipment can benefit from this characteristic. Since corrective maintenance deals with failures that already exist in the industrial equipment, it can give rise to significant safety issues and economic losses, for example, the occurrence of accidents and stoppages in the manufacturing line [26].

Predictive maintenance, on the other hand, arises as an efficient alternative to traditional practices. It is a proactive strategy that uses data analysis techniques and predictive models to anticipate and prevent equipment failures before they occur. This kind of strategy optimizes maintenance activities, minimizing unplanned downtime, reducing maintenance costs, and improving the safety of the manufacturing line [27].

Instead of following a fixed maintenance schedule or waiting for equipment to fail, predictive maintenance uses real-time data collected from sensors, historical maintenance records, and other relevant sources of information to identify patterns that may suggest imminent failures. By analyzing the data, predictive maintenance systems can provide insights and warnings about the condition of the equipment, allowing maintenance teams to take faster and more accurate actions, minimizing downtime and hazards and reducing maintenance costs.

AI plays an important role in predictive maintenance, allowing the development of models, either handcrafted or learned from data, which provide automatic and accurate predictions. Nowadays, the research in AI applied to Industry 4.0 focuses on complex ML-based models like ensemble methods and deep neural networks, given their capability to handle a large amount of information. Among ML methods, those based on deep learning are the most adopted since they achieved state-of-the-art results in many tasks, for example, computer vision, natural language processing, time-series forecasting, and anomaly detection.

Regarding the solutions based on deep learning, the neural network architectures adapted to model time behavior, for example, RNNs, are widely used. This is an important feature because most data in predictive analysis consists of time series obtained from sensor measurements. For example, Pagano and Küfner et al. proposed two solutions based on RNNs. They used the association of neural networks with other algorithms to model the time series of data collected by sensors and to predict the status of plant operation [28].

Despite the advances provided by using AI in predictive maintenance, this field still presents some challenges to be overcome. The existence of noisy or erroneous sensor data, the necessity to collect, transmit, and process large amounts of data on time, and the diversity of operating conditions of industrial equipment are some examples of challenges still faced by Industry 4.0.

### 14.3.4 Optimizing Energy Consumption

The industrial sector is important to the economy of many countries worldwide. Industrial activities, such as metallurgy and recycling, yearly demand large amounts of energy. In 2021, for example, the industry accounted for 38% of the global final energy consumption, following a global increasing trend that has lasted for more than 20 years; that is, this industry percentage in 2000 was 33% [29].

It is still necessary to consume resources such as oil and coal to generate the amount of energy the industry needs to operate. It occurs because the energy generation from renewable sources remains inferior to 15% of the total energy generated. The high consumption of resources such as oil and coal is undesired not only because they are finite resources but also due to their contribution to world pollution and the related problems.

A possible solution to mitigate this problem is to invest in the expansion of renewable sources of energy. In addition, we can reduce the consumption of resources like fossil fuels in industrial processes by decreasing the amount of energy consumed. This goal can be achieved by making those processes more efficient, which can be performed in a number of ways, for example, by operating the industrial equipment properly and by planning suitable maintenance strategies.

Another way to improve the energy consumption of industrial processes is to use ML models, for example, neural networks. These can learn energy consumption patterns in the processes, and we can use those models to identify opportunities for optimization.

To exemplify the use of neural networks in this kind of task, we mention a research developed in the chemical industry that suggests the use of neural networks to optimize the structure of energy resources consumed in industrial processes [30]. The energy resources considered in the study were water, diesel, and coal, among others. The authors employed MLP neural networks to model the relation between the maximum energy consumed in the chemical process and the distribution of energy resources. This way, they could infer the maximum energy consumed in each process regarding different compositions of resources.

The research developed by Elsisi et al. is another example of neural networks applied to optimize energy consumption in Industry 4.0. They proposed to use the YOLOv3, an object-detection algorithm, to count the number of people in a given area. This information was used to manage the operation of air conditioners to optimize energy consumption [31].

Despite the success achieved by neural networks in the optimization of energy consumption in Industry 4.0, using those algorithms still faces some challenges. Besides the challenges typically related to ML approaches, for example, the capability of handling large amounts of data, a noteworthy issue is the aging effects of industrial equipment. The aging changes the equipment's optimal operating parameters over time, eventually making the models obtained by the ML algorithms obsolete and demanding these to be substituted or retrained.

### 14.3.5 Cybersecurity

Cybersecurity is a key issue in Industry 4.0. With the increasing interconnectivity of devices, systems, and data in industrial environments, the risk of cyber threats increases. The potential consequences of cyberattacks comprise disruption of operations, production downtime, theft of sensitive data, compromise of intellectual property, and even physical safety risks [32].

To mitigate these risks, robust cybersecurity measures must be implemented, including secure network architectures, encryption, access controls, intrusion detection systems, and continuous monitoring. In this context, ML-based models, for example, neural networks, are widely employed to improve cybersecurity. As an example, we can train those models to learn the regular patterns of network traffic. By knowing the expected patterns of the network traffic, it is possible to detect intrusion and suspect activities [33].

Also, we can use ML to perform biometric authentication, for example, facial [34] or voice [35] recognition, increasing the security in the access of critical systems and preventing security vulnerabilities.

Another approach to neural networks related to cybersecurity is the concept of federated deep learning [36], which is an approach to training deep learning models, which allows data to be stored on the devices or locations where it is generated, rather than being transferred to a centralized server. Unlike traditional ML methods, where data is collected and stored in a single location, federated deep learning allows deep learning models to be trained on devices or distributed systems such as smartphones, IoT devices, or servers in different locations. In this context, instead of sending the raw data to a centralized server, the processing algorithm is sent to the devices or places where the data is stored. Deep learning models are trained locally on each device using the data available on that device without the need to move the data to a centralized location. After local training, only model parameters, which contain the knowledge learned during training, are sent to a centralized server. On the server, these parameters are securely aggregated from all participating devices. The server combines information from each device without having direct access to the actual data, thus protecting users' privacy.

This decentralized training process allows deep processing models to be continually updated with the latest local data from each device without exposing personal or sensitive data. This approach is especially useful in scenarios where data privacy and security are concerns, such as in Industry 4.0, where a variety of connected devices collect and generate large volumes of data. Federated deep learning offers significant advantages regarding cybersecurity as well as data privacy, latency reduction, and bandwidth savings.

However, federated deep learning also presents challenges [37], such as the need to deal with device heterogeneity and ensure security during the model aggregation process and the constraint of power in some embedded systems. These challenges are being examined through ongoing research in the area of federated deep learning to improve its efficiency and effectiveness.

### 14.3.6 Soft Sensors

Sensors are devices that play a vital role in Industry 4.0. They are heavily employed in industrial plants since they provide valuable information about the various processes. The accelerometer, for example, is a sensor that measures acceleration. It is commonly used to monitor the vibration level of rotating machines [38]. Also, the thermocouple is a sensor used to monitor the temperature of industrial processes [39]. Distance, sound, and pressure are other three examples of physical quantities measurable by sensors. The information provided by those and other sensors can support decision-making and improve maintenance policies and resource allocation [40].

We have plenty of sensors in the market that can measure the various physical quantities. However, measuring certain process variables may be difficult, expensive, or impractical. In such a scenario, an alternative solution is to use the information provided by alternative data sources, for example, other sensors, to infer a desired process variable. The research of Lui et al. exemplifies this scenario in the petrochemical industry context [41]. They use a mechanism called a "soft sensor" to infer the amount of butane content at the bottom flow of debutanizer columns. Since the equipment and the operational environment present limitations, they use alternative information to infer the butane concentration, for example, gas temperature and pressure measurements.

In Industry 4.0, soft sensors are mathematical models that estimate certain process variables using alternative information. The word "soft" in soft sensors comes from "software" since they are often implemented as computer software [40]. Unlike physical sensors, which measure the variables of interest directly, soft sensors use indirect measurements or data correlations to infer the values of desired variables ([42]). This type of solution can be conceived through various techniques, such as statistical models, ML algorithms, or fuzzy logic.

Among those techniques, neural networks arise as a relevant tool due to their capability of modeling complex patterns of correlations among the industrial process variables [43]. We can use neural networks to model soft sensors in many tasks. Yuan et al., for example, proposed a deep learning-based soft sensor consisting of stacked encoders and decoders. This measuring system used process variables measured by hard devices, for example, temperature and pressure, to predict the butane content in the bottom product [44]. Also, Wang et al. used variational auto-encoders based on neural networks to build soft sensors for the industrial zinc roasting process [45].

Identifying an alternative sensor to infer the measurement provided by the desired sensor is one of the main challenges in using this kind of solution since the information provided by both sensors must be correlated. Also, models obtained from available plant measurements may have errors from measurement errors and industrial plant dynamic behavior.

## 14.4 Some Cases of Success Deploying ML in Industry 4.0

In this section, we outline two approaches that employ ML models to address specific challenges of Industry 4.0.

### 14.4.1 Detecting Defects in Sanitary Ware with Deep Learning

The authors proposed an automatic system based on one-dimensional convolutional neural networks (1D-CNN) to support the human test operators in defect detection routines [46]. The case study was a sanitary ware company in the Brazilian Northeast, which produced more than 7,000 products per day.

In this company, the detection of defects in the products relied on the senses of human test operators. One test consisted of analyzing the visual aspect of the product, looking for defects such as external cracks. The other one consisted of two tasks: (i) hitting the sanitary ware with a small hammer and (ii) evaluating the sound emitted, which aimed to identify internal defects. Those tests, although traditionally employed, may have their efficiency compromised due to fatigue and lack of attention from an operator. Also, they are highly dependent on the experience of the operator.

**Figure 14.1** Fault detection pipeline.
*Source:* Monteiro and Bastos-Filho.

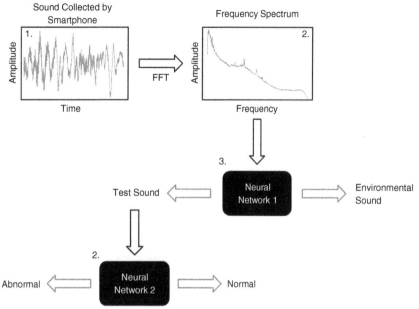

In this sense, the authors proposed a system based on neural networks to support the operator's decision. The proposed solution focused on the sound analysis and worked according to the same principle of traditional routines; that is, the good-quality products presented a typical sound pattern. Sounds not conforming with such a pattern, on the other hand, may suggest the presence of a defect. The fault detection pipeline is presented in Figure 14.1.

The first step presented in Figure 14.1 is to collect test sounds. They were recorded in the manufacturing line with a smartphone since the authors intended to implement the fault detection system in a mobile platform. They adopted this kind of platform due to usability and economic issues. The audio recorded comprised environmental and test sounds regarding faulty and good-quality products.

The second step is to obtain the frequency representation of the audio recording, that is, the frequency spectrum. A frequency spectrum represents the range of frequencies contained in a given signal, as well as their respective amplitudes [47]. This kind of representation allows the identification of typical frequency patterns present in each class of the fault detection problem.

The third step regards identifying whether a sound recorded belonged to the test routine or the environment. The authors modeled this step as a binary classification problem, using the one-dimensional CNN to perform the task.

The fourth step was also modeled as a binary classification problem. On the other hand, the authors used the 1D-CNN to classify test sounds as belonging to good-quality or defective products. If any fault was detected, the system warned the human operator and the product was taken for repair.

The authors trained the 1D-CNNs with data consisting of audio recordings collected in the manufacturing line with the same smartphone. Also, they used the accuracy and classification time to assess the resulting models. In the ML context, accuracy is defined as the ratio between the number of correct predictions and the total number of predictions, suggesting how well the model performs across the different classes. Regarding the challenge of distinguishing between environmental and test sounds, the one-dimensional CNN used by the authors achieved accuracy values superior to 98%, suggesting the model can properly identify most of the test sounds. Once the test has been identified, the second one-dimensional CNN could perform the defect detection task. In this step, the classification model also achieved an accuracy value superior to 98%, suggesting that this kind of approach is a reliable tool to support the test operator in the manufacturing line.

On the other hand, the classification time is defined as the time that the model needs to classify a single input sample, suggesting how fast the responses of the classification model are. The classification time is an important measure since the defect detection system must respond promptly to avoid reducing the production pace. In this sense, the authors achieve time values below 18 milliseconds, suggesting the feasibility of this solution since the testing rhythm is about one test every 30 seconds.

The authors also compared the results achieved by the 1D-CNN with the ones achieved by a traditional ML method: the SVM. The 1D-CNN achieved higher accuracy values on both tasks, that is, distinguishing between environmental and test sounds and identifying defective products. However, the differences between performances were superior on the second task, that is, identifying defective products. The reason was that the sound signals of both classes involved in this task shared more patterns among each other than in the first task since both classes regarded test sounds. This way, the complex architecture of the 1D-CNN allowed the trained model to learn more specific patterns to distinguish finer details in the data of both classes.

In this sense, the results achieved by the authors suggested the feasibility of using ML to build reliable tools to support the decision in industrial processes. The defect detection system based on neural networks provided accurate classifications promptly. Also, the resulting neural networks required small storage capacity, that is, less than 50 MB each, being suitable to be implemented on smartphones and other portable platforms.

### 14.4.2 Detection of Anomalies in Embedded System Using Electrical Signature

In this approach, the authors propose the detection of anomalies in an embedded system using the electrical current consumed by the device under test (DUT) [48]. In this way, it is possible to obtain behavioral information from the system. It is also possible to verify that the DUT works as expected, both from hardware and software points of view. Furthermore, the detection takes place noninvasively, as the current measurement only requires a tap to the reference node (ground) of the power supply itself.

The block diagram of the detection system is illustrated in Figure 14.2. The electrical current consumed by the DUT is acquired by an analog-to-digital converter (ADC) for a predefined time interval $\Delta T$. The spectrogram of the acquired signal is converted into an image (ImaA). The trained auto-encoder reconstructs the input image into its output (ImaB). The reconstructed image and the original image are compared. If the difference between the images is greater than a threshold $\mu$, an anomaly is detected.

A typical embedded system (Figure 14.3) was built to experimentally validate the proposed methods, algorithms, and anomaly detection techniques. A typical embedded system is understood as a computational system implemented in a microcontroller, which executes software (in this context, it is common to refer to software as firmware). The microcontroller, in turn, composes a complete computational system in the form of an integrated circuit. Hence, the DUT consists of two parts: the hardware and the firmware. The hardware was implemented on an evaluation platform for microcontrollers based on ARM Cortex-M4F called "Tiva C" [49]. The development board contains input/output interfaces (general purpose input/output (GPIO)), which we use to emulate typical embedded system tasks, for example, activating devices and transferring data.

The test firmware has the structure traditionally found in such systems: an initialization and hardware configuration block followed by an infinite loop (while (TRUE)). Inside the loop, input/output ports are read and/or triggered, data is

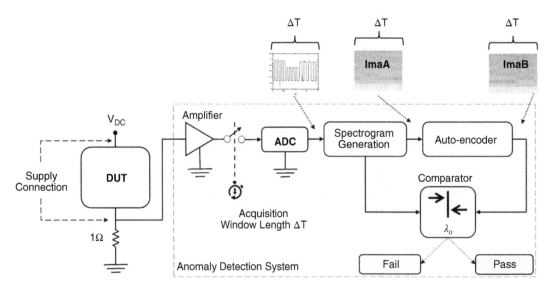

**Figure 14.2** Anomaly detection system architecture. If $|ImaA - ImaB| > \mu$ anomaly is detected. The resistor ($R = 1\ \Omega$) converts electric current ($I$) into voltage ($V$) by Ohm's law $V = R \times I$; ADC – analog-to-digital converter. *Source:* de Oliveira et al.

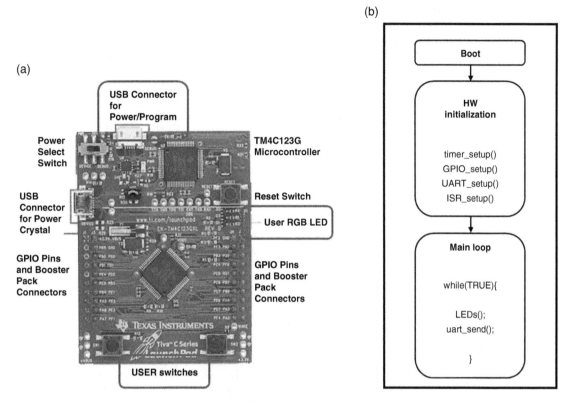

**Figure 14.3** Experimental prototype developed to validate our research. Device under test. (a) The development board used as DUT in our experiment; (b) Test firmware architecture. *Source:* José Paulo G et al. 2022, reproduced with permission from Elsevier.

processed, and information is transferred through an appropriate interface. In that specific implementation, the activities carried out by the system are activation of multicolored LEDs (RGB), interrupt routine handling, and data transfer via serial interface UART (universal asynchronous receiver/transmitter) [50]. The LEDs() and uart_send() functions are responsible for LED operation and serial data transmission, respectively.

To perform experimental tests, multiple firmware versions were implemented. One of them is considered anomaly free. and the rest emulate some type of anomaly. A summary of these versions and their applications can be seen in Table 14.1. Each version represents a simulated system operation.

In addition to multiple firmware versions, samples with different acquisition intervals ($\Delta T$) were also generated. The objective, in this case, is to analyze the performance as a function of this width. Table 14.2 contains information about the generated samples and their quantities, as well as their use in the experiment (training/validation $\rightarrow$ class OK; test $\rightarrow$ class NOK1-5). For shorter windows ($\Delta T \leq 2000$ *ms*), the number of training/validation samples is 12,000. For longer windows, this value is smaller because the total acquisition time is proportional to $\Delta T$.

Regarding the detection of anomalies, the results are presented here. By varying the threshold ($\mu$) and calculating the corresponding error rate, the curves in Figure 14.4 can be plotted. From the curves, the minimum values for the average error rate are below 1% for acquisition intervals equal 1,500 milliseconds and above.

The minimum error rate as a function of $\Delta T$ for all test cases is shown in Figure 14.5. We observe that performance increases with the acquisition interval. This is expected since the spectral content of the sample increases if the observation window is larger. Consequently, it is easier for the auto-encoder to differentiate images from different classes. From

**Table 14.1** Summary of firmware versions used in the experiments: operational description of each version.

| Designation → | OK | NOK1 | NOK2 | NOK3 | NOK4 | NOK5 |
|---|---|---|---|---|---|---|
| **Operation →** | Anomaly free | Delay in LED activation | Delay in transmission over the UART | Inverted order of LED activation | Delay in interrupt execution | Red LED defect simulation |

**Table 14.2** Generated samples for experimental validation and respective amounts used for training, validation, and test.

| Acquisition interval – ΔT (ms) | Designation | #training | #validation | # test |
| --- | --- | --- | --- | --- |
| 500 | $\Delta T\_500$ | 9600 | 2400 | 2000 |
| 750 | $\Delta T\_750$ | 9600 | 2400 | 2000 |
| 1000 | $\Delta T\_1000$ | 9600 | 2400 | 2000 |
| 1250 | $\Delta T\_1250$ | 9600 | 2400 | 2000 |
| 1500 | $\Delta T\_1500$ | 9600 | 2400 | 2000 |
| 1850 | $\Delta T\_1850$ | 9600 | 2400 | 2000 |
| 1750 | $\Delta T\_1750$ | 9600 | 2400 | 2000 |
| 2000 | $\Delta T\_2000$ | 9600 | 2400 | 2000 |
| 2250 | $\Delta T\_2250$ | 8811 | 2209 | 2000 |
| 2500 | $\Delta T\_2500$ | 7767 | 1942 | 2000 |
| 2750 | $\Delta T\_2750$ | 6770 | 1692 | 2000 |
| 3000 | $\Delta T\_3000$ | 6202 | 1551 | 2000 |
| 3500 | $\Delta T\_3500$ | 5080 | 1270 | 2000 |

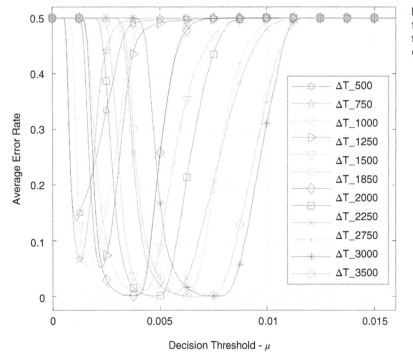

**Figure 14.4** Average detection error rate as a function of the decision threshold for each of the indicated windows – an average of 15 experiments. *Source:* de Oliveira et al.

Figure 14.5, the minimum error tends to be zero for windows greater than 1,500 milliseconds. This result is important because, in practical applications, it indicates the smallest acquisition window width with which the performance is optimal. Thus, time is saved, both in the implementation (data collection and training) and in the testing process.

The authors compared their results with a random forest-based ML model and with an image-processing algorithm called "SIFT" [51], both used to detect anomalies on the same board as the auto-encoder approach. The random forest approach yielded similar results to the auto-encoder proposed by the authors. However, the SIFT algorithmic solution proved to be inadequate in detecting anomalies in the desired manner. Nevertheless, the study highlights that the use of auto-encoders offers the advantage of not relying on data from both classes (OK and NOK) for training. This is a significant benefit since obtaining data from the anomalous class (NOK) is challenging in real industrial environments.

The anomaly detection solution as proposed is carried out with minimal interference imposed on the DUT. Therefore, there are no limitations related to the physical specifications of the system, such as the size and power of components or

**Figure 14.5** Minimum detection error rate as a function of the acquisition window width. Boxplot for 15 tests. *Source:* de Oliveira et al.

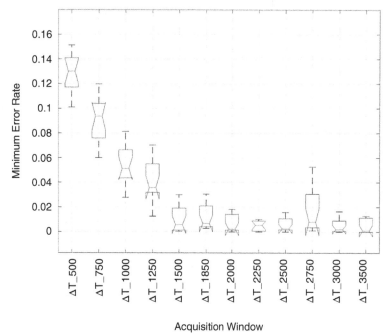

the printed circuit board. Hence, the proposed method can be easily adapted to other embedded systems by training the auto-encoder with spectrograms generated from their electrical current signature.

## 14.5 Conclusions and Final Remarks

A relevant challenge in Industry 4.0 is processing and extracting information from the massive amount of data generated by interconnected devices and systems. Traditional analytical methods may have problems handling this volume and complexity. Also, the variability and unpredictability of industrial processes require adaptive and intelligent systems that can learn and respond in real time. Another challenge is the need for efficient decision-making in complex environments, where multiple variables and factors interact. Addressing these challenges requires technologies capable of handling large amounts of information, adapting to variable operating conditions, and enabling intelligent decision-making.

In this context, neural networks arise as prominent tools to address the challenges faced in Industry 4.0. They can process and analyze complex data patterns, making them well-suited for handling large amounts of data generated in industrial plants. Neural networks can learn from data and adjust their behavior accordingly, allowing them to adapt to changing conditions and optimize processes over time. They are states of art at tasks such as pattern recognition, anomaly detection, predictive modeling, and optimization. By leveraging neural networks, businesses can extract valuable insights from data, improve operational efficiency, and enhance decision-making in real time.

Looking ahead, the future of Industry 4.0 with neural networks appears promising. As neural network models keep improving, they are likely to become more powerful, efficient, and specialized for specific industrial applications. It will enable the development of more advanced predictive maintenance systems that can accurately detect and address potential equipment failures before they occur. Also, neural networks can enhance quality-control processes by quickly identifying defects in products. Furthermore, integrating neural networks with robotics and autonomous systems can lead to more intelligent and adaptive production lines, enabling automated optimization and self-adjustment based on real-time data.

## References

**1** Mazzei, D. and Ramjattan, R. (2022). Machine Learning for Industry 4.0: A Systematic Review Using Deep Learning-Based Topic Modelling. *Sensors* 22 (22): 8641.

**2** Fordal, J.M., Schjølberg, P., Helgetun, H. et al. (2023). Application of sensor data based predictive maintenance and artificial neural networks to enable Industry 4.0. *Advances in Manufacturing* 11 (2): 248–263.

**3** Khan, M., Haleem, A., and Javaid, M. (2023). Changes and improvements in Industry 5.0: A strategic approach to overcome the challenges of Industry 4.0. *Green Technologies and Sustainability* 1 (2): 100020.

**4** Zhou, Z.H. (2021). *Machine learning*. Springer Nature.

**5** Burkart, N. and Huber, M.F. (2021). A survey on the explainability of supervised machine learning. *Journal of Artificial Intelligence Research* 70: 245–317.

**6** Alloghani, M., Al-Jumeily, D., Mustafina, J. et al. (2020). A systematic review on supervised and unsupervised machine learning algorithms for data science. In: *Supervised and Unsupervised Learning for Data Science*, 3–21.

**7** Botvinick, M., Ritter, S., Wang, J.X. et al. (2019). Reinforcement learning, fast and slow. *Trends in cognitive sciences* 23 (5): 408–422.

**8** Taunk, K., De, S., Verma, S., and Swetapadma, A. (2019). A Brief Review of Nearest Neighbor Algorithm for Learning and Classification. *2019 International Conference on Intelligent Computing and Control Systems (ICCS)*, Madurai, India, pp. 1255–1260, doi: 10.1109/ICCS45141.2019.9065747.

**9** Charbuty, B. and Abdulazeez, A. (2021). Classification based on decision tree algorithm for machine learning. *Journal of Applied Science and Technology Trends* 2 (01): 20–28.

**10** Cervantes, J., Garcia-Lamont, F., Rodríguez-Mazahua, L., and Lopez, A. (2020). A comprehensive survey on support vector machine classification: Applications, challenges and trends. *Neurocomputing* 408: 189–215.

**11** Abiodun, O.I., Jantan, A., Omolara, A.E. et al. (2019). Comprehensive review of artificial neural network applications to pattern recognition. *IEEE Access* 7: 158820–158846.

**12** Yang, S., Yu, X., and Zhou, Y. (2020). LSTM and GRU Neural Network Performance Comparison Study: Taking Yelp Review Dataset as an Example. *2020 International Workshop on Electronic Communication and Artificial Intelligence (IWECAI)*, Shanghai, China, pp. 98–101, doi: 10.1109/IWECAI50956.2020.00027.

**13** Li, Z., Liu, F., Yang, W. et al. (2021). A survey of convolutional neural networks: analysis, applications, and prospects. In: *IEEE Transactions on Neural Networks and Learning Systems*. https://doi.org/10.1109/TNNLS.2021.3084827.

**14** Park, Y.J., Fan, S.K.S., and Hsu, C.Y. (2020). A review on fault detection and process diagnostics in industrial processes. *Processes* 8 (9): 1123.

**15** Tripathi, S., Muhr, D., Brunner, M. et al. (2021). Ensuring the robustness and reliability of data-driven knowledge discovery models in production and manufacturing. *Frontiers in artificial intelligence* 4: 576892.

**16** Liu, Y., Ramin, P., Flores-Alsina, X., and Gernaey, K.V. (2023). Transforming data into actionable knowledge for fault detection, diagnosis and prognosis in urban wastewater systems with AI techniques: A mini-review. *Process Safety and Environmental Protection*.

**17** Carvalho, D.V., Pereira, E.M., and Cardoso, J.S. (2019). Machine learning interpretability: A survey on methods and metrics. *Electronics* 8 (8): 832.

**18** Krishnan, M. (2020). Against interpretability: a critical examination of the interpretability problem in machine learning. *Philosophy & Technology* 33 (3): 487–502.

**19** Herm, L.V., Heinrich, K., Wanner, J., and Janiesch, C. (2023). Stop ordering machine learning algorithms by their explainability! A user-centered investigation of performance and explainability. *International Journal of Information Management* 69: 102538.

**20** Yan, H., Wan, J., Zhang, C. et al. (2018). Industrial big data analytics for prediction of remaining useful life based on deep learning. *IEEE Access* 6: 17190–17197.

**21** Sun, C., Qu, A., Zhang, J. et al. (2022). Remaining Useful Life Prediction for Lithium-Ion Batteries Based on Improved Variational Mode Decomposition and Machine Learning Algorithm. *Energies* 16 (1): 313.

**22** Benkedjouh, T., Medjaher, K., Zerhouni, N., and Rechak, S. (2013). Remaining useful life estimation based on nonlinear feature reduction and support vector regression. *Engineering Applications of Artificial Intelligence* 26 (7): 1751–1760.

**23** Zhao, L., Zhu, Y., and Zhao, T. (2022). Deep learning-based remaining useful life prediction method with transformer module and random forest. *Mathematics* 10 (16): 2921.

**24** de Beaulieu, M.H., Jha, M.S., Garnier, H., and Cerbah, F. (2022). Unsupervised remaining useful life prediction through long range health index estimation based on encoders-decoders. *IFAC-PapersOnLine* 55 (6): 718–723.

**25** Zhang, H., Xi, X., and Pan, R. (2023). A two-stage data-driven approach to remaining useful life prediction via long short-term memory networks. *Reliability Engineering & System Safety* 237: 109332.

**26** Rojek, I., Jasiulewicz-Kaczmarek, M., Piechowski, M., and Mikołajewski, D. (2023). An artificial intelligence approach for improving maintenance to supervise machine failures and support their repair. *Applied Sciences* 13 (8): 4971.

**27** Pagano, D. (2023). A predictive maintenance model using Long Short-Term Memory Neural Networks and Bayesian inference. *Decision Analytics Journal* 6: 100174.

**28** Küfner, T., Döpper, F., Müller, D., and Trenz, A.G. (2021). Predictive Maintenance: Using Recurrent Neural Networks for Wear Prognosis in Current Signatures of Production Plants. *International Journal of Mechanical Engineering and Robotics Research* 10 (11): 583–591.

**29** IEA (2022). Industry. Paris, France: IEA. https://www.iea.org/reports/industry (accessed on 20 December 2023).

**30** Shinkevich, A.I., Malysheva, T.V., Vertakova, Y.V., and Plotnikov, V.A. (2021). Optimization of energy consumption in chemical production based on descriptive analytics and neural network modeling. *Mathematics* 9 (4): 322. https://doi.org/10.3390/math9040322.

**31** Elsisi, M., Tran, M.Q., Mahmoud, K. et al. (2021). Deep learning-based industry 4.0 and internet of things towards effective energy management for smart buildings. *Sensors* 21 (4): 1038.

**32** Bakakeu, J., Baer, S., Klos, H. H., et al. (2021). Multi-agent reinforcement learning for the energy optimization of cyber-physical production systems. *Artificial Intelligence in Industry 4.0: A Collection of Innovative Research Case-studies that are Reworking the Way We Look at Industry 4.0 Thanks to Artificial Intelligence*, 143–163, doi: https://doi.org/10.1007/978-3-030-61045-6_11. Available at: https://ouci.dntb.gov.ua/en/works/7Wa058Zl/

**33** Pawlicki, M., Kozik, R., and Choraś, M. (2022). A survey on neural networks for (cyber-) security and (cyber-) security of neural networks. *Neurocomputing* 500: 1075–1087.

**34** Chidumije, A., Gowher, F., Kamalinejad, E., Mercado, J., Soni, J., & Zhong, J. (2021, May). A Survey of CNN and Facial Recognition Methods in the Age of COVID-19*. *2021 the 5th International Conference on Information System and Data Mining*, 104–109.

**35** Bohnstingl, T., Garg, A., Woźniak, S. et al. (2021). Towards efficient end-to-end speech recognition with biologically-inspired neural networks. arXiv preprint arXiv:2110.02743.

**36** McMahan, B., Moore, E., Ramage, D. et al. (2017). Communication-efficient learning of deep networks from decentralized data. In: *Artificial intelligence and statistics*, 1273–1282. PMLR.

**37** Ferrag, M.A., Friha, O., Maglaras, L. et al. (2021). Federated deep learning for cyber security in the internet of things: Concepts, applications, and experimental analysis. *IEEE Access* 9: 138509–138542.

**38** Fahmi, A.T.W.K., Kashyzadeh, K.R., and Ghorbani, S. (2022). A comprehensive review on mechanical failures cause vibration in the gas turbine of combined cycle power plants. *Engineering Failure Analysis* 134: 106094. https://doi.org/10.1016/j.engfailanal.2022.106094.

**39** Kumari, N., & Sathiya, S. (2022). An intelligent temperature sensor with non-linearity compensation using convolutional neural network. *Proceedings of International Conference on Industrial Instrumentation and Control: ICI2C 2021*, 319–327. Singapore: Springer Nature Singapore, doi: 10.1007/978-981-16-7011-4_32. Available at: https://link.springer.com/chapter/10.1007/978-981-16-7011-4_32

**40** Kadlec, P., Gabrys, B., and Strandt, S. (2009). Data-driven soft sensors in the process industry. *Computers & chemical engineering* 33 (4): 795–814.

**41** Lui, C.F., Liu, Y., and Xie, M. (2022). A supervised bidirectional long short-term memory network for data-driven dynamic soft sensor modeling. *IEEE Transactions on Instrumentation and Measurement* 71: 1–13. https://doi.org/10.1109/TIM.2022.3152856.

**42** Javaid, M., Haleem, A., Singh, R.P. et al. (2022). Exploring impact and features of machine vision for progressive industry 4.0 culture. *Sensors International* 3: 100132.

**43** Perera, Y.S., Ratnaweera, D.A.A.C., Dasanayaka, C.H., and Abeykoon, C. (2023). The role of artificial intelligence-driven soft sensors in advanced sustainable process industries: A critical review. *Engineering Applications of Artificial Intelligence* 121: 105988.

**44** Yuan, X., Gu, Y., Wang, Y. et al. (2019). A deep supervised learning framework for data-driven soft sensor modeling of industrial processes. *IEEE Transactions on Neural Networks and Learning Systems* 31 (11): 4737–4746.

**45** Wang, C., Li, Y., Huang, K. et al. (2022). VAE4RSS: A VAE-based neural network approach for robust soft sensor with application to zinc roasting process. *Engineering Applications of Artificial Intelligence* 114: 105180.

**46** Monteiro, R.P. and Bastos-Filho, C.J.A. (2019). Detecting defects in sanitary wares using deep learning, *IEEE Latin American Conference on Computational Intelligence (LA-CCI)*, 1–6.

**47** Galar, D. and Kumar, U. (2017). *eMaintenance: Essential electronic tools for efficiency*. Academic Press.

**48** de Oliveira, J.P.G., Bastos-Filho, C.J., and Oliveira, S.C. (2022). Non-invasive embedded system hardware/firmware anomaly detection based on the electric current signature. *Advanced Engineering Informatics* 51: 101519.

**49** Mazidi, M.A., Chen, S., Naimi, S., and Naimi, S. (2017). *TI Tiva ARM Programming For Embedded Systems: Programming ARM Cortex-M4 TM4C123G with C*, vol. 2.

**50** Nanda, U. and Pattnaik, S.K. (2016). Universal asynchronous receiver and transmitter (uart). In: *2016 3rd International Conference on Advanced Computing and Communication Systems (ICACCS)*, 1–5. IEEE.

**51** Lowe, D.G. (2004). Distinctive image features from scale-invariant keypoints. *International journal of computer vision* 60: 91–110.

# 15

# A Smarter Way to Collect and Store Data: AI and OCR Solutions for Industry 4.0 Systems

*Ajay R. Nair*[1], *Varun D. Tripathy*[1], *R. Lalitha Priya*[1], *Manigandan Kashimani*[1], *Guru Akaash N. Janthalur*[1], *Nusrat J. Ansari*[1]*, and Igor Jurcic*[2]

[1] *Computer Science Department, Vivekanand Education Society's Institute of Technology, Mumbai, Maharashtra, India*
[2] *Telecommunications and Informatics Department, HT ERONET, Mostar, Bosnia and Herzegovina*

## 15.1 Introduction

Pattern recognition by machines has become increasingly important in many areas of our lives. From automated optical character recognition (OCR) to face recognition, text recognition, speech recognition, and more, the ability of machines to recognize patterns accurately and reliably has opened new possibilities in fields such as health care, finance, security, and entertainment. With advances in technology and machine learning algorithms, the accuracy and efficiency of pattern recognition systems have significantly improved over the years, making it possible for machines to perform tasks that were once thought to be the exclusive domain of humans.

OCR technology has enabled the conversion of scanned documents, images, and videos into editable and searchable digital formats. This has revolutionized the way we handle documents and has made it easier to store and retrieve information. OCR has also made it possible to extract text from handwritten documents, which was previously a tedious and time-consuming task. With OCR, it is possible to digitize large volumes of documents quickly and accurately, making it an essential tool for businesses and organizations that deal with large amounts of data. Additionally, OCR models also face challenges with handwriting recognition, where the styles of handwriting can vary greatly from person to person, making it difficult for the model to accurately recognize the characters. Video text suffers a lot which makes text detection and text recognition more difficult. Another challenge is dealing with multiple languages and scripts, where the OCR model needs to be able to recognize and distinguish between different scripts, such as Latin, Arabic, and so on. For example, the OCR model could not properly recognize the letter "o" and the digit "0," especially in a dark and noisy environment. Today, to solve these problems, there are several OCR models such as rule-based OCR models, template matching OCR, feature-based OCR models, deep learning OCR models, and many more. This chapter attempts to elaborate on the methods for recognizing numerous kinds of display units with several design patterns and font sizes.

## 15.2 Background

Here, we discuss the various challenges that are faced in Optical Character Recognition (OCR) and how it saves countless hours of manual effort and reduces human errors. This has proven particularly beneficial in industries such as banking, insurance, and logistics, where large volumes of paperwork need to be digitized and processed efficiently. OCR is a process that uses advanced algorithms and pattern recognition techniques to convert scanned or photographed text into editable and searchable digital content. By analyzing the shapes, patterns, and structures of characters, OCR software identifies and translates text from various sources such as documents, books, or images. OCR enables efficient document retrieval and archiving. With the ability to convert physical documents into searchable digital formats, OCR has facilitated quick and accurate access to information, improving productivity and decision-making processes. Despite its numerous advantages, OCR technology does have its limitations. One of the primary challenges lies in accurately recognizing handwritten or poorly printed text, which can result in errors during the conversion process. Complex formatting, such as tables or columns, may also pose difficulties for OCR algorithms, potentially leading to inaccuracies in the output. Factors like low-resolution scans, faded ink, or smudges can hinder the OCR software's ability to accurately interpret the text, reducing its overall efficiency.

Karez Hamad et al. [1] explore how OCR may be used to convert any type of text or text-containing material, such as handwritten text and printed and scanned pictures, into a digital format for deeper and more extensive analysis. They also highlighted the difficulties encountered in obtaining the correct digital format for the text materials. Among them are scene complexity, uneven lighting, skewness, blurring, aspect ratios, typefaces, and wrappings. To locate a better digitally prepared picture, OCR includes several processes, including preprocessing, segmentation, normalization, feature extraction, classification, and template matching. Statistical techniques, neural networks, a combiner classifier, and a post-processing phase are all used. Handwriting recognition, receipt imaging, the legal industry, and more applications employ OCR. The power law technique can be used to adjust the brightness and contrast of an image.

Haojin Yang et al. [2] explore how video text may be utilized as a beneficial source for automatic video indexing in digital video libraries. They discuss the methods necessary to obtain the video's text. The chapter described the two procedures involved, which are text detection and text recognition. It initially detects probable texts with a high recall rate during the text detection step. The identified texts are next processed using picture entropy-based filtering and lastly Stroke Width Transform (SWT) algorithms. These approaches are used to eliminate non-text blocks with the inclusion of the Support Vector Machine (SVM) for sorting out non-text patterns, which helps to enhance detection accuracy. The binarization technique is used for text recognition to extract text from the complicated backdrop so that OCR is utilized with greater accuracy.

Narendra Sahu et al. [3] show how OCR is applied in pattern recognition with a range of practical applications. Object Character Identification models, both theoretical and numerical, are addressed. OCR and Magnetic Character Recognition (MCR) methods are utilized for text pattern recognition. These strategies are employed in a variety of commerce and banking applications. The network components are either 1 or 0. When the constituents have more than 50% coverage, the vector element's value is 1; otherwise, it is 0. The Main Neural Network contains 10 numbers. This fund will contain 10 rows of this type of vector, with each row representing 0 or 1. Noise elimination, skew recognition/improvement, and binarization are the key preprocessing stages performed during OCR. Following the completion of OCR, Sentence Boundary Detection and Natural Language Processing methods such as tokenization and part-of-speech tagging can be used to reduce noisy inputs and increase the model's accuracy.

Chirag Indravadanbhai Patel et al. [4] describe automatic number plate recognition, which is one of the most important uses of OCR. This chapter demonstrates numerous approaches and tactics that assist the traffic police department in locating persons who are speeding and breaching traffic regulations. Several issues, such as fast vehicle speed, nonuniform vehicle number plate, the language of the vehicle number, and varied lighting conditions, are obstacles that impact the overall identification rate of the system. The usual technique of an automated number plate recognition system is to capture a picture of the vehicle, followed by number plate detection, character segmentation, and character recognition. Character recognition can be accomplished using an Artificial Neural Network and template matching, with a character recognition rate of roughly 85%. From 1,176 photos, the character recognition percentage was around 95.7%. Tesseract is another character recognition (OCR) technology maintained by Google. Based on the image size of the number plate, the success rate of character recognition is also achieved.

Ravneet Kaur et al. [5] proposed a novel Industrial Internet of Things approach that can be applied in various fields. A manufacturing shop floor will be outfitted with video cameras to capture images of the Human-Machine Interface (HMI) controller continually. Various OCR engines, such as Google Vision and Tesseract, may be used to recognize streaming data and test it for common factory HMIs and realistic lighting conditions. Image preprocessing, where the Region of interest (ROI) is formed, is one of the functional blocks involved in data extraction. Based on the ROI, OCR modules and data post-processing techniques are used, with the edge analytics module calculating the confidence level of the OCR result acquired. It is then saved on the Internet of Things (IoT) Gateway and utilized as input data for analytics apps.

Shashank Shetty et al. [6] propose a method for detecting and recognizing text in a video frame. It is made up of three key processes. The video is first separated into individual frames, followed by a text detection phase that consists of two steps: text localization and text verification. The text recognition step includes text verification and OCR. The result is the detection phase from the video frames in a Word file, and the final table displays the OCR accuracy level regarding extraction from the video frames. The proposed technology reduces noise by about 97.08%. The extracted and recognized proportion is around 91.60%.

Datong Chen et al. [7] discuss the Gabor filter, which can be efficiently used to estimate the scales of the stripes, which is a common substructure of text characters. The authors presented an algorithm that enhances the stripes in the three preselected scale ranges. The resultant enhancement gives a better performance than the Binarization process. Work-related text detection is classified into region-based, texture-based, and edge-based methods. The proposed algorithm uses Gaussian scale space as a texture feature and classifies the input image pixels into text pixels and backgrounds by using the k-means process. The enhancement method is used to enhance the contrast between the background and the text pixels, which are segmented easily from the background using the Binarization algorithm. The approach showed significant results in improving the recognition rate of superimposed text. Various frame enhancement techniques, such as the power law technique and inverse law technique, can be applied for accurate results.

David B. Wax [8] et al. describe a method that may collect data values from hospital anesthesia machine ventilators for research and decision assistance. They have employed the principle of extracting and storing data from a picture rather than storing the image itself. Collecting data and storing it in a relational database has been shown to be less expensive than vendor-sourced alternatives. They began by extracting data from anesthesia machine ventilators. They can mirror the ventilator display to the other slave machines through an external Video Graphics Array (VGA) connector. This output is routed through a VGA to the National Television Standards Committee (NTSC) converter and then to a USB video digitizer. These photos are subsequently sent to the Electronic Health Record workstation computer. The picture from the digitizer is recorded using open-source software, and the image bitmap is then sent to our departmental server through the local network. As the images arrive at the server, they are utilized to extract data using open-source image-processing tools. To retrieve data from sub-images, OCR is performed. Extracted data is then saved with the original image's location and time of capture. The process is controlled by an applet, which uses shell commands for image-processing programs and database functions to store the data. Since the ventilator machine has a 7-segment display unit, the general OCR model can help extract data from the unit. This also works if they vary in font size as well as font style. This can improve the efficiency of data extraction.

Igor Jurčić and Sven Gotovac [9] present and describe a comprehensive techno-economic (CTE) model for assessing the telecom operator potential. Telecom operators will have a key role in the following period for data transmission, and it is important to adapt to the changes brought by the Industry 4.0 era. This model will assist managers in making several key business and strategic decisions. The CTE model is a newly created model, based on research of existing models of analysis and experience from the telecommunication segment. This model is fast, reliable, and efficient in assessing the potential of a particular telecom and provides accurate potential assessment results. In addition, this model is modular and can be used to estimate individual parts of telecoms.

## 15.3 Architecture of Wireless Extraction of Display Panel

### 15.3.1 Block Diagram

The block diagram illustrated in Figure 15.1 depicts the various stages or components of the system. Initially, a stable video is captured using the IP Webcam application. The device with the application should have the same IP address as the server for data transfer. Text detection is performed on the collected video using the KerasOCR model. The model attempts to

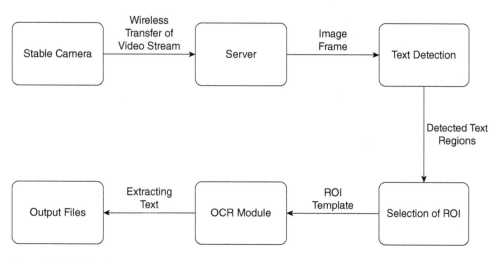

**Figure 15.1** A block diagram.

detect all words from the video. However, regions with improper detection or undetected regions can be selected by the user to create their own Region of Interest (ROI). Multiple ROIs can be created to detect the text areas in the video. Once the ROIs are selected, an ROI template is created. Finally, the OCR model is used to extract text from the frame using the ROI template.

Cropped images are created from the frames of the video using the ROI template. The EasyOCR model is applied to these cropped frames to recognize the words. An array is generated with various changing values of the words.

### 15.3.2 Modular Diagram

The system described in the modular diagram in Figure 15.2 follows a series of steps to perform OCR on video frames. First, display units with various font sizes and styles are used as the source of information. Stable video recording is done to ensure that the results are not affected by any camera movement. To capture stable video, an IP Webcam application can be used where the IP address of the device running the application should match the IP address of the server or the device that is receiving the video feed. This ensures a stable connection and reduces the chances of any lag or disruptions in the video. The KerasOCR model is used for text detection on the image frames, but due to noise and other factors such as brightness or blurriness, some parts of the display units may not be detected properly. In such cases, the user can create their own ROI to focus on the specific area of interest. Multiple ROIs can be created as well. ROI template will be created from both text detection ROIs and user ROIs.

Cropped images are then created from the video frames using the ROI template. In the next phase, the EasyOCR model is used for text recognition, generating results that are stored in an array.

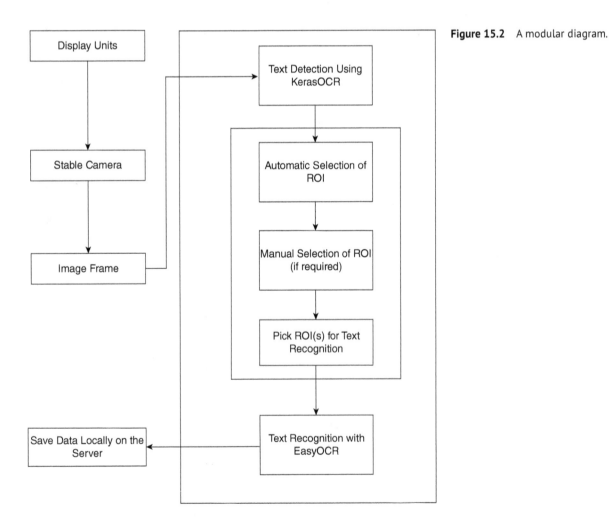

**Figure 15.2** A modular diagram.

### 15.3.3 Equations Applied

For text detector evaluation (TedEval) metrics:

$$\text{Recall} = \frac{True\ Positive}{\left(True\ Positive + False\ Negative\right)}$$

$$\text{Precision} = \frac{True\ Positive}{\left(True\ Positive + False\ Positive\right)}$$

Recall measures the proportion of actual positive cases that are correctly identified by the model. Precision measures the proportion of positive predictions made by the model that are correct.

$$\text{Hmean} = \frac{2*\left(Precision * Recall\right)}{\left(Precision + Recall\right)}$$

The harmonic mean is a commonly used metric for evaluating the overall performance of a text detector.

$$\text{Average Precision}\left(AP\right) = \left(R1*P1\right) + \left(\left(R2 - R1\right)*P2\right) + \ldots + \left(\left(Rn - Rn - 1\right)*Pn\right)$$

The AP metric is a measure of how well the detector can identify all relevant instances of text in the image while minimizing false positives.

Rn is the recall at the nth precision level.

Pn is the precision at the nth recall level.

Character Error Rate (CER) is the evaluation metric of the EasyOCR model for text recognition.

$$\text{CER} = \frac{S + D + I}{N}$$

S = Number of substitutions

D = Number of deletions

I = Number of insertions

## 15.4 ESP32 Cam Module

The ESP32-CAM is indeed a small-sized camera module based on the ESP32 chip, as shown in Figure 15.3, which is manufactured by Espressif Systems. It is designed to be a compact and versatile solution for various IoT applications.

The module itself is quite small, measuring only 27*40.5*4.5 mm in size. It is capable of working independently, acting as a standalone system. Additionally, it has a low deep-sleep current consumption of just 6 mA, making it suitable for battery-powered applications where power efficiency is crucial.

However, it is important to note that the ESP32-CAM requires a relatively higher power supply to function properly. It needs at least a 2A 5V power supply to ensure the board boots up correctly. This power requirement is higher compared to other ESP32 development boards, mainly due to the power demands of the camera module.

The ESP32-CAM offers several GPIO pins, as shown in Figure 15.4, allowing you to connect various peripherals and expand its functionality. It also supports the OV2640 camera module, which provides a resolution of up to 2 megapixels for capturing images and videos. Additionally, it includes a microSD card slot for storing captured media or other data.

**Figure 15.3** ESP32-CAM module. *Source:* Ajay Nair (Book author).

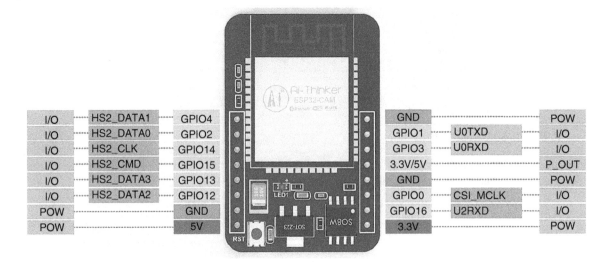

| | | | | | | | |
|---|---|---|---|---|---|---|---|
| I/O | HS2_DATA1 | GPIO4 | | GND | | | POW |
| I/O | HS2_DATA0 | GPIO2 | | GPIO1 | U0TXD | | I/O |
| I/O | HS2_CLK | GPIO14 | | GPIO3 | U0RXD | | I/O |
| I/O | HS2_CMD | GPIO15 | | 3.3V/5V | | | P_OUT |
| I/O | HS2_DATA3 | GPIO13 | | GND | | | POW |
| I/O | HS2_DATA2 | GPIO12 | | GPIO0 | CSI_MCLK | | I/O |
| POW | | GND | | GPIO16 | U2RXD | | I/O |
| POW | | 5V | | 3.3V | | | POW |

**Figure 15.4** ESP32-CAM pin diagram [10].

With its compact size, GPIO pins, camera support, and microSD card slot, the ESP32-CAM can be used in a wide range of IoT applications. It is particularly suitable for home automation devices, industrial wireless control systems, wireless monitoring applications, and more. The module is designed with a DIP package, which means it can be easily integrated by plugging it into a compatible bottom plate, enabling efficient product development and production.

The ESP32-CAM module does not come with a built-in USB connector for direct programming. Instead, there are two common methods to program the ESP32-CAM module.

FTDI Programmer Method: This method involves using an external FTDI programmer to upload code to the ESP32-CAM module.

ESP32-CAM-MB USB Programmer Method: This method involves using a dedicated USB programmer board called "the ESP32-CAM-MB USB programmer" as shown in Figure 15.5.

Both methods allow you to upload code to the ESP32-CAM module, but the specific steps may vary slightly depending on the programming software and hardware you are using.

### 15.4.1 Why Is the ESP32-CAM-MB USB Programmer Preferred over the FTDI Programmer?

The ESP32-CAM-MB USB programmer is specifically designed for programming the ESP32-CAM module. It is a dedicated programmer board that eliminates the need for additional wiring and connections. This simplifies the programming process and reduces the chances of making errors in wiring. The ESP32-CAM-MB USB programmer is designed to work seamlessly with the ESP32-CAM module. The pin mappings and connections are predefined and optimized for programming the module. This ensures better compatibility and reduces the likelihood of compatibility

**Figure 15.5** FTDI programmer and ESP32 camera module. *Source:* RandomNerdTutorials.com, https://randomnerdtutorials.com/upload-code-esp32-cam-mb-usb/, last accessed September 19, 2023 [11].

issues that may arise when using a generic FTDI programmer. The ESP32-CAM-MB USB programmer typically comes with its programming software, which provides a user-friendly interface for uploading code to the ESP32-CAM module. This software often includes additional features, such as firmware updates and configuration options specific to the ESP32-CAM module. ESP32-CAM-MB USB programmer includes a reset button and a boot mode jumper for easy programming while the FTDI programmer requires a manual reset button and boot mode selection using external wires and jumper cables. The ESP32-CAM-MB USB programmer often includes an integrated power supply that can provide sufficient power to the ESP32-CAM module during programming. This eliminates the need for an external power supply and simplifies the setup.

### 15.4.2 So How Is the ESP32-CAM-MB USB Programmer Used?

To begin, it is necessary to establish a connection between the ESP32 camera module and the ESP32-CAM-MB USB programmer, which should then be connected to a laptop using a USB data cable. It is important to verify the presence of the appropriate CH340C drivers and select the suitable ESP32 board and COM port while leaving the remaining options in the Tools menu at their default settings. The next step involves writing the code and adding the necessary libraries to the sketchbook in the Arduino IDE. To complete the process and flash the ESP32 cam module, simply click on the Upload button. Two streaming options are available: web server and real-time streaming protocol (RTSP) server. When the ESP32 initiates a web server, it transmits images from the camera to a web browser upon connection, resembling a continuous image stream. For RTSP server functionality, a streamer for video streaming and a session to handle RTSP communication is required alongside the server. The Wi-Fi client acts as the interface for the connected RTSP client. Finally, open the Serial Monitor and copy the stream links for use in the application.

## 15.5 Wireless LAN Network Setup

For setting up the LAN router, we need to connect the router with an Internet cable. Switch on the router. Further, go to the login page of the router by entering 192.168.4.100 in the browser. Enter the network administrator credentials (i.e., username and password). After successful login to the page, note down the IP address and subnet mask. If the IP address and subnet mask are not set, then they need to be configured, and the router needs to be rebooted. In the "DHCP" menu, configure the "DHCP settings." Note down the starting and ending IP addresses and the default gateway. You now are provisioning that range of addresses for your devices to connect to this router. You can obtain the Media Access Control (MAC) addresses of your ESP32-CAM modules in the Serial Monitor in Arduino IDE itself. In the "DHCP" menu, click on "Address Reservation" and add new static IP addresses to the list and enable them. In the "IP & MAC Binding" menu, add your ESP32-CAM's MAC address and bind it to an IP address permanently by enabling the "ARP Binding" option. You can view your ESP32-CAM modules in the "DHCP Clients List" in the "DHCP" menu if they are connected to the given router, as shown in Figure 15.6. Note that the "Lease Time" for your module is permanent here as it has been assigned a static IP address.

### ARP List

| ID | MAC Address | IP Address | Status | Configure |
|----|-------------|------------|--------|-----------|
| 1 | D0-51-62-38-B2-A7 | 192.168.0.101 | Unbound | Load Delete |
| 2 | CC-AF-78-91-7E-81 | 192.168.0.102 | Unbound | Load Delete |
| 3 | 50-B7-C3-58-8F-03 | 192.168.0.103 | Unbound | Load Delete |
| 4 | 00-16-17-BA-2B-06 | 192.168.0.104 | Unbound | Load Delete |
| 5 | 00-16-17-BA-2B-06 | 192.168.0.150 | Bound | Load Delete |

**Figure 15.6** IP MAC binding.

## 15.6    Optical Character Recognition for Text Detection and Text Recognition

### 15.6.1    KerasOCR's Text Detector – CRAFT Model

A brand-new text detector called "CRAFT" [12] locates certain character regions and connects the characters it finds to a text instance. The character region score and affinity score are produced by a convolutional neural network in the framework, known as "CRAFT" for character region awareness for text detection. Each character in the image is localized using the area score and grouped into a single instance using the affinity score. They suggest a weakly supervised learning method that estimates character-level ground truths in current real-world datasets to make up for the dearth of character-level annotations. Their approach has been thoroughly tested on International Conference on Document Analysis and Recognition (ICDAR) datasets, and the results demonstrate that the suggested method outperforms cutting-edge text detectors. Additionally, testing on the MSRATD500, CTW-1500, and TotalText datasets demonstrates the proposed method's high adaptability in challenging situations like long, curved, and/or arbitrarily formed texts.

Its foundation is a fully convolutional network architecture based on VGG-16 with batch normalization. In the decoding section of our model, skip connections aggregate low-level characteristics similar to U-net. The region score and the affinity score are the two score maps that make up the final result. Figure 15.7 depicts the network architecture schematically.

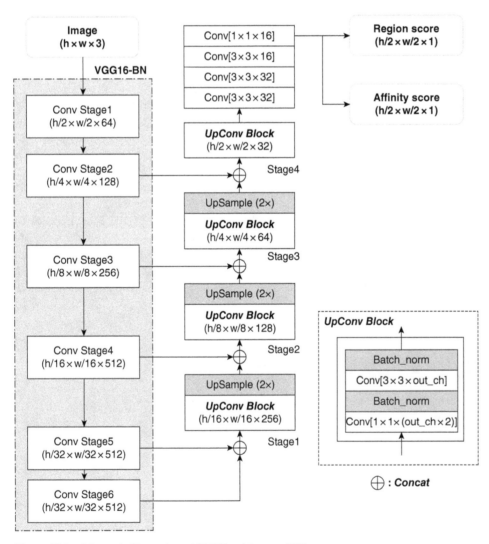

**Figure 15.7**    Schematic illustration of CRAFT architecture [13].

### 15.6.2 Training Custom KerasOCR Models

One folder containing text images for your required task and another folder containing its corresponding annotations must be provided for fine-tuning this KerasOCR's text detector. Using label studio, 41 images were selected, and 9,831 character annotations were done on them. This dataset was then used for fine-tuning the KerasOCR detector model, for customizing it to our OCR tasks. For training, training and validation sets were created in the ratio 4:1, and the model was trained for 1,000 epochs. The best model weights were then saved, which was then loaded for further use. It was tested on another set of images relevant to our task whose results were found satisfactory.

Here are the results obtained, including recall, precision score, harmonic mean score, and average precision score in the terminal's screenshot, as shown in Figure 15.8.

**Recall:** 0.8666219839142091
**Precision:** 0.81217277486911
**Harmonic Mean:** 0.8385143898689302
**Average Precision Score:** 0

### 15.6.3 EasyOCR Architecture

EasyOCR is a Python-based package for employing an OCR model that is ready to use [14]. The PyTorch library and Python are used to implement it. All common writing scripts and 80+ languages, such as Latin, Chinese, Arabic, Devanagari, and Cyrillic, are supported by EasyOCR.

The architecture of EasyOCR shown in Figure 15.9 is as follows:

**Text Detection:** EasyOCR's initial step is to find text within a picture. An already-trained text detection model is used for this. The deep learning model used to recognize text in photos was trained using a sizable dataset of text-containing images. By scanning for pixel patterns that correspond to the text, the text detection model can recognize text in photographs.

EasyOCR's text detection model is a deep learning model that was trained on a sizable dataset of text-containing photos. By searching for pixel patterns that are consistent with writing, the model can recognize text in photos. The following layers make up the text detection model:

**Convolutional Layers:** These layers are in charge of taking features out of the image. The convolutional layers scan the image using a filter to find pixel patterns that are consistent with the text.
**Pooling Layers:** The feature map's size is decreased as a result of the pooling layers. The pooling layers merge nearby features into a single feature using a pooling technique.

```
(tedval) D:\TedEval>python script.py -g=gt/gt.zip -s=test/gt.zip
Calculated!{"recall": 0.8666219839142091, "precision": 0.81217277486911, "hmean": 0.8385143898689302, "AP": 0}
(tedval) D:\TedEval>
```

**Figure 15.8** Evaluation metrics.

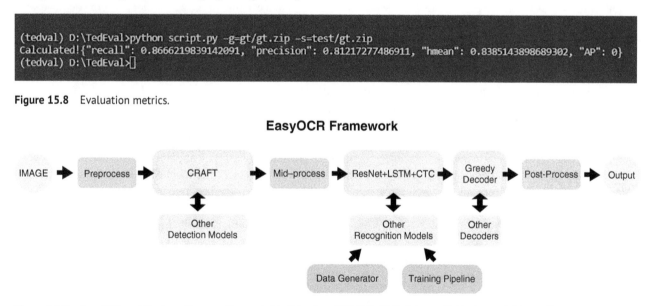

## EasyOCR Framework

**Figure 15.9** EasyOCR architecture diagram. *Source:* JaidedAI – EasyOCR [15], GitHub, https://github.com/JaidedAI/EasyOCR/blob/master/examples/easyocr_framework.jpeg.

**Fully Connected Layers:** The fully connected layers use a softmax layer to put the features into a probability distribution and are in charge of identifying the characteristics as text or non-text.

**Text Recognition:** After the text has been located, the text must then be identified. An already-trained text recognition model is used for this. The deep learning text recognition model was developed using a sizable text dataset. The text recognition model can recognize each character in the text and translate it into text.

Another deep learning model that was trained on a sizable text dataset is the text recognition model in EasyOCR. The text's characters may be recognized by the model, which can then be translated into text. The following layers make up the text recognition model:

**Embedding Layer:** This layer turns the characters into vectors. A lookup table is used by the embedding layer to convert each character into a vector.

**LSTM Layers:** The LSTM layers are in charge of foretelling the text's subsequent character. The LSTM layers make predictions about the following characters in the text based on the preceding characters using a long short-term memory network.

**Output Layer:** The output layer is in charge of producing the finished text. The output layer creates a probability distribution over the potential characters using a softmax layer.

**Post-Processing:** EasyOCR's final step entails post-processing the identified text. Spelling corrections, punctuation removal, and formatting of the text are examples of tasks included in this.

An effective OCR tool is EasyOCR. It is simple to use and supports OCR text in many different languages.

The post-processing step in EasyOCR is responsible for correcting spelling errors, removing punctuation, and converting the text to a desired format. The post-processing step is made up of the following steps:

**Spelling Correction:** The spelling correction step is responsible for correcting spelling errors in the text. The spelling correction step uses a dictionary to find the correct spelling of the words in the text.

**Punctuation Removal:** The punctuation removal step is responsible for removing punctuation from the text. The punctuation removal step uses a regular expression to remove punctuation from the text.

**Format Conversion:** The format conversion step is responsible for converting the text to a desired format. The format conversion step can be used to convert the text to a different font, size, or color.

### 15.6.4 Training of Custom EasyOCR Model

To create a dataset for training the EasyOCR model, you can either prepare your dataset with scene text and images, or you can use any kind of text generator. EasyOCR needs word-level annotations for training the recognition model, so you need to crop all the words in the images and store the words in the images in a csv file. In the same folder where you have your images dataset, you need to create the labels.csv (The name for the file can be changed in the configuration file for the training script) file. The labels.csv file will only have two fields, which are "filename" and "words." The filename column will have the name of the images, and the words column will have the actual word in the image. You need to make sure that each image in the dataset should be of the dimension $(600 \times 64)$, or at least they do not exceed the threshold. All these default values can be changed accordingly in the configuration file.

Once you have prepared the dataset, you need to split the dataset into two parts: the training and validation sets. Both the sets will have their labels.csv file, where the words in the image and their location will be specified. Now, you can feed both the folders to the training script provided in the EasyOCR repository. Once you have stored the training and validation set in their appropriate places, you can now change the default configurations provided by EasyOCR and start the training of the model.

The training has 3,00,000 iterations, and after each 20,000th iteration, a weight file will be saved to the saved_models folder. In case the training stops midway, you can specify where the training should resume from in the configuration file and resume the training from saved_weight. Once the training is finished, you can use the best_weight file to load it into your device, and you can use the custom_easyocr model for recognition. The mean CER (Character Error Rate) a metric which is used to evaluate the performance of the recognition system is shown in Figure 15.10.

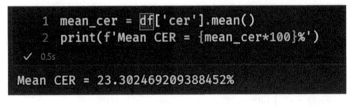

**Figure 15.10** Character error rate.

**Mean CER:** 23.302469209388452%

### 15.6.5   Real-Time Streaming Using IP Webcam

Using the IP camera application IP Webcam from Google Play Store, the real-time data stream of the display panel can be wirelessly transferred to our application using a secured RTSP link provided by the application. Any camera that provides an IP link for wireless transfer can be used here. Using multithreading, a separate daemon thread runs in the background, which reads the next available frame from the stream. Because of the huge number of incoming frames, a queue data structure is maintained of length *1* so that latency is reduced and the latest frame is obtained. The main thread can now run our main processing application.

### 15.6.6   Processing the Data

KerasOCR's detector model and EasyOCR's text recognition model are loaded with the weights obtained from their training process. The RTSP (h264_pcm) link provided by the IP Webcam application is provided to OpenCV's VideoCapture method as stream ID. The daemon thread captures frames, and the main thread does processing simultaneously.

OpenCV shows the video stream for the user to capture a screenshot for the further text detection process. This captured frame goes through KerasOCR's detector model, and most of the textual regions are detected in this phase itself. Figures 15.11 and 15.12 show the regions where texts are detected from the power supply display unit. These regions are then drawn on the image frame with their corresponding ROI number for the user to choose from. Suppose some text region is not

**Figure 15.11**   Detecting of ROI on power supply display unit. *Source:* Ajay Nair (Book author).

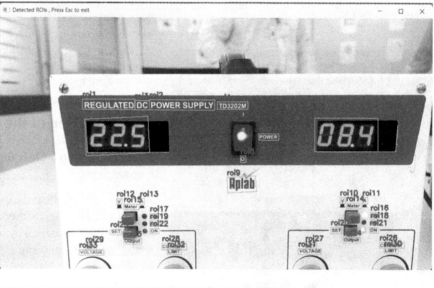

**Figure 15.12**   Detecting of ROI on power supply display unit. *Source:* Ajay Nair (Book author).

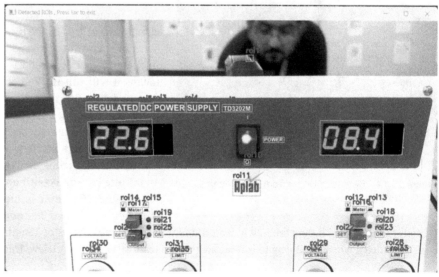

**Figure 15.13** Detection values from display unit stored in csv format.

**Figure 15.14** Detection values from display unit stored in csv format.

detected; unfortunately, an additional feature is provided, which allows the user to draw their ROI on the frame. ROIs created from both these options are combined to create an ROI template, which can then be used for further text recognition processes.

The video streaming process is started again to obtain real-time frames. The ROI template is then used to crop images from these frames based on their coordinates. If ROI coordinates have a length of 2, then the frame can be cropped by slicing, whereas if it has a length of 4, a four-point transformation is done initially to obtain a consistent order of coordinates, which are then applied with perspective transformation to get cropped images.

Finally, these cropped images of each frame are passed to EasyOCR's recognition model to extract the text present in them. A dictionary of these extracted texts with their timestamp when OCR is performed is created, which is finally fed into a Pandas DataFrame, which logs all the changing values. This DataFrame is then converted to a CSV file, which is saved at the end with the final timestamp (Figures 15.13 and 15.14).

Figures 15.11 and 15.12 show the Region of Interest (ROI) creation on the video frame. Figures 15.13 and 15.14 show the Results stored in Results (.csv) file. Figure 15.15 shows the detection of texts in function generator display, and its associated OCR results are shown in Figure 15.16.

## 15.7 Working of the Model

### 15.7.1 Creating a Class for Handling Web Camera Stream

A class called "WebcamStream," which allows access to a webcam stream and provides methods for reading frames from it, was written. The class initializes by opening the video capture stream and setting up a queue to store the frames. If the stream fails to open, an error message is printed, and the program exits. It also retrieves the frames per second of the webcam hardware/input stream. A single frame is read from the video capture stream to check if the camera is working in the method mentioned earlier. It also starts a thread to continuously read frames from the stream and put them in the queue. The thread is marked as a daemon thread, which means it will run in the background while the program is being executed. The "start" method is used to begin the next available frame-reading process. The read method is used to get the latest read frame from the queue. It waits for a frame to be available in the queue. Once a frame is available, it is retrieved and returned. The update method is the target function for the thread. It runs in an infinite loop and continuously reads frames from the video capture stream. If a frame is successfully read, it checks if the queue is not empty. If it is not empty, it discards the previous (unprocessed) frame. Then, it puts the newly read frame into the queue. The "stop" method stops reading frames. Overall, the code provides a convenient way to interact with a webcam stream and retrieve frames in real time.

**Figure 15.15** Detection of ROI on function generator display unit. *Source:* Ajay Nair (Book author).

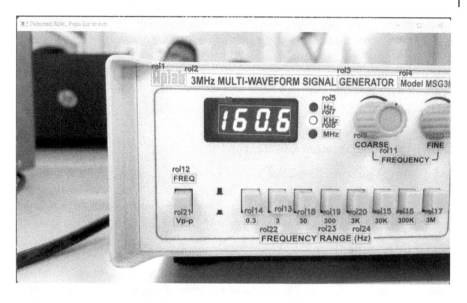

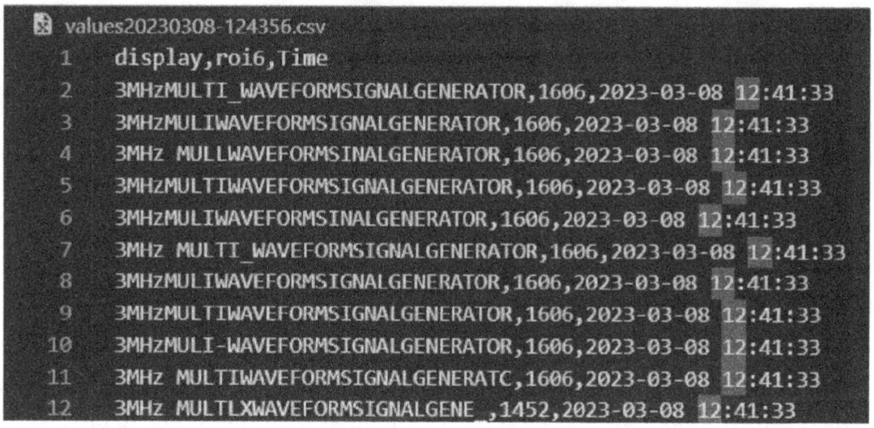

**Figure 15.16** Detection values are stored in csv format from function generator display unit.

```
values20230308-124356.csv
1    display,roi6,Time
2    3MHzMULTI_WAVEFORMSIGNALGENERATOR,1606,2023-03-08 12:41:33
3    3MHzMULIWAVEFORMSIGNALGENERATOR,1606,2023-03-08 12:41:33
4    3MHz MULLWAVEFORMSINALGENERATOR,1606,2023-03-08 12:41:33
5    3MHzMULTIWAVEFORMSIGNALGENERATOR,1606,2023-03-08 12:41:33
6    3MHzMULIWAVEFORMSINALGENERATOR,1606,2023-03-08 12:41:33
7    3MHz MULTI_WAVEFORMSIGNALGENERATOR,1606,2023-03-08 12:41:33
8    3MHzMULIWAVEFORMSIGNALGENERATOR,1606,2023-03-08 12:41:33
9    3MHzMULTIWAVEFORMSIGNALGENERATOR,1606,2023-03-08 12:41:33
10   3MHzMULI-WAVEFORMSIGNALGENERATOR,1606,2023-03-08 12:41:33
11   3MHz MULTIWAVEFORMSIGNALGENERATC,1606,2023-03-08 12:41:33
12   3MHz MULTLXWAVEFORMSIGNALGENE ,1452,2023-03-08 12:41:33
```

### 15.7.2 Using the Web Camera Stream Class

An instance of the WebcamStream class is created, passing the URL of the webcam stream as an argument. This URL could be an RTSP stream or any other supported format. The start method is called on the WebcamStream class object to start reading frames from the stream in a separate thread. A while loop is set up to continuously read frames from the stream and perform operations on them. Inside the loop, the read method is called on the object to retrieve the latest frame from the stream. The frame is resized by dividing its height and width by 2. This halves the dimensions of the frame. The resized frame is saved as an image. The resized frame is displayed in a window so that the user can then check the orientation of the camera and if the display panel is visible correctly. Then, he can select that frame, which will further be used for the text detection process.

### 15.7.3 Detecting Texts or Selecting Regions of Interest

The saved resized image is then passed to another function that detects text and performs a series of operations to create ROIs in an image and obtain user input for selecting specific ROIs. The resized frame is given as input to KerasOCR's text detector module based on the CRAFT text detection model. The detector model is loaded with the trained weights initially. Then, all the text regions detected by the model are returned where each rectangular ROI has eight coordinates for all four vertices. The orientation of the detected rectangular ROI may be horizontal or rotated. Then, we assign sample names for all ROIs in incremental order. Then, we use KerasOCR's tools library to draw bounding boxes of the detected ROIs in the resized image. We also draw text above the bounding boxes so that the user can select their desired ROI for further processing. The user is also provided with another option to interactively draw ROIs themselves if the detector model is somehow not able to detect the text. The user can also name these ROIs, which have four coordinates for two vertices. Each bounding box is represented by a tuple of four integers, $(x, y, w, h)$, where $(x, y)$ is the top-left corner of the bounding box, $w$ is the width of the bounding box, and $h$ is the height of the bounding box. Both these final ROIs are returned by the function for further processing.

### 15.7.4 Cropping Selected ROIs from an Image

While looping through the selected ROI coordinates, if they have a shape of four, then we simply crop the ROI from the original image using these extracted coordinates. This is done by slicing the appropriate region from the image array. Else, when they have the shape of eight, we crop them using a four-point transform function. This function takes an image and a list of four points representing the corners of a quadrilateral in the original image. It calls another order function that takes a list of four points representing the corners of a quadrilateral and orders them in a specific manner: top-left, top-right, bottom-right, and bottom-left. It returns the ordered coordinates. It computes the width of the new transformed image by calculating the Euclidean distance between the bottom-right and bottom-left points, and the top-right and top-left points. It computes the height of the newly transformed image by calculating the Euclidean distance between the top-right and bottom-right points, and the top-left and bottom-left points. The maximum width and height are determined by selecting the larger values between both width values and between both height values, respectively. It defines a destination array with four points representing the top-left, top-right, bottom-right, and bottom-left corners of the transformed image. It computes the perspective transformation matrix, which maps the coordinates of the quadrilateral to the destination points. It applies the perspective transformation to the original image to return the warped image. These cropped regions from both methods are saved to a folder.

### 15.7.5 Performing Optical Character Recognition on the Cropped Images

EasyOCR's text recognition module is based on Convolutional Recurrent Neural Network (CRNN), which is composed of three main components: feature extraction (using Resnet) and Visual Geometry Group (VGG), sequence labeling (Long Short Term Memory [LSTM]), and decoding (Connectionist Temporal Classification [CTC]). The model, loaded with our trained weights, is used for the further recognition process. The OCR function is defined, which takes a list of our cropped images as input. A loop is initiated to process each image in the list. Inside the loop, the current image is passed as input to the recognition model to perform OCR on the image and obtain the OCR results. Then we use a dictionary to map these results to the ROI name. The code checks if the OCR results for the current image's label exist in the dictionary. If not, it initializes an empty list for that label and appends the last recognized text from the result to the list (if available) or None otherwise. If the label already exists in the dictionary, it appends the last recognized text from the result to the list (if available) or None otherwise. After processing all images, the current timestamp is also added to the dictionary. The OCR results stored in the dictionary are converted into a Pandas DataFrame. This DataFrame is then saved as a CSV file with a current timestamp in its name for further analysis or processing.

## 15.8 Application GUI

Figures 15.17 and 15.18 shows the dashboard UI of the application. Figure 15.19 shows the Dashboard UI when the ROIs are being created. Figure 15.20 shows a video of a controller that we have obtained from industry for our project, and Figure 15.21 shows the ROIs detected by our application on the same video. Figure 15.22 shows the option for the user, where he/she can draw the ROIs themselves. Figure 15.23 shows the final recognized OCR results from our application.

**Figure 15.17**  Home screen.

**Figure 15.18**  Entering stream link.

**Figure 15.19**  Waiting message before showing the stream.

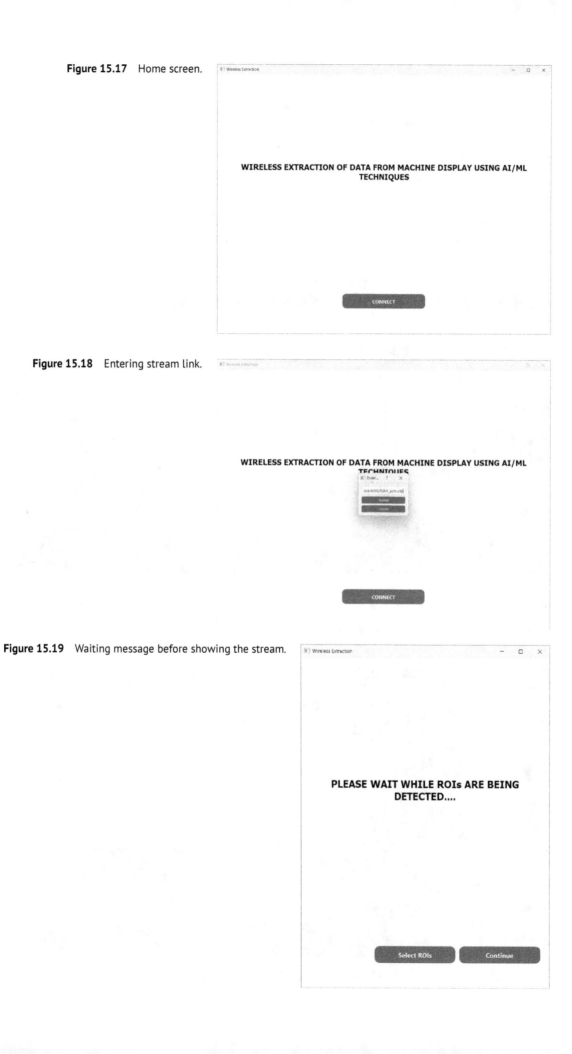

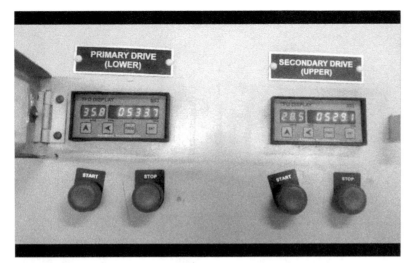

**Figure 15.20** Machine display. *Source:* Manigandan Kasimani (Chapter author).

**Figure 15.21** ROI detection by algorithm. *Source:* Manigandan Kasimani (Chapter author).

**Figure 15.22** Manual selection of ROI. *Source:* Manigandan Kasimani (Chapter author).

| Primary ▲ | Secondary | RoI8 | Time |
|---|---|---|---|
| 533 | 5291 | Start | 2023-06-26 14:27:35 |
| 533 | 5291 | Start | 2023-06-26 14:27:36 |
| 533 | 5291 | Start | 2023-06-26 14:27:42 |
| 5337 | 5291 | Start | 2023-06-26 14:27:34 |
| 5337 | 5291 | Start | 2023-06-26 14:27:34 |
| 5337 | 8529 | Start | 2023-06-26 14:27:35 |
| 5337 | 5291 | Start | 2023-06-26 14:27:35 |
| 5337 | 5291 | Start | 2023-06-26 14:27:36 |
| 5337 | 5291 | Start | 2023-06-26 14:27:36 |
| 5337 | 5291 | Start | 2023-06-26 14:27:37 |
| 5337 | 5291 | Start | 2023-06-26 14:27:37 |
| 5337 | 5291 | Start | 2023-06-26 14:27:38 |
| 5337 | 5291 | Start | 2023-06-26 14:27:38 |
| 5337 | 5291 | Start | 2023-06-26 14:27:38 |
| 5337 | 5291 | Start | 2023-06-26 14:27:38 |
| 5337 | 5291 | Start | 2023-06-26 14:27:39 |
| 5337 | 5291 | Start | 2023-06-26 14:27:39 |
| 5337 | 5291 | Start | 2023-06-26 14:27:40 |
| 5337 | 5291 | Start | 2023-06-26 14:27:41 |
| 5337 | 5291 | Start | 2023-06-26 14:27:41 |
| 5337 | 5291 | Start | 2023-06-26 14:27:41 |
| 5337 | 5291 | Start | 2023-06-26 14:27:42 |
| 5337 | 5291 | Start | 2023-06-26 14:27:42 |
| 5337 | 5291 | Start | 2023-06-26 14:27:42 |
| 5337 | 5291 | start | 2023-06-26 14:27:43 |
| 5337 | 5291 | Start | 2023-06-26 14:27:43 |
| 5337 | 5291 | Start | 2023-06-26 14:27:44 |

**Figure 15.23** Detected results.

## 15.9  Conclusion

In this research, we provide a full method for text detection and recognition on video. Text detection and recognition are handled by models such as KerasOCR and EasyOCR, respectively. The text recognition model supports many sorts of display unit systems that vary with font size and font style. Users can also construct an ROI for the region that is not detected. Text recognition of video in real time is also possible. Text detector evaluation (TedEval) and CER evaluation metrics are used for text detection and text recognition, respectively. The accuracy of the model can be improved further by adding more training datasets.

## References

1 Hamad, K.A. and Kaya, M. (2016). A detailed analysis of optical character recognition technology. *International Journal of Applied Mathematics Electronics and Computers* https://doi.org/10.18100/ijamec.270374.

2 Yang, H., Quehl, B., and Sack, H. (2012). A framework for improved video text detection and recognition. *Multimedia Tools and Applications* https://doi.org/10.1007/s11042-012-1250-6.

3 Sahu, N. and Sonkusare, M. (2017). A study on optical character recognition techniques. *The International Journal of Computational Science, Information Technology and Control Engineering* http://dx.doi.org/10.5121/ijcsitce.2017.4101.

**4** Patel, C.I., Shah, D., and Patel, A. (2013). Automatic number plate recognition system (ANPR): a survey. *International Journal of Computer Applications* http://dx.doi.org/10.5120/11871-7665.

**5** Kaur, R., Acharya, J., and Gaur, S. (2019). Non-invasive data extraction from machine display units using video analytics. https://www.researchgate.net/publication/334001257_Non-Invasive_Data_Extraction_from_Machine_Display_Units_using_Video_Analytics (accessed 13 January 2024).

**6** Shetty, S., Chakkaravarthy, S., Varun Kumar, K. A., and Devadiga, A.S. (2014). Ote-OCR based text recognition and extraction from video frames. http://dx.doi.org/10.1109/ISCO.2014.7103949

**7** Chen, D., Shearer, K., and Hewe, B. (2002). Text enhancement with asymmetric filter for video OC. https://doi.org/10.1109/ICIAP.2001.957007

**8** Wax, D.B., Hill, B., and Levin, M.A. (2017). Ventilator data extraction with a video display image capture and processing system. *Journal of Medical Systems* https://doi.org/10.1007/s10916-017-0751-2.

**9** Jurčić, I. and Gotovac, S. (2022). A comprehensive techno-economic model for fast and reliable analysis of the telecom operator potentials. *Applied Sciences* 12: 10658. https://doi.org/10.3390/app122010658.

**10** SHENZHEN AI-THINKER TECHNOLOGY CO., LTD, *ESP32-CAM*. https://loboris.eu/ESP32/ESP32-CAM%20Product%20Specification.pdf (accessed 23 June 2023).

**11** Santos, R. and Santos, S. (2021). Upload Code to ESP32-CAM AI-Thinker using ESP32-CAM-MB USB Programmer (easiest way). https://randomnerdtutorials.com/upload-code-esp32-cam-mb-usb (accessed 29 August 2023).

**12** Morales, F. *GitHub - faustomorales/keras-ocr*. https://github.com/faustomorales/keras-ocr (accessed 23 June 2023).

**13** Baek, Y., Lee, B., Han, D. et al. Character Region Awareness for Text Detection. (2019). https://doi.org/10.48550/arXiv.1904.01941.

**14** CLOVA AI RESEARCH, NAVER CORP. *GitHub - clovaai/CRAFT-pytorch*. https://github.com/clovaai/CRAFT-pytorch (accessed 23 June 2023).

**15** JaidedAI – EasyOCR. https://github.com/JaidedAI/EasyOCR/blob/master/examples/easyocr_framework.jpeg (accessed 29 August 2023).

# Index

*Topics in Artificial Intelligence Applied to Industry 4.0*, First Edition. Edited by Mahmoud Ragab AL-Refaey,
Amit Kumar Tyagi, Abdullah Saad AL-Malaise AL-Ghamdi, and Swetta Kukreja.
© 2024 John Wiley & Sons Ltd. Published 2024 by John Wiley & Sons Ltd.

Printed in the USA
CPSIA information can be obtained
at www.ICGtesting.com
CBHW082054300524
9218CB00051B/10

9 781394 216116